Praise for *A Life Decoded*

"For decades now, our infinitely more vulgar media has called Venter many things: maverick, publicity hound, risk-taker, brash, controversial, genius, manic, rebellious, visionary, audacious, arrogant, feisty, determined, provocative. His autobiography shows that they are all justified."

—*Nature*

"*A Life Decoded: My Genome: My Life* will attract a broad audience of scientists and science watchers. It offers a window into the life and mind of a scientist who . . . has indisputably become an extraordinary figure. *A Life Decoded* . . . like a sociological tracing agent, charts a course through the world of genomics, highlighting structural features and tensions."

—*Science*

"A personal glimpse into one of the most monstrously successful scientific lives of our time." —*Seed*

ABOUT THE AUTHOR

J. Craig Venter is one of the leading scientists of the twenty-first century. A pioneer in the world of genomic research, he is recognized for his visionary contributions to the field. In February 2001, Venter published the completed sequence of the human genome. He is the founder and president of the J. Craig Venter Institute. He lives near Washington, D.C.

J. CRAIG VENTER

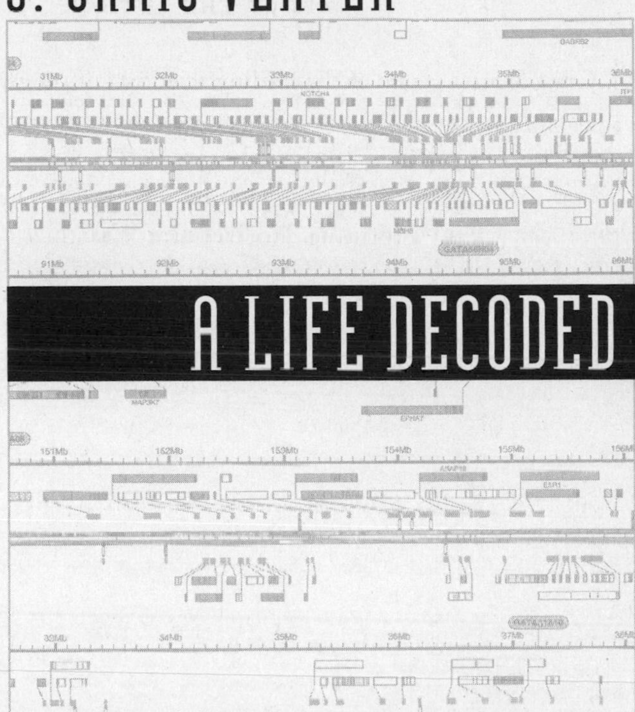

A LIFE DECODED

MY GENOME: MY LIFE

PENGUIN BOOKS

PENGUIN BOOKS
Published by the Penguin Group
Penguin Group (USA) Inc., 375 Hudson Street, New York, New York 10014, U.S.A.
Penguin Group (Canada), 90 Eglinton Avenue East, Suite 700, Toronto,
Ontario, Canada M4P 2Y3 (a division of Pearson Penguin Canada Inc.)
Penguin Books Ltd, 80 Strand, London WC2R 0RL, England
Penguin Ireland, 25 St Stephen's Green, Dublin 2, Ireland (a division of Penguin Books Ltd)
Penguin Group (Australia), 250 Camberwell Road, Camberwell,
Victoria 3124, Australia (a division of Pearson Australia Group Pty Ltd)
Penguin Books India Pvt Ltd, 11 Community Centre, Panchsheel Park, New Delhi – 110 017, India
Penguin Group (NZ), 67 Apollo Drive, Rosedale, North Shore 0632,
New Zealand (a division of Pearson New Zealand Ltd)
Penguin Books (South Africa) (Pty) Ltd, 24 Sturdee Avenue,
Rosebank, Johannesburg 2196, South Africa

Penguin Books Ltd, Registered Offices:
80 Strand, London WC2R 0RL, England

First published in the United States of America by Viking Penguin,
a member of Penguin Group (USA) Inc. 2007
Published in Penguin Books 2008

10 9 8 7 6 5 4 3 2 1

ISBN 978-0-670-06358-1 (hc.)
ISBN 978-0-14-311418-5 (pbk.)
CIP data available

Printed in the United States of America
Designed by Nancy Resnick

To my son, Christopher Emrys Rae Venter, and
my parents, John and Elizabeth Venter

CONTENTS

A LIFE DECODED

INTRODUCTION

DNA neither cares nor knows. DNA just is. And we dance to its music.

—Richard Dawkins

DNA provides the music. Our cells and the environment provide the orchestra.

—J. Craig Venter

Traditional autobiography has generally had a poor press. The novelist Daphne du Maurier condemned all examples of this literary form as self-indulgent. Others have quipped that autobiography reveals nothing bad about its writer except his memory. George Orwell thought that an autobiography can be trusted only "when it reveals something disgraceful." His reason? "A man who gives a good account of himself is probably lying." Sam Goldwyn came to this conclusion: "I don't think anybody should write his autobiography until after he's dead."

Because I have been fortunate enough to take part in one of the greatest, most exciting, and, potentially, most beneficial scientific adventures of all time, I believe my story is a story well worth telling—even more so because it became so controversial, for political, economic, and scientific reasons—but I am well aware of the research that shows that memory is notoriously malleable. I flinch at the thought of claiming that this side of my story is the only truth, because it depends on chance, on many other people, on my recollections and my partial experience of the events that came to influence my life. Since this is also the first biography to benefit from having six billion base pairs of the author's genetic code as an essential appendix, new interpretations of Craig Venter, based on my DNA, will continue to be made long after life has left my body. I have no choice but to leave the ultimate interpretation to you and to history.

This is a tale of seemingly impossible quests and grand objectives. It features great rivalries and bitter disputes, and clashes of egos involving some of the biggest figures in biology. My adventure has swept me from peaks of incredible exhilaration as I marshaled a relatively small but dedicated army of scientists, computers, and robots to achieve what seemed almost impossible, and then plunged me into black pits of depression as I faced opposition from Nobel laureates and senior government officials, my colleagues, and even my wife. Even today some of these memories remain painful. But I still have a great deal of respect for my critics. My opponents were for the most part honorable as we engaged in a battle of ideologies, morals, and ethics. Everyone felt passionately that he had right on his side.

Apart from the sheer excitement of the science that is at the heart of this book, I hope, too, that my story will serve as an inspiration. My early years were hardly a model of focus, discipline, and direction. No one who met me as a teenager could have imagined my going into research and making important discoveries. No one could have predicted the arc of my career. No one could have seen me rise to head a major research effort. And certainly no one could have foreseen the war of words in which I became embroiled, let alone held out much hope that I could ever beat the establishment.

One of the fundamental discoveries I made about myself—early enough to make use of it—was that I am driven to seize life and to understand it. The motor that pushes me is propelled by more than scientific curiosity. For many years I have been trying to make sense and meaning out of the lives I saw destroyed or maimed due to the government policies that involved us in the war in Vietnam. I have struggled to understand the deaths of two men who were briefly under my care, an eighteen-year-old who should not have been alive at all as a result of his wounds and a thirty-five-year-old who should have survived but gave up.

Decades later, with the benefit of hindsight, perhaps it was inevitable that such experience would have compelled me to understand life in its most intimate detail. Although the two men were ultimately victims of war, the manner of their dying has remained with me. I had witnessed the power of the human spirit, which can be stronger than any drug. There are still so many questions to answer about the workings of the human body and, most mysterious of all, how it is influenced by our state of mind. By facing such basic questions, I was transformed during my short time in Vietnam from a risk taker who shunned the conventions of the 1960s to one who took a major gamble on a career that was far removed from his upbringing. I willed myself through a junior college to a university and ultimately a Ph.D. I became a scientist, focusing initially on the protein molecules that mediate our responses to adrenaline, and then switching to molecular biology to gain access to the tools that would enable me to read the DNA code that determined the structure of the protein molecules for which I was searching. In this way I was ultimately led to the biological instructions—the genetic code—that the cells use to guide the way proteins are made. My first glimpse of the code of life gave me an appetite to see more. I wanted to look upon the bigger picture: the entire set of genes within an organism, what we call the genome. After nearly a decade of work I developed new techniques that led me to decode the first entire genome of a living species, and the challenge grew to the ultimate, to sequence the human genome. And what greater challenge than to try to understand one's own life in the context of being the first person in history to be able to gaze upon his own genetic legacy, to focus in detail on those segments, regions, and genes that provide a genetic context for his own life and his own complex, unique blend of nature and nurture?

Although it will take decades to interpret completely what my DNA reveals, I have already made out some hints and whispers of its message. From time to time you will come across boxes containing these insights, which I use to illustrate our current, extremely limited abilities to interpret what the code of my life actually means. While the interpretation of DNA is very much a work in progress, and sheds only a weak light on my destiny, we have arrived at a uniquely fascinating point in human history as we develop the ability not only to connect our own existence to our evolutionary past but, for the first time, to begin to see what the future has in store for us. Still, one of the most profound discoveries that I have made in all my research is that you cannot define a human life or any life based on DNA alone. Without understanding the environment in which cells or species exist, life cannot be understood. An organism's environment is ultimately as unique as its genetic code.

I suspect many people decide to write an autobiography to help make sense of their lives. It is all too human to dwell on our mistakes, our triumphs, and those electric moments when a life-changing decision was made. *A Life*

Decoded might be thought of as an extreme example of this, as the sum of six billion base pairs of my DNA struggles to understand itself. Now that we have a DNA replicating machine (me) reading his own DNA for the first time, new opportunities inevitably beckon. As we begin to interpret its contents, we humans may perhaps begin to transcend our own DNA and can perhaps even begin to modify it. We might change the story of life and even create synthetic and artificial versions. But that will be the subject of my next book.

1. WRITING MY CODE

We must, however, acknowledge, as it seems to me, that man with all his noble qualities . . . still bears in his bodily frame the indelible stamp of his lowly origin.

—Charles Darwin

Of all my early memories, the most vivid is my total and absolute freedom. Today's soccer moms schedule activities for every minute of a child's day. They give their kids mobile phones so they can stay in touch. Some even want to track their children using GPS, or keep an eye on them from work using a webcam. But half a century ago the life of the average kid was unstructured and often unobserved.

I was lucky that liberty was a family tradition. As a child my mother loved to climb barefoot on the cliffs in Ocean Beach near San Diego. My father had grown up fly-fishing on the Snake River in Idaho and spent summers working on his uncle's Wyoming cattle ranch. During my childhood in California, my parents simply told me to "go play." Given this heady license, I discovered that I loved taking risks and facing challenges—a side of me that I have not yet outgrown. I especially loved to race all those years ago and still do.

One of my favorite destinations was the local airport, where I'd watch from the long grass near the runway as a plane's propellors disappeared into a blur. Then, as the DC-3 edged closer, I became giddy with anticipation. Once it had taxied into position for takeoff, my time had come. The race was about to begin. I crouched down close enough to count the gleaming rivets that studded its aluminum skin. As the two supercharged engines roared I leaped up and straddled my bicycle. My muscles tensed, and I leaned hard on the pedals. The plane began its eastward dash for the blue Californian skies. With my head down and my heart pounding I began to pedal the bike as hard as I could down the runway.

Close to the reclaimed land where the airport was built, my parents owned a $9,000 bungalow, in Bayside Manor. This was a lower-middle-class neighborhood of Millbrae, a city of fewer than eight thousand inhabitants constructed on land that had once belonged to the Mills family, some fifteen miles south of San Francisco. Highway 101 ran to the east and the railroad to the west, while north and south of me were pastures grazed by cows, a rural setting that was gradually erased as the airport grew—from San Francisco Municipal Airport to San Francisco International Airport in 1955—and then continued its relentless expansion. The roar of the turboprops flying over our little house would eventually give way to the scream of the jet.

San Francisco Airport was very different when I was a boy. There was no security, no cameras, no barbed wire fences. All that separated the main runway from the road was a drainage ditch and a small creek. With a few friends I would cycle down the incline, through the water, and up the other side. At first we would just sit in the grass and watch with amazement as the planes taxied out and then took off, surprised by how slowly the great birds began their passage down the runway. I don't remember who came up with the idea, but one day we decided that we could probably ride our bikes faster than the airplanes were traveling. So we waited until a plane started to taxi in preparation for its ascent, and then we jumped onto our bikes to race it for as long and as far as we could. We even pulled out in front of the planes for a brief but exhilarating moment before they accelerated and overtook us.

Today I often fly in and out of San Francisco International Airport, and whenever I am on that same east-west runway, I think back to my childhood. It is not hard to imagine the anxiety a pilot must have felt seeing a gang of little boys furiously pedaling beside his plane. Some of the passengers did stare out the windows, either waving or openmouthed or just plain aghast and appalled. Occasionally pilots shook their fists at us or even notified the control tower, which would dispatch the airport police. But because the runway was such a long way from the terminal, we could easily see them coming and escape across the creek. Then one day we rode to the airport and found that our racing days were over: A new fence had been erected around the runway.

Every day of my childhood was a day of play and exploration that had a bigger impact on my development than anything I was taught in school and, although I can't know for certain, at least as much as my DNA. I think that one reason I was able to become a successful scientist was that my natural curiosity was not driven out of me by the education system. I discovered even then that competition—competition as simple as a bunch of kids trying to outrun great, lumbering airplanes—can generate long-term benefits as much as short-term thrills. Today, whenever I see that fence around the runway, I feel proud of my contributions to airport security.

My DNA, My Life

My genetic autobiography can be found throughout my body. Each of my one hundred trillion cells (with the exception of my sperm and red blood cells) packages my DNA in forty-six chromosomes, a number typical of humans and of no particular significance, save that chimpanzees, gorillas, and orangutans have forty-eight. (It was once thought we did as well, until a diligent scientist bothered to count them properly in 1955.) Along my chromosomes are distributed around twenty-three thousand genes, a much lower number than we once thought. They do not seem to be organized in any particularly clever way, and genes of similar function are not necessarily grouped together.

The linear code in genes is a three-letter one, with a triplet of DNA "letters" coding for a particular amino acid that, when joined with a string of other amino acids, folds into a protein, one of the basic units that build and operate my cells. With a repertoire of only twenty different amino acid building blocks, my cells can create a bewildering array of combinations to produce proteins as different as keratin in my hair and hemoglobin, the red pigment that transports in my blood. Proteins can carry signals, such as insulin, or receive them, such as our visual pigments, neurotransmitter receptors, and taste and smell receptors, all of which are structurally similar. There is no chromosome that codes for the heart or the brain; each cell has the entire set of genetic information to build any organ, but they do not. We are only at the earliest stages of understanding how potent cells in the embryo, stem cells, end up playing various combinations of genes to form about two hundred types of highly specialized cells found in the body, such as nerve and muscle cells that in turn form the organs such as the brain and heart. But overall we do know that the order of the letters in the DNA provides a recipe (translated into action in cells through the medium of a more ancient genetic molecule, RNA—ribonucleic acid) to make, in my case, one Craig Venter.

My particular edition of the human genome was written in January 1946 in the married student housing of the University of Utah, Salt Lake City, where my parents, John and Elizabeth Venter, lived with my older brother in spartan university housing that had once been occupied by the U.S. military. My mother and father were both familiar with barracks life, however, having served on different shores of the Pacific as part of the Marine Corps during World War II. They first met at Camp Pendleton in California but ended up in Salt Lake City to be near my father's parents. While my paternal grandmother was a devout Mormon, my grandfather was anything but.

One visitor recalled being invited into the garage by my grandfather to meet "Malcolm," an old friend. Good ol' Malcolm turned out to be a bottle of scotch. My grandfather had never agreed to a temple wedding, so my grandmother held the ceremony after he died, with her brother filling in for him. Following in my grandfather's footsteps, my father ended up being excommunicated by the church. I had always thought it was for drinking coffee and smoking, though it may have been because he objected to the Mormon tithing system (giving one tenth of one's income to "God's work"). Excommunication, in any case, did not seem to bother him, other than the effect it might have on his mother, because he was not particularly religious. But his antipathy hardened at my grandmother's funeral when he had words with the church people because of the way they, and not the family, had dominated the proceedings.

I was born innocent enough on October 14, 1946. At the time my father was attending school in Salt Lake City, thanks to the GI Bill, studying for his degree in accounting and barely making ends meet as he struggled just to look after my mother and my fourteen-month-old brother, Gary. I imagine that he must have perceived me as an added burden, one that would make his life more difficult. My mother says that of his children I am the most like my father. But we were never very close.

My mother sold real estate for a while, though my dad resented her working, which was considered a sign of lower status in those days. It did not help that she was too honest: She couldn't sell a house she didn't like. Art was her creative outlet, expressed through endless paintings of the ocean, but for most of my childhood she was a homemaker. Among my earliest memories are of her clipping coupons from newspapers and going from one supermarket to the next to get the best deals. We would use powdered milk to save money, even though many dairy farms were nearby. Summer holidays meant camping or staying with my maternal grandparents in San Diego, a city that would come to play a central role in my life. I can also recall only too well how my folks constantly rewarded my better-behaved older brother, a mathematical prodigy. I, on the other hand, was threatened with visits to the juvenile detention home to scare me into better behavior.

The Y and the Wherefore

One major feature of my genome would have been obvious to my mother, the delivery team, and in fact anyone else from the very first moment I entered the world, let alone a Nobel Prize–winning geneticist, genome guru, or sequencing czar. Rather than having a pair of each and every chromosome (as women do), I have one X chromosome (women have two) and a lonely Y chromosome. The X and Y are called sex chromosomes to distinguish them from the rest, which are called autosomes.

My Y chromosome, like that of every man, is responsible for the characteristics that make men male, particularly a gene called SRY (the Sex-determining Region of the Y). Although it is just fourteen thousand base pairs long, the power of SRY was first demonstrated when a British team transplanted one to turn a female mouse into a male, nicknamed Randy. This male genetic endowment is, relatively speaking, unimpressive. The recipe for the human body consists of at least twenty-three thousand genes, but only one thousandth of these—a mere twenty-five genes—lie on the Y. Nonetheless, this little group of genes has proved highly influential. For people cursed with a Y chromosome, life is hard from the very start and only gets tougher. Look at the oldest residents of this planet and you will see they usually lack a Y chromosome. From fertilization to death, those who bear this chromosome are in relative decline compared with those blessed with two Xs. The Y confers many peculiarities, from a greater risk of committing suicide, developing cancer, and becoming rich, to having less hair on the top of the head.

Even by the age of two, the one trait that has perhaps most characterized my success had become apparent: taking risks. I don't remember anything about the incident, but I am told that I almost drowned as a result of an encounter with a high diving board. In later life one of my mentors, Bruce Cameron, joked that I liked to do high dives into empty pools. ("He tries to time it so the pool is filled by the time he hits the bottom.")[1] After my very first leap into unknown waters, my parents insisted that I learn to swim at the local YMCA. I am glad they did, for swimming would build my confidence in years to come and, one way or another, would help me survive in Vietnam.

As I grew up in Millbrae, more evidence emerged that risk-taking was in my blood. I had many early adventures on the railway tracks, which seemed to dominate our lives. The clatter of trains was in the background all day and all night, while the trains also took my father to work in—what seemed to me—the big, distant city of San Francisco and brought him home each evening. We lived on the "wrong side" of the tracks. The ritual of food shopping involved my older brother, Gary, and me following our mother and pulling our little red wagon, a Radio Flyer, over the tracks to haul the groceries back home.

The tracks were another forbidden playground. I would hang out with Gary and our friends in a large drainage pipe that ran beneath them. From the perspective of a small boy, the steam engines that passed by almost every hour of every day were unbelievably powerful, fascinating, and thrilling machines. As a train slowed in its approach to the station in Millbrae, the engineers would blow steam out the side at us. We were enveloped and enthralled as one of the giant mechanical objects of our affection rumbled past.

We often played on the tracks, as well. One game involved putting your ear to the rails to see who could remain listening the longest after the first tinny vibrations signaled that a train was approaching. We would put pennies on the track so the train would flatten them. (I tried this on a recent visit to Millbrae and was disappointed to find that modern pennies, not being copper, are too hard to flatten; they merely leave a metallic ghost on the track.) As I got older, around age seven, our adventures became more daring, and we began to jump on the freight cars as they passed by. This is not as easy as it sounds for a small boy, even with a slow-moving train, and often a conductor in the caboose would yell at us to chase us off.

By then my father had been made a partner in John F. Forbes and Co. Although he was doing well, he had to pay for the privilege. His regular payments to the company limited our family income, and his heavy workload meant he often fell asleep in front of the television at the end of the day while watching boxing matches, or *The Dinah Shore Chevy Show* with her famous farewell kisses to the audience. But by 1953 we had saved enough to buy a car from my grandfather, a sturdy four-door machine, a 1949 Studebaker. We also needed somewhere bigger to live. I now had a sister and another brother, Susie (Suzanne Patrice Venter) and Keith (Keith Henry Venter). We moved to a house on the other side of the train tracks, in the hills of Millbrae. Our new home was much farther from the railway, but we still felt the influence of the steel tracks. Each morning my mother would drive my father to the train, and each evening she would pick him up again. These little journeys became defining events in an average day of the Venter family as my mother briefed my father on the mischief and other troubles that I had caused. Today my brother Gary argues that we were not that different in terms of behavior; it is just that I wanted to be caught. "Maybe you were after a bad-boy image even then!"

Our new home and neighborhood did not change me in that respect. My mischief making was now inspired by a friend whose father was a railroad worker and whose family lived in a railcar on a side track near the station. I thought he had the coolest life and visited him as often as possible. With his inside knowledge he showed me useful tricks, such as how to release the brakes to stop a train and how to couple and uncouple railcars. At the end of the school year his family had to move to another station, to the deep relief of my parents. I watched as his railcar clattered off, taking him out of my life.

My risk taking accordingly shifted away from the tracks to the Old Bayshore Highway, which followed the edge of the bay and into the developing San Francisco Airport. In the 1950s this area was pristine, just open fields and no buildings. My friends and I would often ride our bikes across the overpass to the old Bayshore road. Years later, in high school, this was one of the places we would race cars and even eventually play chicken in the way described by the book *Hot Rod,* with its romantic view of young speedsters.

While my upbringing showed that I loved my freedom and could be reckless, there was more to me than that. If I had to state another key trait that was obvious early on, it would be my insatiable urge to build things, from crystal radio sets to forts. I was always happiest when I was lashing something together, relying on my imagination to fill out what my crude materials and tools could not provide. While my peers were getting the creativity bashed out of them at school, I would do anything I could to lay my hands on material to build things from scratch.

Many of my early efforts—often aided by my little brother, Keith (who is now paid to continue in the same vein as an architect with NASA)—took place in our backyard, which we called the "back-back" yard because it lay beyond our gardens and a three-foot-high fence. At one end were a compost heap, a swing set, and a crab apple tree. At the other was my domain: an apricot tree, blackberry bushes, and the best feature of all, plenty of dirt where I could build.

My construction efforts were modest in the beginning, mostly small but elaborate tunnels and forts. Each month I also managed to save a dollar, enough to buy a plastic model of a ship or warplane. We discovered, thanks to boredom, lighter fluid, and matches, that we could have more realistic battles by setting the models ablaze. Toy soldiers burned well, too, and dripped hot plastic in a satisfyingly dramatic way. As my tunnels progressed in size and scale, my pyromania advanced to setting off firecrackers in them.

My rapidly expanding underground tunnel fort was hidden with plywood, two-by-fours, and dirt. But after several weeks it was discovered by my father who, fearing a lethal collapse, ordered me to fill it in. My construction efforts now moved aboveground. With vast amounts of scrap lumber obtained from construction sites, I began to build a combined fort and clubhouse. We would spend hours pulling and then straightening old scavenged nails so we could use them again. The result was a two-story structure that would have impressed even Spanky, that take-charge kid from 1930s vintage movies. We had an endless supply of apricots, berries, and crab apples—handy ammunition for fights. Eventually, a neighbor complained that the fort was an eyesore, and once again my creation was torn down.

All this building took place between the ages of seven and ten. Afterward, my activities moved out of the yard and into the street, where I managed to combine my two first loves—construction and risk taking—in a series of vehicles, from wagons and soapbox racers to crude coasters, all of which took advantage of how our house sat on the crest of a hill. We even made an early version of a skateboard by screwing the two halves of a skate at either end of a two-by-four. Keith and I would race it down the hill until a moratorium was declared on skateboarding when my brother Gary tried it and broke his arm.

Inspired by *Popular Mechanics,* my ambitions later moved up a gear. The magazine contained plans to build an eight-foot hydroplane from marine plywood, without the need for a complex underlying frame. Living in the San Francisco Bay area we were surrounded by water, but as a family of terrestrial interests—tennis and golf—we took little advantage of it. In these drawings I saw my first chance to enjoy the waves. From money earned by mowing lawns and delivering newspapers I had enough to buy the necessary materials. My friend Tom Kay, who had a talent for art, showed me how to translate the plans into wood, and using basic tools, it took me months to build the boat. My father was certain I would never succeed, and for good reason: I had no outboard engine. Eventually, a mechanic friend of his offered to sell me a broken-down, late 1940s outboard engine for $14. I was forced to learn how to strip the engine and put it together again to get it working once more. I used a fifty-gallon oil drum as a test tank and was so pleased when I succeeded in rebuilding this antique that I let my refurbished engine chug away in the backyard for hours.

Once the engine was installed, I painted my boat jet black with orange lightning bolts on its bow. I was surprised and pleased that my father was proud to help me with the ultimate test. The hydroplane was hoisted onto the roof of his new 1957 Mercury station wagon, and the family drove to the harbor in Coyote Point, south of the airport. There was no ramp into the bay or bottle of champagne to break on her bow. We had to wade through the mud to launch her. After flooding the engine, a result of nerves, I got it going and did several high-speed laps around the harbor, blinded by the spray from the bow. The day ended well, but I remember being dissatisfied with the temperamental motor. I dreamed of doing better.

For a seventh-grade science project I discovered that I had an interest in doing something practical with my knowledge. After a trip to Candlestick Park with my father to watch the San Francisco Giants, I was inspired to build an electronic scoreboard for the junior high school baseball field. It was a satisfying feeling when it displayed the scores for the first time and was the most obvious of a number of signals to that point that I would end up doing something practical for a living.

In September 1960 I started attending Mills High School but I don't remember much about this milestone because school was neither my interest nor my strength. I still have problems with spelling because of my early hatred of rote learning and tests. I felt that I had been going downhill ever since I peaked in kindergarten, and in high school my only A's were in PE, swimming, and woodshop. My woodshop instructor was unimpressed when I said that instead of making furniture I wanted to build a hydroplane, based on a new design that had set a 60 mph speed record. I sent off for the plans and found that it was a complicated affair with a skin of mahogany marine plywood that had to be molded over a frame of steam-bent members. The $300 plus cost for the materials was well beyond my means, but my project was saved by Ted Myers, who was privileged enough to drive a shiny Thunderbird to school each day. Ted was a fellow swimmer, and while very good academically, he had little interest in working with his hands, much to the annoyance of his father, who owned a construction company and wanted a more macho son. Ted's father offered to buy enough material to build two boats if I would get Ted's hands dirty. I agreed, as did Ted.

While Ted did occasionally come to help and watch me work, only one boat was ever completed. In the process of building the hydroplane, I annoyed my own father by taking over his garage. (I eventually found a way to make room for his car by using a complex pulley system to raise my boat to the ceiling each night.) I also made a discovery that has largely held true ever since: The sweetest moment of any project is not when it is finished but when the end is in sight. I loved this particular project best when I had built its frame—when my imagination could fill in all the details of the completed boat. Today, after having been responsible for building new houses, laboratories, and other projects, I still find that the final version often falls short of the one glimpsed in my head during construction.

Education is, of course, among the biggest influences of nurture, along with friends and family. But December 1961 saw events unfold on the world stage that would also come to have a huge bearing on my life: President John F. Kennedy committed the United States to suppressing a Communist-led rebellion and to preserving South Vietnam as a separate state governed by an American-sponsored regime in Saigon. For a teenager in Millbrae, though, the war still seemed far away, especially one whose role models were not lantern-jawed stars of war movies but cinematic antiheroes such as James Dean and Marlon Brando. My friends and I aspired to look like hoods. With my blond hair it was hard to look especially greasy, but I did manage a ducktail, combed to a point in front, a haircut that was inspired by a Bay area gang member who liked to ride his motorcycle to middle school.

Rebellious and disobedient, I was constantly in trouble during my freshman year and more often than not grounded or in detention. My mother would sometimes even check my arms for needle marks. Despite my freedom to wander as a child, I was not even allowed to go to high school football games at night without being chaperoned. On one occasion I snuck away from my family to hang out with my friends under the bleachers, where they drank beer and smoked. The gang left the stadium for downtown San Mateo, and I found myself in the backseat of a stolen car that was being hot-wired. We set out on a joy ride. When a police car began to follow with flashing lights and siren, I panicked. We swerved into an alley close to the stadium and braked hard. I got out and ran as fast as I could back to the bleachers, my family, and normal life. However adventurous I might be, I was not a criminal. I began my second year of high school with a new outlook. I didn't know what I wanted or where I was going, but I was not going to be a teen hoodlum. A big chance to change my direction came during the summer of 1962 when I met a girl with a slight build and light blond hair who was a violinist in the school orchestra and had also toured abroad. Linda began to introduce me to a world far beyond the narrow confines of California, swimming, and boat building. We drank coffee in the local bookstore and discussed literature. We listened to classical music and went to the College of San Mateo to hear Bob Dylan. When the summer ended, so did the romance, but by then she had set my life in a new direction.

Blame It on My Genes

Attention deficit/hyperactivity disorder (ADHD) is marked by inattention, excessive motor activity, impulsivity, and distractibility—an accurate description of the teenage Venter. Recent research has linked ADHD with a genetic stutter that involves ten repeats of a section of a gene called DAT1, the dopamine transporter gene. This gene is responsible for the reuptake of the messenger chemical dopamine in the brain and is also the target of amphetamines and cocaine. Perhaps it is no accident that variants of this gene also influence how children respond to methylphenidate (Ritalin), a stimulant that is used to treat ADHD. My genome reveals that I do indeed have these ten repeats. Hence, my behavior—if, that is, you really believe a simple genetic stutter can trigger such a complex trait. Not everyone agrees.

Another turning point came with an English literature class taught by Gordon Lish, a twenty-eight-year-old with thick blond hair who was a follower of the Beat Scene. His first assignment was, appropriately enough, J.D. Salinger's *The Catcher in the Rye*, whose hero I could easily identify with. Until then I had been uninvolved, bored, and surly in most of my classes. I would tune out, staring out the window or over the teacher's head—but only if the teacher was lucky. Just as often I would talk to my friends, disrupting the class and driving the teacher to distraction. Today, of course, a kid who behaved as I did would be put on Ritalin. But Lish had an inspiring intellectual energy that could engage a young mind so much more powerfully than drugs. He also seemed to have a real interest in me, and we spent time outside of class discussing literature and life. I was attempting to absorb ideas for the first time, reading, learning, and thriving at school at long last.

English literature class with Lish was the first of the day, and like every other student in the country, we had to recite the Pledge of Allegiance. We would give Lish grief about this, because it seemed to contradict the liberal ideas we had been discussing in class. Lish acknowledged our resistance by sometimes rushing through the pledge. Then one day the principal announced that Lish had been fired for being un-American.[2] We were stunned, and some of the girls even cried. I was outraged that the one teacher with whom I had connected was now gone.

I convinced some other students to join me in what had to be one of the first high school sit-ins. The protest snowballed until we shut the school down. We continued the demonstration the following day, and by then the local press picked up on the story. As ringleader I was summoned to present our demands to the principal. That was simple: Lish had to be reinstated. Given my far from stellar academic record, the principal asked if I was upset because I would lose my only A grade from Lish. The truth was that I was destined to fail even that class because I had not done the specific assignments Lish had asked for. The demonstration ended quietly the next day after the students were threatened with suspension.

For my loyalty to Lish I was suspended from school for a week. My parents grounded me for a month, longer than ever before. I missed Lish, and, not surprisingly, a few years later another English teacher would become my mentor. Lish himself later recalled how he had been forced from his teaching post "as a function of the usual kinds of distresses aroused by persons who create any kind of stir in the classroom."[3] Other crimes he admitted to at Mills High School included, fortunately for me, "letting pupils get so excited about ideas their voices can be heard in the hall."[4] Almost four decades later, Lish saw me on television and remarked to a friend, a Nobel laureate who also happened to be my biggest critic, "I think I know that guy. I did not forget—in Yiddish we say the *punim* (face)—that God gave Craig."

Today, on reflection, my early rebelliousness showed all the hallmarks of a second child who lived in the shadow of a successful older sibling. I suspect that while my parents rewarded Gary for being a superachiever, I found I could get their attention by being a super*under*achiever. Gary was, granted, a hard act to follow. But I tried. I chased him into cross-country running but decided that it was both painful and painfully uncool. In the spring I joined the swim team, only to be in Gary's wake once more. At first it was hard to adapt to the discipline of three hours of swim practice every afternoon, and I stank in these sessions.

But I really liked the actual races and surprised everyone, especially myself, by winning. Recognizing that I could now beat him at backstroke, Gary switched to butterfly. Winning was a new experience for me, and an addictive one. Despite my poor technique and ungainly flip turns, I still managed to be the fastest in the B division. Gary and I joined a club run by former Olympic-eligible swimmer Raymond F. Taft, and I improved my speed substantially over the summer. Gradually, swimming began to dominate my life. I was invited to join the varsity team and soon became the fastest in the 100-yard backstroke. The distance suited me perfectly: I could go all-out for a minute because I thrived on adrenaline. I would still often lose in practice but, thanks to a hormonal surge, never in an actual race.

During the last summer of my high school swimming career, I finished a 400-yard medley relay race in which the four of us, including Gary, set a new American record. I ended my senior year by winning the league championship, defeating a longtime rival, and setting new school, county, and league records. Gold medals and local newspaper coverage had a substantial impact on my self-confidence and self-worth. I was a hit with the girls, too.

But while my sporting achievements became increasingly impressive, my academic skills were lagging far behind. My grades were so dismal that they threatened to sabotage not only my eligibility for the swim team but my graduation. Fortunately, I wrote a glowing paper about the presidential bid of the extreme Republican Barry Goldwater, whose slogan was "In your heart, you know he's right." The teacher who marked my paper seemed to be a Goldwater conservative and was sufficiently impressed to give me a D minus instead of an F, which would have ended my chance of graduating.

Ray Taft felt that I had strong Olympic potential. He told me that I had more guts than anyone he had ever known, but the problem was that I was winning on them alone. He wanted me to unlearn all my bad habits in the pool and develop new techniques. But I liked winning too much to change, and by then, in any case, I did not want to have anything more to do with competitive swimming. Despite my grades I had also been offered a swimming scholarship to Arizona State University. But at the age of seventeen I turned my back on swimming, on school, and on Millbrae. Tired of being grounded and craving my freedom, I headed for southern California.

My Y Chromosome and Sex

At the age of sixteen, when I still had long blond hair, I was fully introduced to the genetic programming on my Y chromosome. I had a girlfriend named Kim, whom I had met earlier that year after she transferred to Mills High School. My Y came into its own when Kim had a Sweet Sixteen party while her parents were away. This moment was particularly sweet for me, too. She wore baby doll lingerie—see-through—and encouraged me to explore. Before then my only experience had been adolescent dreams, fantasies, and heavy petting with various girls at school, including Kim's best friend. The endorphin high of my first love affair came after the sex. All this was ultimately due to a chromosome consisting of around twenty-four million base pairs, the human Y, which has roughly twenty-five genes and gene families. As I mentioned before, one of them is SRY, the gene that drives the development of the testes.

We had more opportunities after Kim ended up moving with her family to a house in Burlingame, only ten minutes away. I constructed a rope ladder so I could escape my own house after bedtime, drive to Kim's, and climb through her first-floor bedroom window. This went on for weeks that summer of 1963 until one morning I returned and discovered that my rope ladder had vanished. Keith was playing a joke, I thought. But when I sneaked in, I was confronted by my father, who was sitting on the stairs in the hallway. If he caught me again, he warned, he would call Kim's father to tell him I was having sex with his daughter.

After a few weeks, my nocturnal escapades resumed. Once again I soon found my ladder removed and the door locked, and my father and I had a blazing argument. The next time I visited Kim, her father was waiting for me. He pulled out a gun and held it to my head. Six months later Kim's family moved out of Burlingame and my life. I never fully got over what I viewed as my father's betrayal, which I considered even worse than having a gun pointed at me. I can blame it all on the Y, which also plays a key role in the production of the androgen hormones linked to aggression.

In Newport Beach I could surf in 70-degree water rather than the 50-degree waves of northern California. In those days the beach had a small boardwalk and the atmosphere was like something out of the *Gidget* films. My pursuits in that period involved drink, girls, and bodysurfing at the Wedge, a spot with direct shore break waves that can be spine-crushing. I lived in a small house with four roommates. To make money I worked nights at a Sears, Roebuck warehouse, where I put price tags on toys. I also tried night clerk, airport fuel truck driver, and baggage handler (where I was motivated by my peers' points system, which rewarded crushed, dripping suitcases and other damage from tossing bags onto the airplanes).

Although I had a few dollars, my days were free, and there was endless surf to enjoy, I realized even then that I had to do more with my life than body surf the Wedge, hang out, and work at menial jobs. I enrolled in the Orange Coast Junior College in Costa Mesa, which was conveniently only a few minutes away from some beautiful beaches. But by now the developing situation in Southeast Asia had begun to catch up with me. I was too late to apply for a student deferment and received my draft notice. Like hundreds of thousands of other young men, I would be swept far from the sheltered environment of sixties suburban America.

I was very conflicted. I was personally against the war but had a long family history of military service. One ancestor was a fifer and medic during the Revolutionary War. My great-great-great-grandfather served in the cavalry during the War of 1812. My great-grandfather was a sharpshooter in the Confederate Army during the Civil War. My grandfather was a private in World War I, serving in France, where he was badly wounded and had to crawl for miles to safety. And, of course, both my parents had been Marines.

As a consequence, my father was disturbed that I had been drafted into the army. He persuaded me to talk to a Navy recruiter—probably the most useful piece of advice he ever gave me. Thanks to my swimming records, I received what I thought was a great offer: a three-year enlistment instead of the usual four, plus a slot on the Navy swim team and a chance to compete in the Pan American Games. Despite my qualms about the war, I looked forward to serving my country and swimming for it too. This did not seem a bad option, and it never dawned on me that I might end up in Vietnam, even though by the time I reached boot camp in San Diego, the war had escalated and the U.S. presence would rise to 185,000 troops by Christmas 1965. I began my military career with my long blond hair being shaved off. I found myself milling around behind barbed wire with tens of thousands of young men, from wannabe professionals to farm boys and escaped prisoners. I was at the start of the long, traditional process of having my spirit broken so I would become an efficient and pliant sailor. Misbehavior meant running around all day carrying a yoke saddled with sand buckets. (The guards would beat you if you slowed down or stopped.) I was miserable; boot camp was the closest thing imaginable to prison.

I even considered going AWOL with a fellow recruit who was equally despondent. Once again, the ocean would provide my escape. All we had to do was swim down a creek that ran through the base and out into the sea. I worked out each day with a two-mile swim, so I thought it would be easy. I didn't realize that when I discussed my plan with a friend, someone was standing on the other side of the wall. The night before we were to make our joint bid for freedom, the company commander had an important announcement to make: Two idiots were planning an escape, and he wanted to remind them that desertion during wartime was a capital offense.

My actual naval swimming career likewise ended before it had started. August 1964 saw the Gulf of Tonkin incident, involving two alleged attacks by North Vietnamese gunboats on American destroyers. (A report released in 2005 by the National Security Agency indicates that the second attack did not occur.) Lyndon Johnson ramped up the war effort and canceled all military sports teams, which meant I had to find a new career in the Navy.

An IQ test revealed a respectable score of 142, among the best of thousands of draftees and recruits. I was amazed, and it was high enough to give me the choice of any Navy career. Of all the interesting options, from nuclear engineering to electronics, only one did not require any enlistment extensions and extra time in the military. My career choice was, I thought, a no-brainer: I would go to hospital corps school. (With foresight that perhaps revealed my hidden interests, I had said in my seventh-grade yearbook that I was going to be a doctor.) No one explained to me at the time that the reason the military did not bother to extend the enlistment for corpsmen was that the casualty rate was high.

After corps school I was posted for medical training to nearby Balboa Navy Hospital, then the biggest military hospital in the world. I soon became the senior corpsman and was allowed to live off base. My grandparents had a shack in the back of their old house near Ocean Beach, which they let me live in. I commuted to the hospital on my 305cc Honda Dream (the largest motorcycle Honda made at the time). The corpsman I bought it from had seen so many scrambled brains during neurosurgery that he was desperate to get rid of it.

Work gave me a remarkable overview of human disease. I discovered that I had a talent for performing spinal taps on meningitis patients, taking liver biopsies from hepatitis patients, and so on. After a short while I was placed in charge of the large infectious disease ward, where I ran a team of more than twenty corpsmen on three shifts around the clock tending hundreds of patients with diseases ranging from malaria to tuberculosis to cholera. Decades later I would decode the genomes of the infectious agents that caused many of these illnesses.

The hospital became my refuge from the rule book, military discipline, and the 7:00 a.m. daily dress inspection. I rarely wore my uniform, preferring jeans or surgical scrubs. Each afternoon, once my shift ended at 3:00 p.m., I revved up my bike to go surfing. I grew my hair as long as I could get away with, since the girls at the beach stayed as far away as possible from Navy boys.

Compared with high school, it was always a struggle during this period of my life to find female companionship. I was surrounded by Navy nurses, but as an enlisted corpsman I was technically prohibited from dating them since the nurses were officers. That did not stop me, of course. First it was the head nurse. Then I became more interested in her friend—so interested, in fact, that I began to date her instead. This would turn out to be a big mistake.

Each month corpsmen faced a potential second draft from the Navy into the Marines, who used Navy medics. My hospital training was slated to be only six months at most before I was packed off to Vietnam. Most corpsmen served as medics in combat, where they did not last long. The Vietcong would pay a bonus to any soldier who could show that he had killed a corpsman, usually by bringing back a trophy such as his ID card. After six weeks in the field, a corpsman had only a fifty-fifty chance of survival.

But because I was highly valued by the doctors in San Diego, each month when the draft list came up, my name was absent—sometimes removed at the last minute. I ultimately avoided the draft for fourteen months before my name was finally posted. But the posting included a footnote: I was to be sent to the naval station in Long Beach, where I would run the emergency room. I was stunned, relieved, and delighted. The head doctor had a big grin and was pleased with his last-minute save.

The head nurse whose friend I had been dating was annoyed when she found out that, once again, I had managed to avoid Vietnam. As I was leaving, she told me to get a haircut. I had two weeks of surfing ahead of me before reporting to Long Beach, so a crew cut was the last thing I wanted. I answered with something flippant. Before I had even gotten to my motorcycle, two MPs arrested me and told me I was to be held for a court-martial. I was quickly found guilty of disobeying a direct order—my long blond hair was damning enough—and sentenced to three months in the brig at Long Beach. I faced hard labor, a criminal record, and a certain posting to Vietnam or a dishonorable discharge from the Navy.

The MPs took my orders and records, which were contained in a thick manila envelope that had the original copy of my orders taped to the outside, and returned a while later to send me on my way with my modified orders in bright red type over the original ones. One of the peculiarities of naval bureaucracy was that I still had my two weeks of leave before going to the brig. I returned to my grandparents' home feeling lonely and upset. I sulked and worried about telling them, let alone my ex–Marine Corps parents, about my court martial. My promising start in medicine had been derailed.

I stared at the manila envelope for what felt like days, brooding on what it contained. This was 1966. Computerized records did not exist, so when military personnel moved to new assignments, they carried all their records with them. I thought about my high-speed court-martial and began to wonder if the copies of the orders inside the envelope were the same as the original on the outside. Did they change all my orders or only the original? I decided to tell my uncle Dave about my predicament and ask whether I should risk tampering with the envelope. Though concerned, he was also amused and intrigued, and brought my grandmother in on the problem. On examining the envelope she ordered Dave to boil some water on the stove. I could not believe it when she held the envelope over the steam. After a few minutes it was open, and she handed it back to me.

When I pulled out my records, I found untouched copies of my original orders to report to the medical station in Long Beach. After making certain all the paperwork was in order, my grandmother helped me reseal the envelope. All I had to do now was "lose" the revised order attached to the envelope and come up with a plausible reason that it was missing. Since I would be driving to Long Beach by motorcycle, I had a perfect excuse: I had dropped the envelope while flying down the highway, and the paperwork became detached. My uncle thought it was a good plan. For extra realism we went out in the street and began throwing the envelope around and sliding it along the pavement. We stripped off the remains of the original order so that all traces of red were missing.

There was always a chance that a second copy of the orders to put me in the brig had been sent on ahead of me. But I reasoned that since I was being incarcerated anyway, it seemed like a straight choice between a chance of freedom and an extended time in the brig. By then I had faith in the incompetence of military bureaucracy. Nonetheless, dark thoughts consumed me during the two-hour motorcycle ride to Long Beach. At the base gate I was directed to the check-in office, which I anxiously approached.

I handed my tattered envelope to the imposing figure behind the desk. The chief tore it open, examined my orders, and scowled. "Son, you are in deep shit." My heart sank, and I began to panic. I told my sad motorcycle story, but he was having none of it and just repeated, "Son, you are in deep shit." He briefly disappeared, then returned to chew me out and tell me that I was to be dealt a severe punishment: I had to go to the transit barracks, where I would be confined for one week on cleaning detail. Only when I had learned my lesson about negligence of Navy property would I proceed to the emergency room and resume my duties as head corpsman. I had indeed learned an important lesson: It really pays to take risks and to take control of your own life.

While I was running the emergency room in Long Beach, my love life picked up once again. Through a friend I met Kathy, an art student living in Pasadena, and we were soon spending most nights together. Despite my good fortune in life and love, it soon became clear that I could stay in Long Beach for only a few months, because Vietnam was now unavoidable. I felt a knot in my stomach as I began to psych myself up for what was to come. However, my future—or at least my chances of living—would be changed by a chance encounter with a young officer who had just returned from Vietnam, where he had done research. He told me that if I could get sent to the Navy hospital in Da Nang, the odds of my surviving the year would increase dramatically. But there was a big problem: Only a tiny fraction of corpsmen in my position got that posting. How was I going to finesse it? The head of the medical center suggested that I write to the U.S. Navy Surgeon General and volunteer to go to the hospital before I received my official assignment.

Everyone thought this attempt to influence the slowly grinding wheels of naval bureaucracy was a long shot. But once again I felt I had nothing to lose. My letter described my extensive experience in the infectious disease ward in

San Diego and my emergency room experience. As a final flourish, I argued that my medical skills would be put to much better use in Da Nang than if I stayed in Long Beach. As the weeks went by I became more and more pessimistic. Then, after a month, I was given my orders: I was to report to the Navy hospital in Da Nang. Seizing the initiative in this situation probably saved my life.

I had thirty days' leave before being sent to the war zone, and they turned out to be one of the best times of my life. Kathy had a little British sports car, a Triumph TR4, that we drove up the coast from L.A. to San Francisco. We stopped at my parents' house in Millbrae, and I spent time waterskiing with my brother Gary and my school friends behind my hydroplane on Lake Shasta. Kathy and I went to San Francisco and hung out in the Haight-Ashbury district, where I worked at the free medical clinic for a week. The streets around the intersection of Haight and Ashbury were at the heart of the hippie movement, where life was just one big hazy marijuana party and where antiwar sentiment and flower power were at their peak. Everyone I met told me to go to Canada to avoid being sent to Vietnam, but I felt somehow that it was wrong. Perhaps it was because I did not want to throw away my one chance of a career in medicine; perhaps it was the uncertainty about the war; perhaps it had something to do with the military background of my family.

I read what I could about the Vietnam conflict but found no middle ground. The government line—that the war was the only way to stop Communism—was bought by the older generation, including my strict Marine Corps parents. I wanted to believe that argument, but I found the doubters' position more plausible. In some strange way, however, I also believed that the war could transform me personally. I had met servicemen who fought in Vietnam, and it was clear they were different in some indefinable way from those who had not. I wanted to experience the adventure and I thought Vietnam could offer answers to some fundamental questions about life. Even before I arrived in the theater of war, I encountered the understanding that exists between all servicemen, the brotherhood of arms. I was driving with Kathy in the TR4 along Highway 1, the coastal route between San Francisco and Washington State, taking the tight curves on the road at high speed, when a motorcycle policeman began to give chase. When he eventually caught up, I explained that I was on my final leave before shipping out to Vietnam. As it turned out, he was a veteran whose life had been saved by a corpsman, and for a moment, we formed a bond. He told me to slow down and not get myself killed before I even got to Vietnam.

Before shipping out I was sent for counterinsurgency training. My first two weeks of the month-long course were spent in Little Creek Amphibious Base in Virginia Beach. There we were fed an extreme version of the government position by some hapless officer. As he dutifully presented the official line, I pestered him with uncomfortable questions. Although I resisted the political indoctrination and had initially refused weapons training, I did discover that I was a good shot and, inevitably, came to enjoy target practice. Perhaps this was just as well. I had been told that in Vietnam most corpsmen ended up being more heavily armed than their buddies.

For our last week, we were divided into teams and dropped off in a swamp with no food. We were to rely on our survival training as we were hunted by troops with live ammunition. If caught, we would be put in a POW camp for the remainder of the week. One of my team, a southern guy, grew up eating collard greens, which constituted the main part of our diet along with wild strawberries and tea made from roots. Others ate frogs and swamp fish. We got by somehow.

My team was the only one to avert capture. When we finally turned ourselves in, we were threatened, intimidated, and marched into the POW camp. By then the other teams had been subjected to various humiliations. We found them squatting in the mud and dirt in their underwear. I was brought up in an area of California populated by few minorities and felt a long way from the civil rights movement, Mississippi freedom, and the rise of Black Power. But I was appalled when all my prejudices about white southerners appeared to be confirmed. Blacks were singled out for particularly harsh treatment. One was hit with a rifle butt. I stepped up to treat his head wound, which was bleeding badly. When I was ignored, I demanded the right to attend to him. I, too, was hit with a rifle and forced into a small cell with my "patient." On returning to the main camp, I filed a protest.

The camp commander summoned me on the day before graduation. He went to great lengths to explain that blacks needed to be subjected to this brutal treatment to prepare them for what would happen in Vietnam if they were captured; they could be reeducated, tortured, or killed. Then I was offered a "choice" that was typical of military justice of the day: withdraw my complaint and graduate the very next day or remain, with a guarantee of "special treatment" in the POW camp, until I agreed to his demands. I was forced to sell out and drop my protest. (A few years later, after a congressional investigation into alleged abuses to soldiers, the camp was closed.) Now there was nothing left to keep me from the country where only a decade earlier Ho Chi Minh had told the French, "You can kill ten of my men for every one I kill of yours, yet even at those odds, you will lose and I will win." Vietnam would teach me more than I ever wanted to know about the fragility of life.

2. UNIVERSITY OF DEATH

I am tired and sick of war. Its glory is all moonshine. . . . War is hell.

—General William T. Sherman

Each [organic being] at some period of its life, during some season of the year, during each generation or at intervals, has to struggle for life, and to suffer great destruction. When we reflect on this struggle, we may console ourselves with the full belief that the war of nature is not incessant, that no fear is felt, that death is generally prompt, and that the vigorous, the healthy, and the happy survive and multiply.

—Charles Darwin, *The Origin of Species*

I wanted to escape. I was determined to get away from the living, the dead, and the dying: The ones who wanted to live but couldn't. The amputees who would survive but wanted to die. Those who were so badly injured that they hardly knew if they were alive or dead. The stream of bodies that were being medevaced from the jungles, the bomb-scarred rice paddies, and the ruins of thatch-and-mud-wattle hamlets. I thought about the "Dear John" letter from my girlfriend, Kathy, who could not bear to hear any more about the grisly things that I had seen and what I had felt. Now all I could sense were the warm waters of the sea off China Beach. After five months, I was determined to swim away from the bullshit, the madness, and the horror.

My plan was to carry on until I was exhausted and then sink into dark waters and oblivion. This would not be easy, because I was a strong swimmer and in great shape. More than a mile from the beach, as I saw venomous sea snakes surface nearby to breathe, I began to have doubts about what I was doing. But I still swam on through the emerald green—until, that is, I made contact with reality when a shark began testing me, prodding me in a "bump and bite" attack. I kept swimming but now more slowly and less determined. I began treading water and looked around. The air was hazy, and I could no longer see the shore. For a moment I was angry that the shark had disrupted my plan. Then I became consumed with fear. *What the fuck am I doing?* All thoughts of dying departed. I wanted to live, more than I had ever done in the previous twenty-one years of my life.

I turned around and swam for the shore in a state of panic, driven on by sheer adrenaline. I was more afraid than I had ever been before, not of the shark or of the venomous snakes but that I might not make it back safely because of my stupidity in wanting to die, in wanting to take the easy way out of the brutality and loneliness of Vietnam.

Getting back to shore seemed to take forever; I could not believe that I had swum out so far, and wondered if I was even heading in the right direction. All I could think of was how much I wanted to live and how foolish I had been. Then I came upon the breakers and knew I had a chance of making it if I could bodysurf to the beach. I caught a wave and rode it as long as I could, and then caught a second and a third. Then came a final wave. My feet could now touch the sandy bottom. I swam on for a yard or two and then ran through the water until I collapsed.

Endurance

My ability to swim great distances rests in part on the fact that I do *not* have a mutation in one of my genes, the one responsible for the enzyme adenosine monophosphate deaminase 1, or AMPD1, which plays a key role in muscle metabolism. A mutation here—one of the most common—causes a deficiency in the enzyme and, as a consequence, can lead to aches, cramps, and early fatigue. All it takes is for a single letter change, from C to T, to stop the enzyme from being manufactured, sapping endurance as a result. Fortunately for me I am a C/C and not a T/T.

I lay on the sand, naked, for what felt like hours. I was exhausted and relieved. I was grateful to be alive and to have escaped the fatal consequences of my flawed view of life. There was now no doubt in my mind that I wanted to live. I wanted my life to mean something; I wanted to make a difference. I felt pure; I felt energized. I felt I had the strength to carry on.

I walked up the beach to the dirt trail that would take me past the Army Special Forces' camp. There, on the edge of a sand dune ridge, stood a series of bamboo cages, each about three feet high and a square yard or two in area. In every one a man was squatting—clearly Vietnamese, presumably Vietcong. Their situation was hard for me to identify with, yet I knew that my predicament would seem like heaven to any of the figures imprisoned before me. I felt even angrier and more disappointed with myself at my willingness only moments earlier to take my own life.

I passed the Marine airbase and walked on to the main highway, "Highway 1," a two-lane paved road that ran from Da Nang to Monkey Mountain and past the Navy hospital. I passed through the rear gate, up some wooden steps, and then opened the door of a hut that was in permanent twilight, a space filled with shadow and darkness. These were my sleeping quarters, the only place where I could escape Vietnam and stay alive. This was now home. I had come a long way from being a freedom-crazy kid in the San Francisco Bay area, a southern California surfer, and an antiwar protester to my latest incarnation: Navy corpsman.

My first real glimpse of life in Vietnam had come on the packed charter plane that transported me there. Its lights were blacked out, and when it began its descent in a steep dive, I could see the flashes from the gunfire aimed at us as we approached the runway. From the moment I landed in Vietnam on August 25, 1967, life was traumatic. I found myself in the port of Da Nang, bordered by Quang Nam Province and the South China Sea, in a job that was *M*A*S*H* without the jokes and pretty women. Alongside us in the battle against Communism was the Army of the Republic of Vietnam (ARVN), whom we knew colloquially as the "Arvin." Two years earlier, millions of American viewers had been shocked by a *CBS Evening News* report about the first Vietnamese peasant homes to be torched by U.S. troops in hamlets near Da Nang. By the time I arrived at the Da Nang naval hospital, such "Zippo jobs" were notorious only for being routine.

Like almost everyone else, I was assigned the ubiquitous cheap housing found at military airstrips, barracks, and hospitals: the Quonset hut. This row of semicircular steel ribs covered with corrugated sheet metal was named after its town of manufacture, Quonset, Rhode Island. My bunk and locker were situated right inside the door so they would be easy to find—the interior was always dark because the hospital ran two twelve-hour shifts: seven in the morning to seven at night and the opposite.

I preferred the night shift. Working during the hours of darkness gave me the chance to go to the beach nearly every day to run, swim, or surf. That schedule had other tangible benefits, since after dark there was usually a rocket attack on the Marine airbase across the street, with machine gun emplacements around us firing almost every night. The later shift also meant that I could avoid the rats, which hunted when it was dark. Your only real escape from military life—dreaming—would often be disturbed by traps snapping or flipping around, even by rats scurrying over your face. It is a hell of a way to wake up.

Night Genes

Everyone knows instinctively if they are owls or larks—"evening" or "morning" people. I have always burned the midnight oil and, as a consequence (or perhaps it is the cause), find it hard to get up in the morning. The answer must lie in my "body clock." In fact, there is no single chronometer in the human body but a superclock: Timepieces probably reside in every cell that consist of interlinked cycles of waxing and waning proteins.

Working together, the protein cogs made by these clock genes create so-called circadian rhythms, which help control the timing of a variety of biological changes, including hormone production, blood pressure, and the slowdown of metabolism during sleep. What is it about this mechanism that makes me hate getting out of bed?

One study has linked a mutation in the gene Per2 with "advanced sleep phase syndrome" (wanting to go to bed early in the evening and wake in the small hours). Here my genome is consistent with my lifestyle: I lack this particular mutation.

More promising, a correlation between a difference in the length of another clock gene—Per3—and being a night owl, has been uncovered by a team at the University of Surrey, at St. Thomas's Hospital in London, and in the Netherlands. The longer variant of this gene has been linked with being a lark (though that correlation is still debated). The shorter variant is significantly more common in people with an evening preference, with its most extreme version known as "delayed sleep phase syndrome" (DSPS). But because my version of Per3 suggests that I am unlikely to be an owl, more work needs to be done to understand the idiosyncrasies of my body clock.

Although the hospital had been set up as a buffer between the base and the Vietcong, rockets often fell short, detonating somewhere in the hospital compound. As if to guide the aim of our attackers, red crosses were painted on the Quonsets. (We presumably believed that the Vietcong had accepted the terms of the Geneva Convention.) One night while I was working a rocket exploded directly in front of my hut, close to my bunk. The wall was full of holes from shrapnel, and a large piece was embedded in the mattress on which I had been sleeping only hours earlier. The air raid siren was near my hut and, when it gave the warning, often evoked more fear than the rockets themselves. I have since learned this was a classical case of Pavlovian conditioning, and its effect was more powerful when people were in a deep sleep.

Genes and Addictions

I hated Vietnam, but it did not drive me to addiction. One reason could be linked to dopamine, a messenger chemical that is active in pathways related to the brain's reward and pleasure centers. A gene that codes for a class of protein where this chemical acts, the dopamine receptor 4 gene (DRD4), contains a stretch consisting of forty-eight base pairs that is repeated between two and ten times. There have been claims—though the evidence is not consistent—that the longer forms of this gene are linked with schizophrenia, mood disorders, and alcoholism. Some variations of the (D2) DRD2 gene, which codes for another dopamine receptor, are also linked to substance abuse. Like that of other sensation seekers who resort to drink and drugs, their genetic makeup means they get more out of using, a direct way to activate the brain's pleasure centers.

I do enjoy a drink even though there is a history of alcohol abuse in my family. The complications of alcoholism claimed the life of my grandfather at the age of sixty-three. His father died while drunk, run over while racing a horse and buggy. Could the susceptibility lie in our dopamine genes? Could my own destiny have been shaped by a genetic repetition? In fact, I have four copies of the repeated section of DRD4, which is about average.[1] Other genes are linked with dopamine, so DRD4 does not give the whole picture. Although I have examined other types of my dopamine receptor genes (DRD1, 2, 3, 5, and IIP), I have found nothing of particular note there, either.

Making friends was not easy. Few were interested in much beyond getting high to escape the horror and hopelessness. Marijuana was available on every base as tons of grass easily entered Vietnam. At the hospital gate I could buy a bag of two hundred prerolled joints of high-quality Bangkok gold for just $2. (It was only later in the war that troops switched to heroin and harder drugs that gave them a fast high.) Getting drunk was the only alternative. The hospital had a night club of sorts, a dingy bar where liquor was cheap and available twenty-four hours a day. Vietnamese bands and singers belted out cover versions of the Beatles, the Animals, and the Stones. Most corpsmen went there when off duty and drank themselves into oblivion. I did that on occasion, along with smoking joints, but spent much of my free time doing exercise.

I would run almost every day on My An Beach, now called China Beach. The sand swept in a great arc from Monkey Mountain, which rose above the bay of Da Nang, out to Marble Mountain, where battles raged almost every night. Although it was hard to do, in the rare moments that I could forget the war I realized that I was in a stunning setting. The mountains were riddled with mysterious caves and tunnels, some containing Buddhist and Confucian shrines. One cathedral-like cave only a few miles from my Quonset hut served as a Vietcong field hospital.

Running the three miles along the beach was an adventure in its own right as I made my way past the barriers of concertina wire and guard towers that had been set up at half-mile intervals. For entertainment the Marines would shoot at me with either a 50-caliber machine gun or their M16s. I learned to keep my pace constant during these strafings. I would swim after each run and bodysurf for hours on end until I got a surfboard.

Da Nang had rough surf and a powerful riptide: an undercurrent or "river" that takes the water brought in by waves back out to sea again. Given what I faced in the camp on a daily basis, I had no qualms about diving into the rip current. The tide would carry me quite far into the ocean, much like a chair lift on a ski slope. I had to hold my breath sometimes for a minute or so as I was pulled underwater and out to sea, but once I mastered it, it was a thrilling ride. The water was teeming with marine life, including sharks, six-foot barracudas, and sea snakes. The latter were the most worrisome. *Pelamydrus platurus* and *Aipysurus laevis,* snakes common to the South China Sea, would travel in large herds measuring miles long and up to a half-mile wide. The snakes were not typically aggressive but if disturbed would strike and bite, releasing a neurotoxin venom that can be rapidly lethal. There were routine reports of Vietnamese fishermen being killed from sea snake bites when they attempted to remove them from their nets.

One afternoon while out bodysurfing, I felt something hit my leg. I reached down to push away the intruder and knew immediately that I had grabbed hold of a sea snake. My hand circled the rounded part of its body near the head, not the flattened tail, and I knew I could not let go. Its jaws were wide open, and it was trying to bite. Sea snakes are strong swimmers, and it was all I could do to hang on. Swimming with one arm while being tumbled by ten- to twelve-foot waves and holding on to a writhing snake is not something I would recommend. Finally I was able to stand and started to run but was knocked down by a wave once more. Stumbling breathlessly toward the beach I saw some driftwood and used it to hit the snake on the head until it stopped moving. A friend took a photo of me holding my trophy, recording one of those crossroads in life that, with the wrong luck, could easily have led to death. I did not want to forget what had just happened. I took my knife, skinned my attacker, and back at the hospital, pinned it with hypodermic needles to a board to dry in the sun. I still have the snake skin hanging in my office as a reminder of the encounter.

I was the senior corpsman in the intensive care ward, a single Quonset hut with no windows, two doors, and twenty beds. Conditions were oppressive thanks to the heat and humidity, and during monsoon season, which at its height saw a steady cold rain called the crachin (French for *drizzle*), the floor often became so flooded that we moved around on planks. Most nights we faced the nearby explosions of rocket attacks, but because our patients were immobilized, we stayed with them, talking to ones who were awake to calm their fears as much as our own. There was not much opportunity to sleep, in any case, for across the highway was the Marine airbase and the constant noise from the H21 Shawnees, the "flying bananas"; the tadpole-shaped H-34 Choctaws; cargo-carrying CH-47 Chinooks; Hueys; and other "great iron birds," as the Vietnamese called them. Helicopters touched down in the zone at the back of our hut, and every landing brought more victims of mines, punji stakes, bullets, grenades, shells, mortars, high explosives, napalm, and white phosphorus.

The hut was lined with stryker frame beds, which rotated on a circular frame; these enabled us to place a second thin mattress on a paralyzed patient so he could be flipped end to end. The beds were never empty, and I saw their occupants tested in ways that I hoped I would never have to be. We usually had several double amputees, victims of land mine explosions. The fact that they were there at all, after having both their femoral arteries severed, was a tribute to the skill of the corpsmen in the field and the use of helicopters to evacuate casualties. (The Vietcong would shoot seriously wounded men rather than take them prisoner.) These patients were often vividly aware of their predicament and not infrequently would scream in pain or horror at the realization that they no longer had any legs, feet, hands, or arms. The brain surgery patients, the "veggies," usually had no idea of who they were or what they had lost. Between these grim extremes were the chest and abdominal injuries.

On my ward the patients faced one of two fates: They could either improve sufficiently to survive being medevaced to Japan or the Philippines for more sophisticated care, or they would breathe their last there. I witnessed several hundred soldiers die, more often than not while I was massaging their hearts (at times with my bare hand) or attempting to breathe life into them. A few of these men are deeply etched in my memory. There was the eighteen-year-old Marine, whose plight baffled me. He looked normal and healthy and had no obvious wounds, but he remained unconscious. On closer examination I found a small gauze pad on the back of his head stained with a spot of blood. Before we could investigate further, he went into cardiac arrest. This was a common occurrence, and as head of the arrest team, I started a well-drilled routine. We had a good success record because our patients were young and strong, but this young man was an exception. We tried defibrillation and then injections of adrenaline into his heart and continuous cardiac massage, but only after more than an hour did we finally accept that he was dead.

His death made no sense, and an autopsy was scheduled. Since I was clearly disturbed by the demise of this teenager, I was asked if I wanted to assist. The next morning I went to the pathology hut where my patient lay naked on the table. The young pathologist noted no wounds other than the small hole at the back of the head. I found the autopsy hard to bear even before the first incision: Once you have smelled formaldehyde oozing from a cadaver in a hot and humid hut, you will never forget it. I fought back the waves of nausea as the pathologist made a horseshoe-shaped cut from one side of the chest to the other and pulled back a large skin flap, resting it on the face of the corpse. With large shears he cut the ribs on the center line, exposing the heart that we worked so hard to start a few hours before. There were no obvious injuries.

The pathologist pulled the flap down and took a scalpel to the soldier's head, revealing his skull, the top of which he cut through with a bone saw.

After he removed the brain, he sliced it open. There we saw a small track the length of a pencil that culminated in a bullet. I was stunned how damage to less than one percent of his brain had been fatal, and I asked the pathologist for an explanation. "The bullet must have hit something important" was all he could answer. Although I had now reached the end of my shift I could not sleep. Each and every one of the cells in our bodies can grow indefinitely in a petri dish, and yet the entire community of 100 trillion cells that made up that young man perished because a tiny fraction of them had been destroyed.

Two other casualties had a great impact on me because they provided a remarkable testament to the human spirit and the will to live or to die. Both men had suffered extensive abdominal wounds: One, a Caucasian around thirty-five years old, had been shot in the abdomen with an M16, either one captured by the enemy or in an incident of "friendly fire." Because the small round of this weapon was extremely unstable—even a layer of tissue paper could send the bullet tumbling in any direction—it could inflict ugly wounds. I have seen an entrance hole in one part of the body and a corresponding exit wound in a seemingly impossible location. This patient was no exception: Some parts of his intestine were shredded, and others had tears. The surgeons had removed the damaged section of his bowel and were certain he would survive. He was not a combat soldier but was in some type of support position and had been unexpectedly caught in a Vietcong attack. He was soon awake and, surprised to be alive, was in good spirits. But the pain of abdominal surgery can be intense, and soon his mood began to turn.

Three days later the ward door burst open as a new patient was wheeled in: an eighteen-year-old African American with bullet wounds to the abdomen from a machine gun attack. He did not have much intestine left, and what remained was mostly heaped on the stretcher. The surgeon and his team had done all they could, but because the man had also lost his spleen and part of his liver, and his intestine was still oozing blood, he was not expected to live until morning. Amazingly, when I arrived for my shift, the patient was awake and alert. He had a warm personality and wanted to talk. He described how his squad had been ambushed and how he was concerned about them. He had a magnetic life energy that I was drawn to in the midst of such chaos and tragedy. Given his grim prognosis I spent as much of the night as possible talking to him about his family, his friends, and the surprise attack, but mostly about going home and his dream of playing basketball. He finally drifted off to sleep as I went off watch. I did not expect to see him again. The next evening when I went on duty, there he was, still talking up a storm, defying his prognosis and his biology and capturing all of our attention.

While all this was going on, the thirty-five-year-old M16 victim asked if I would do him a favor as I changed his dressings. He wanted to dictate to me a letter to his wife, to tell her that he had lived a good life and that he loved her, but that he could not take the pain and did not think he would see her again. I thought his chance of recovery was high, and I expected him to be medevaced in a day or so, and I assured him that with his prognosis, he would soon be out of the war. In any case, my poor spelling and handwriting and my brittle twenty-year-old emotions were not up to the task, so I asked one of the other corpsmen to do it. I was also angry that he was giving up. We were doing all we could within the limits of medicine in 1967 in a war zone in horrible circumstances. I think I was annoyed because I had thought so many times myself that death would be the easy way out. When I returned for my next shift, he was gone. He had died midday; the postmortem indicated that "he gave up." What an incredible contrast. The man who should have lived did not, while the man who was supposed to die immediately lived beyond all conceivable odds because he wanted to. People don't usually give up on life; it is torn from them.

Although the latter patient died a few days after being flown to the Philippines, he showed me the effects of the human spirit and of sheer willpower, which can be stronger than any drug. The effort we had all put into giving him a few more days of life was far from wasted, because he had bestowed on all of us, but most particularly me, a wonderful gift: He had won our respect and had given us a thirst for life, which I have craved almost every day since I first met him. I have spoken of and thought of both these men often and feel that they partly helped drive much of my future career. They helped turn me from a young man without purpose into one compelled to understand the very essence of life. And life was so cheap in Vietnam that my mission had real urgency.

It took only a few weeks of enduring the conflict—and having to deal with hundreds of casualties—for me to feel my opposition to the war crystallize. I was not alone; it was unusual to find anyone in the military in Vietnam who supported the war. Like many of my peers, I was unimpressed when I heard we were to receive a visit from two very important officials: Vice President Hubert Humphrey and General William Westmoreland. Humphrey seemed to support the policies of President Lyndon Johnson to expand the war effort, while Westmoreland, *Time*'s Man of the Year in 1965, was the gung-ho military man who cultivated the war's insane body count philosophy. The idea had brutally simple and sinister ingredients: attrition by more bombs, more shells, and more napalm. The American war machine was going to kill Vietcong guerrillas and the troops of the North Vietnamese army faster than Hanoi could send men down the Ho Chi Minh Trail to fight in the south. Curtis LeMay, the Air Force chief of staff who was satirized by George C. Scott in the movie *Dr. Strangelove,* vowed that he was going to "bomb 'em back into the Stone Age." In this war with no fronts, one measure of progress was the Vietcong "body count." The other was the number of young Americans killed and wounded.

On their arrival with reporters in tow, the general and vice president were introduced to the 150 medical personnel at the Da Nang hospital. I had sometimes fantasized about making a headline-grabbing protest to embarrass the top brass; but all I could muster for the occasion was a refusal to shake hands with either of them and mutter, "We are making a terrible mistake being in Vietnam." While this led only to an awkward moment, a short while later one of my patients, a double amputee, made a more effective gesture. As photographers and reporters looked on, and General Westmoreland prepared to pin a medal on his chest, the amputee told him to "take [his] Purple Heart and shove it up your ass." Westmoreland glared at me and stomped off. The vice president kept his composure, grasping the hand of the man and saying: "I understand why you feel that way." Later in life my opinion of Humphrey improved substantially.

Corpsmen were the link between life and death and often became confessors as well as healers. I came to learn the many shapes and forms of protest and rebellion made by our troops. There were numerous desertions while in Vietnam or on R&R. There were men who refused to go into battle, particularly toward the end of the war. Then there was "fragging," an insidious and lethal form of protest. Despite the mystical comradeship of the Marine Corps, which meant they never abandoned their dead, some would use a single shot to execute the lunatic second lieutenant who was interested only in body counts and promotions rather than whether he was shooting a villager or a Vietcong. Many times fragging involved the use of various types of land mines to eliminate a commander who, in the view of the troops, would get them all killed for no reason.

One of several stories to come out of Khe Sanh illustrates not only the cheapness of life but also the callousness and desperation of those days. A group of three wounded Marines came to the hospital and told me how they had killed their commander. Some land mines explode when they are stepped on while others, like the one they had used, are designed to detonate when the victim steps off, relieving the pressure. This trio described how they had observed the behavior of their commander, noticing how he drank from a bottle of scotch each night in his tent. Driven by thoughts of self-preservation, they placed the bottle of scotch on the land mine. Later that evening the commander went for his last drink. Incidents like these were not isolated occurrences.

Everyone was bending the rules in some way or other to cope with the extraordinary pressure. I found an unlikely ally in Bill Atkinson, who, before he was drafted, had lived off the land in the mountains of Montana near the Canadian border in a log cabin with no power, just oil lamps and fire. He even had a wolf as a "pet." Bill worked in the medical records/patient transfer office at the hospital. One day I asked if he could help me with an Army captain who was rabidly antiwar. Though he had been wounded, he was due to return to the field. Bill had released many men on medical discharges and was glad to help. We soon developed a system to send home those who were at mental breaking point or who had a compelling reason, in our view, to escape.

Toward the end of my first six months in Da Nang, the Navy decided that it should follow the Army's lead and send female nurses to Vietnam. Women were always welcome. But while the Army nurses were young, vivacious, and compassionate, the Navy sent more senior commanders and captains who were far removed from the realities of medical care and brimming with bureaucracy. The combination of hard-pressed staff making the best of trying conditions and the rules and red tape of petty bureaucrats was combustible. For me the conflict between the pragmatists and the fundamentalists led to an explosive encounter during one night shift when the ward was packed with recent casualties and we were short-staffed.

Among the new arrivals was a Korean soldier on a respirator with a severe head wound and a body shredded by shrapnel. Nearby were two prisoners of war. One was Chinese, badly wounded but conscious, and the other a Vietcong, who was on a respirator and on whom I was working with another corpsman. Both had guards, but neither prisoner was capable of going anywhere. As I was on my way to help the Korean, who was thrashing in agony,

I overheard one of the new nurses order one of my corpsmen to clean the fingernails and toenails of the Vietcong patient. He certainly could have done with a thorough cleaning, since he had probably lived for months in an underground bunker. But he did not need one now, given that he could hardly breathe because a tube was not draining the blood from his chest cavity.

As the senior corpsman I did not want my men to be distracted from the more important task of trying to save a life. There was a heated exchange when I told the corpsman to continue working on the chest tube and the nurse-officer "to go screw herself." Early that morning the Vietcong patient died. I finished my shift and went to my bunk to sleep, only to be awakened by an MP who escorted me to the base commander's office. I was told that although I was considered one of the best corpsmen, I could not return to the ward because the nurse had filed court-martial charges for disrespect to a superior officer and for disobeying a direct order. I was relieved of duty. Even though escape was hardly an option in the midst of a war zone, I was also confined to my barracks until the appropriate punishment could be determined. It was a disturbing echo of what had happened months before in Long Beach, as now another nurse wanted to increase the odds of my being killed by sending me into the field.

After two days I was paid a visit by an officer with a distinctive five o'clock shadow. Ronald Nadel was the disheveled doctor in charge of the dermatology and infectious disease clinic, and he was looking for a talented corpsman to work with him. He knew of the charges against me and, if anything, seemed amused by them. I warmed to him immediately and accepted his offer, in part because of my interest and experience with infectious disease and in part because I was relieved to be getting out of the intensive care ward or going to the brig or going into the jungle. That night I pondered how I had been saved once again from a similar fate in a very similar way.

When I joined the clinic, my life improved immediately. Ron Nadel was a wonderful mentor who liked teaching as much as I liked learning. We became an effective team, and he began to trust independent operations to me while he tended to others. He knew that I respected my limits and that if I saw a new condition, I would not attempt to deal with it without input from him or others. Even so, often when I did not recognize something—say, an unusual tissue while surgically removing a cyst—it was new to Ron as well.

We saw two hundred patients or more every day and had to deal with an amazing range of disorders, from malaria and jungle rot to tumors and venereal diseases, the latter almost epidemic in Vietnam. Dotted along Highway 1 from Da Nang were bordello shacks called skivvy houses, presumably named after "skivvies" (underwear).

Prostitution was an industry that involved an army of women, even young children, who were often a collateral effect of the devastation in the countryside. The atmosphere in the shacks where they plied their trade was a blend of acid rock, smoke, dope, beer, and *nhoc mam,* a pungent sauce made from fermented fish. They also congregated in noisy bars with names such as A-Go-Go and The Bunny. Colored water—"Saigon tea"—had to be bought to enjoy the company of bar girls, with after-hours sex costing extra. The pimps would shout, "You want bomb-bomb" or "You want to buy my sister?" My answer was always no. As a corpsman who had treated the men after their visits to the skivvy houses, I could never bring myself to use them, despite enduring long periods without sex.

Cases of antibiotic-resistant syphilis and gonorrhea were mounting daily. To diagnose syphilis I had to ask the Marine to drop his pants and then, with gloved hands, squeeze the primary lesion and blot the pus onto a microscope slide. Under a dark field microscope the microbial universe of venereal disease could be seen on the glass: corkscrew-shaped bacteria, *Treponema palladium.* Thirty years later, when I had to think about which organisms to decode, I placed a high priority on *Treponema* because of the damage it has caused throughout history. I was linked to the story of that organism in 2002, when I was honored with Germany's Paul Ehrlich award, named after the father of modern chemotherapy who was the first to come up with an effective treatment for syphilis.

My favorite time in Da Nang came every Wednesday when we treated children at a local orphanage. Ron and I would fill a Jeep with medical supplies and drive to the small village where the orphanage was located. The Vietcong were ever present deep within Marine lines, and although assassinations of officials did occur on the outskirts of Da Nang, we were clearly on a humanitarian mission and were left alone. With us came a Vietnamese nurse-interpreter nicknamed Bic, a Catholic from North Vietnam who had fled south after the French collapse in 1954. (The Catholics had fought alongside the French in exchange for autonomy.) She treated all of us, Americans and Vietnamese alike, with equal contempt. I would provide complicated instructions on how to use a particular medicine and ask her to explain them to the patient. She would bark a short sentence as a translation. While discussing simpler treatments, she would talk for an age. We dealt with everything from pregnancies to impetigo and insect infestations, and an occasional major wound or broken bone. But the war was never far away. We moved from antibiotic pills to injections when we discovered that the drugs we dispensed were being collected outside the orphanage by the Vietcong.

Practicing medicine at the orphanage was one of the highlights of my time in Vietnam. I found that basic hygiene and soap could often do as much to improve the quality of life for many as advanced drugs. The children had bright, innocent, and eager faces, and it was easy to develop relationships with them since we came back each week. Using my knowledge to do a little good in the midst of so much death and misery, I became convinced of the direction my life should take. If I made it back home and could get into a university and then into a medical school, I would practice medicine in the developing world. But sitting in a hut outside of Da Nang in 1968, after barely graduating from high school four years earlier, the very acts of surviving the war and going home seemed distant, let alone attending a university. But I was fortunate. Ron Nadel, one of the few people whom I respected, helped convince me that I could do it.

My work in the clinic also triggered my lifelong love of sailing. It all came about when a Navy chief walked in one afternoon to inquire about having some tattoos removed and was turned away because elective surgery is not permitted in a war zone. After several attempts, the chief became desperate and told me his sad story. Navy personnel were given two one-week R&R trips out of Vietnam after being in the country for at least six months. The chief, like many others, chose to go to Bangkok on his first R&R. And, as most did, he became dead drunk on his arrival, waking the next morning with three things he did not have the day before: a hangover, a young girlfriend, and a tattoo across the fingers of both hands spelling out the name Mary.

Now back in Da Nang, he was eligible for his second R&R and was scheduled to go to Hawaii, where he would meet up with his wife. Of course she was not a Mary. There was something both innocent and tragic at the same time about this sailor, so I decided to help. After the clinic had closed for the day, I replaced his tattoos with what by today's standards were crude skin grafts from his thigh. Weeks later it was clear that not only were the grafts successful, but he now also had some impressive battle wounds to show off to his wife.

Working in the harbormaster's office in Da Nang, my patient had access to small craft of various types and offered as payment the use of a 25-foot Boston Whaler and nineteen-foot fiberglass Lightning sailboats, centerboard boats that had a single aluminum mast and two sails. On a free day with little going on, Ron and I drove the whaler out of Da Nang harbor, around Monkey Mountain, and up the coast. It was odd to be in the middle of a war zone, bobbing in the ocean in a small powerboat. What a fantastic relief to be out on the open water, away from the realities of everyday life and death in Da Nang. Ever since, I have taken to the sea to reactivate my brain and refresh my senses.

I wanted to try the sailboats too. I had often dreamed about achieving a state of freedom, and sailing became a key part of this fantasy because I read constantly about ocean voyages and adventurers. But a condition of getting my hands on a sailboat was that one of us had to be competent enough to get the boat out safely through the surf and back again. I consulted with my Marine buddies, but no one had the slightest idea how to sail. So I became my own instructor, figuring that sailing could not be that hard to master. On each trip, my motley crew of Marines and I would venture farther and farther off shore. Being near the Marine airbase we were subject to the usual "fun" from our comrades, which usually involved their dropping smoke grenades from their helicopters, but sometimes hand grenades, too.

On days with light winds we would attempt to fish with makeshift tackle made of surgical gear. One day we hooked something and our primitive rig was pulled, and the boat rocked. A few minutes later the line went taut again, and the next thing we knew we were being pulled backwards at a fast clip. I was certain we had caught a shark and did not want to join him in the water. An hour or so later we began to enjoy the ride, but he did not tire. A few times he would stop and swim near the surface, and we could see that we were in over our heads: Our "engine" was an eight-foot Mako, a species noted for their speed and agility. Fortunately, some of the Vietnamese fishermen whom we often saw on our trips noticed us sailing backwards. They motored up in their sampans and offered to trade fish and tackle for our line and its thrashing contents. Deal. We worked to transfer the line to them and watched as four of the wooden flat-bottomed boats worked together to land the shark.

Even in a war zone, curiosity can often overcome fear. The coast of Vietnam is beautiful, with small coves of heavy green vegetation draping down to the water. Near the beach there was often a small wooden pagoda, which served as a shrine or temple, consisting of stepped pyramidal structures, usually adorned with beautiful carvings. We wanted to use the sailboat to explore some of the coves on Monkey Mountain, which was not completely under U.S. control. My crew decided to take their M16s with them on our little adventures, and on more than one occasion the serene scene was shattered by gunfire from the Vietcong. The Marines would return fire and we would try to lie low in our fiberglass boat while sailing away as fast as possible. I think I have had the only Lightning with bullet holes in its sails.

After six months of enduring Vietnam, I was eligible for my first R&R. I had read about the flying doctors in Australia and became fascinated by the country, so I opted to go to Sydney. It is head-reeling to be plucked out of a war zone and placed back into normal society within a few hours, sipping coffee and listening to the Lovin' Spoonful on a hotel radio. I felt as though I had stepped out of a time machine. My first morning I headed in the direction of the beach from my hotel, and after two blocks, I had a fateful encounter. A cute girl was walking toward me, carrying two bags of groceries, and as I passed, she dropped both, their contents spilling across the sidewalk. As I helped her pick them up, I asked for directions to the best beach for bodysurfing. Bronte Beach was a half mile or so down the road, she said, adding that her name was Barbara. We talked for a little while before I set out for the beach and we seemed to click, and Barbara gave me her number.

That day the surf was unusually high, and few people braved the waters. Given my experiences in Vietnam, I was not bothered and began to bodysurf in the fifteen- to eighteen-foot waves. I was there for hours, having the time of my life, and felt free. Then I noticed a crowd on the beach. People were pointing anxiously toward the water not far from me. A girl was struggling against the rip current, so I swam to her. I ended up pulling her onto the beach.

Members of the Bronte Beach lifesaving club helped her ashore and were blown away by the fact that an American had not only bodysurfed their beach in tough conditions but had rescued a swimmer as well. They invited me to their clubhouse and made me an honorary member of the club. That night they took me on a "pub crawl" going by scooter from bar to bar. They urged me to find the girl that I met earlier because she was from New Zealand, and Australians knew that all New Zealand girls were "easy."

Barbara and I got together that night and stayed together for the rest of the week. When I had to return to Vietnam, we agreed to meet again and write each other frequently. Three months later I was eligible for my second R&R, and at the recommendation of one of the doctors I worked with, I chose Hong Kong. To surprise me, he had made reservations under the name Dr. Venter at the Peninsula Hotel, where Barbara had agreed to stop on her way to Europe. After I stepped off the military transport plane direct from Da Nang, I was met by hotel representatives and driven to my grand suite in a Rolls-Royce. Barbara met me at the hotel and was as overwhelmed as I was by the suite and the three servants who went with it. Having no expenses in Vietnam other than booze and weed, I felt that I could splurge, and we spent most of our time either shopping or in bed. At the end of a short week I found it very difficult to return once again to Vietnam and the war. Barbara and I tentatively agreed to meet again in London after my service was up.

When I arrived back in Da Nang, I considered myself a pacifist and was determined to work only to save lives and not take any. Many corpsmen held this philosophy until they were drawn into combat. My own thinking began to change similarly when it became likely that the hospital would be overrun in a series of attacks that became known as the Tet Offensive. (Tet is the week-long lunar new year holiday that to the Vietnamese is Thanksgiving, Christmas, and New Year's rolled into one.) During the campaign, which began before dawn on January 30, 1968, our situation became so grave that Navy nurses were evacuated from Da Nang, and the Marine guards asked that corpsmen be armed to help repel any attack.

The hospital was subjected to near constant shelling, and while I was mentally prepared to use my M16 to avoid being one of the fifty thousand or more young men who were eventually killed in Vietnam, I was never called upon to do so. Instead, I along with every available person dealt with thousands of casualties in long days of death and dying. The Tet Offensive changed public attitudes back in America, and it changed me, as well. I learned more than any twenty-year-old should ever have to about triage, about sorting those you can salvage from those you cannot do anything for except ease their pain as they died. I was not studying at the university of life but of death, and death is a powerful teacher.

My last three months were especially difficult. Too often I had seen first-hand in Vietnam how men began to nurture hopes that they might make it out alive, only to have them dashed. One day I found myself working on a grunt my own age who had taken a mortal wound from a sniper on his way to the airport and a flight home. Around that time I remember one particular night on the airbase I spent with friends after a day of sailing. We were all lying on our backs, feeling mellow after spending an evening smoking joints, when rockets began to explode nearby. Instead of diving into the bunker, we lay there mesmerized by the weird psychedelic light show, which was like a scene from *Apocalypse Now*. The next morning when I thought about the stupidity of the previous night, I decided that if I wanted to survive my remaining time in Da Nang, I had better act a lot smarter. I did not smoke another joint in Vietnam and drank only occasionally. I ran harder and surfed every free moment. Finally it came my turn to say good-bye to colleagues and to leave on a chartered 707 aircraft.

The plane was vulnerable, having no armor. We had learned to sit on our helmets to avoid being shot through the ass while in the medevac helicopters, but there were no helmets on the planes. All takeoffs and landings were scheduled at night for added protection, but the flash of weapons still made the situation frightening. Despite the flashes and the cramped conditions, no one complained. I held my breath, and one could sense a wave of relief wash through the plane when we cleared the sniper zone. At last there was a mood of celebration: We now knew for certain what had been in doubt, that we had survived our tour in Vietnam. I felt overwhelmed, a combination of relief and trepidation about the future. So much had happened, so much had changed. I had been in the military for only two years and eight months, but I was not the same young man who had been drafted off his surfboard. Now I was headed home to a very different life.

Life was my gift. I had seen thousands of men my age killed or maimed in unthinkable ways. I did not feel survivor's guilt, but I did want to do something with my life to honor all those who were now beyond my help. I was back in charge of my destiny again. I recognized that if I had said good-bye to death and destruction, I had also said good-bye to being taken seriously by doctors and practicing a level of medicine that I knew I could return to only after a decade or so of intense study and training in civilian life. I might never even get to that level, given my poor educational record, one that left me unable to spell the most basic words and that had resulted in my being sent to Vietnam in the first place. I was about to greet a life of uncertainty.

Civilian life came closer by the hour as we traveled first to Japan, and then to Guam, Hawaii, Alaska, Seattle, Travis airbase in northern California, and finally to an Air Force base near Los Angeles. On each leg of the trip, I became more apprehensive. When we finally landed, I was briefly elated, and some of the men with me stooped to kiss the ground as we came off the plane. No brass band and flags welcomed us, and while groups of friends and family were milling around to greet many of the returning soldiers, no one had come for me. Back in the U.S.A., I felt incredibly alone.

After two days in the transit barracks, my wait was over. My final orders read: "Released from Active Duty and Transferred to Naval Reserve." It was August 29, 1968; I had $2,800 in the bank after saving most of my pay for the year. I had some serious medical skills, three military medals—the National Defense Service Medal, the Vietnam Service Medal with Bronze Star, and the Vietnam Campaign Medal with Device—an honorable discharge, and, most important, my life. I took my sea bag and boarded the flight for San Francisco.

Life in America had carried on as though there was no war, but it was no longer the same place for me. The family home that I had left four years earlier now felt empty and unfamiliar. Three out of four children had left home, leading to what I call my younger brother, Keith's, "only child" period. My father had finally finished paying for his partnership in John F. Forbes and Co., now a major West Coast accounting firm. For the first time in his life he had disposable income and the time to develop a close relationship with Keith, and they often played golf together. My father now drove a Cadillac, of which he was very proud, and had become a full member of Green Hills Country Club. So this, of course, was where my mother and father had decided to celebrate their son's homecoming from Vietnam.

Fresh out of the University of Death, I felt something like Benjamin Braddock at the welcome home party of his southern Californian suburbanite parents in *The Graduate*. At dinner I came close to losing it when a group of Republican fat cats who were sitting around, drinking and smoking cigars, went on about how great it must have been to kill Communists and gooks. The "gookification" of the Vietcong and the North Vietnamese made it easier for our men to kill men, women, and children, or to undertake absurd acts such as throwing dead buffaloes into wells to poison the locals' water. I wanted to scream about the endless civilian casualties, to tell them about the thousands of young men who were maimed or scarred or killed to achieve nothing—nothing other than proving to our enemies that we were more than willing to sacrifice our youth. I left the dinner claiming I had jet lag and went home.

I booked a ticket for a flight to London the next day. My plan—or hope—was to get out of the United States as quickly as possible and meet up with Barbara. Arriving at Heathrow I was made to feel very unwelcome. The year 1968 had seen antiwar demonstrations increase in frequency and violence, and I was a twenty-one-year-old American with a backpack and sleeping bag and a U.S. passport that had been issued in Saigon. For reasons that even now are hard to fathom, the British officials decided that this tanned and very blond young man had come to the United Kingdom to help stir up protest about the war. I was searched; every item I had was rigorously examined, I presume, to find drugs. Finally, after half a day, I was released and entered the United Kingdom, the country of my ancestors, for the first time. I had no place to stay, and since it was the peak of the summer season, there were no vacancies. One hotel suggested that my only chance was the YMCA, where I found a room for the night.

The next day I found a small room in a cheap hotel and connected with Barbara. She had spent the last three months hitchhiking around Europe and convinced me that it was a great way to travel. We headed to Dover by train, caught a ferry, and began our gypsy existence by camping on a beach in Calais. I soon became tired of hitchhiking: what worked well for a young single female was less efficient for a couple with backpacks, and in Frankfurt I bought a used Volkswagen. As we toured France and Spain, I still wanted to get away from big cities and society, to find a private, peaceful spot to decompress, get my mind in order, and adjust to not being in a war zone. I rented a chalet in the Swiss Alps, high in the mountains near Lausanne at the far end of Lake Geneva. We hiked, cooked, rested, made love, read, and adjusted. But as the gray days and first snows of winter approached, I became more and more restless about starting my new life. I had been encouraged to stay in the Navy on a rapid promotion track, but I could not deal with the blind submission to authority. I could have taken a physician's assistant or a registered nurse exam, but I wanted to try for more. I knew from talking to Ron Nadel and others that to get into medical school I not only had to go to a leading university but also to get top grades there. Since I was not likely to be accepted by that kind of school with my academic record, my only choice was to go to a community college and transfer to a university for the third and fourth years.

The College of San Mateo near where I grew up was among the best of its kind and had transfer programs designed for both Stanford University and the University of California. I could start in January. Even better, Barbara, who had only a certificate in computer programming, liked the idea of going back to school with me. We checked with the American embassy and were told that there was a long waiting list for visas for New Zealanders to go to the United States. Worse, they could not even guarantee that Barbara could eventually get one. If we were married, however, she would be granted a visa immediately.

While I liked the idea of living with Barbara and going to school, I was not keen on the commitment of marriage at this stage of my life. I was young and immature and just wanted sex and companionship. I called my brother Gary for advice. "This is the 1960s," he told me. "Most people will marry four or five times in their lives, so don't worry about it." Barbara and I were married in a civil ceremony in Geneva, which seemed an appropriate geographic compromise for a couple who came from America and New Zealand.

Barbara soon got her U.S. visa, and we headed to England on the way back to the United States to pick up something that I had dreamed about in Vietnam, one of the many fantasies that had helped sustain me through the conflict: the crème de la crème of motorcycles, a Triumph Bonneville 650. This was the classic sixties hot road bike, associated with Hollywood stars such as James Dean, Steve McQueen, and Marlon Brando, and it had a unique sound that ascended from a low growl to a throaty exhaust note, then an open-pipe snarl. I had placed an order while still in Da Nang and said I would take delivery in the United Kingdom—that way I could import it into the United States as a used bike and save a lot of money.

I returned to Millbrae with a new motorcycle and a wife, neither of which was particularly welcomed by my parents. But I was in much better mental shape now than when I had first arrived from Vietnam. I was going to try to understand the events that I had witnessed in Da Nang while taking perhaps the hardest step I had ever taken in going back to school. To find out about life, I was anxious to start my education over, even if it was to be from scratch.

3. ADRENALINE JUNKIE

These changes—the more rapid pulse, the deeper breathing, the increase of sugar in the blood, the secretion from the adrenal glands—were very diverse and seemed unrelated. Then, one wakeful night, after a considerable collection of these changes had been disclosed, the idea flashed through my mind that they could be nicely integrated if conceived as bodily preparations for supreme effort in flight or in fighting.

—Walter Bradford Cannon, *The Way of an Investigator*

It is difficult to underestimate the confidence that I felt in early 1969 about returning to school. At the same time it is difficult to overestimate my motivation to advance myself. Like anyone, I feared failure, particularly given the lackluster state of my academic record. But having seen real poverty of mind, body, and soul in Vietnam, I knew the value of education—in this case, my own education. Fortunately, at least, there were many job opportunities in California. Based on my medical experience I quickly got a position as a respiratory therapist at Peninsula Hospital in Burlingame, on the San Francisco Peninsula. Within a short time I was appointed head of the cardiac arrest team, just as I had been in Da Nang.

With the help of the same GI Bill that partially funded my father's college education after war, I enrolled in classes at the College of San Mateo. At my side was Barbara, who had to attend many of the same basic classes in English, math, and chemistry because California bureaucracy could not make sense of her New Zealand high school record. I was prepared to work hard, and had to learn how to learn and how to study for the first time. However, like so many people who have succeeded in life, I had some great teachers who encouraged and inspired me, taking a real interest in my education.

One was Bruce Cameron, who taught my first class, English literature. Bruce was a forty-something graduate of Hunter College in New York City, where he had funded his education by driving a yellow cab. With his newly minted master's degree, he had moved to California the year before for a teaching position at the college. Bruce would often tell the class that if anyone did not like or was bored by a writing assignment, that they could take a chance and submit something based on their own inspiration. I decided to ignore one of the assignments and, inspired by Cameron's love of nicotine, to tell the tragic tale of Harry Boggs, a heavy smoker, and his succumbing to lung cancer. My amateurish three-act story, and the fact that I was the only one ever to rise to his challenge, seemed to impress him. So started a friendship that soon progressed to regular dinners with him and his wife, Pat. Bruce not only befriended me but challenged my thinking and encouraged me to excel. My writing and my self-assurance improved every week, not just in English but in subjects that I had been quite hopeless at before, such as math.

Chemistry still struck particular fear into my heart. This subject was crucial for a career in medicine, and yet high school had left me allergic even to the thought of grappling with the world of atoms and molecules. My instructor was a recent Ph.D. named Kate Murashige, whose dedicated teaching kindled my enthusiasm. To my surprise I found myself enjoying chemical detective work—using different techniques to discover unknown compounds. Now a patent lawyer (and with the benefit of hindsight), Kate remembers me as someone who looked as if he were going to succeed as an "A student." I began to enjoy my six classes, and my changed attitude was reflected in my grades—I managed to get straight A's even while working full time at the hospital every evening. By taking extra classes I planned to finish the first two years of college in eighteen months and then would try to transfer to the University of California. (I could not afford Stanford University's fees.)

One day as I was sitting in my French class feeling somewhat intimidated by those who not only spoke French but liked to act out its shoulder-shrugging mannerisms, a student burst in to tell us that there had been a military massacre of protesting students at Kent State University. By that time Richard M. Nixon had been elected president on a promise to end the war in Vietnam, but instead the conflict had expanded into Cambodia. The peace movement had responded to the escalation, and across the country campuses had erupted with demands to end the war. On Monday, May 4, 1970, four students were shot and killed by National Guardsmen and nine others were injured at Kent State University.

I could understand the deep antipathy between the protesters and the government. I had seen the senseless killing and brutality firsthand in Vietnam, where my own antiwar feelings had been stirred. But I also felt strongly for the servicemen and -women still in Vietnam. Most of those men had been drafted, like me, or had joined up to escape a grim home life or for an adventure. And there were still many who felt a patriotic duty to our country. The protesters did not take such factors into account. Whatever a soldier's motivation, whatever his reasons for being there, he was just another "baby killer"—a reference to events such as My Lai, a U.S. massacre of hundreds of unarmed civilians in which old men, women, and children were killed. Ironically, many of the "pigs" who had joined the National Guard, and presumably many of those involved with the shooting of the Kent State protesters, had done so to avoid Vietnam.

My antiwar feelings won out on that particular day and I believed that my experience as a medic in Vietnam could help influence opinion. My first thought was that we had to shut down the school with a demonstration to protest the killings. Thousands of students gathered nearby, and since it was a spontaneous rally, the feelings of outrage and frustration were overwhelming. Person after person took the microphone to address the crowd, and when my turn came, I urged that we plan a huge but peaceful march into the City of San Mateo. The next day the local newspaper carried a front-page photograph of me with the headline: "It's Our School, Let's Take It Over."

I eventually organized the demonstration, and through Bruce, the college president and his staff contacted me to see if they could negotiate a peaceful outcome. They seemed encouraged by my nonviolent approach, expressing unofficial support for the march and outrage concerning the student shootings. On the day of the march, more than ten thousand people took part. I led the group, followed by a symbolic coffin.

Because of the sheer adrenaline high, the events of that day are a blur, save one detail. A white van had slowly followed us. With its sliding door open, the men inside were continually photographing me and the other student leaders. I thought that they were members of the press, but I learned later that they were police and FBI agents. The march ended peacefully. The Kent State massacre had led to the only nationwide student strike in history, in which 4 million protested and more than nine hundred American colleges and universities were closed. Amid all the uproar, tear gas, and billy clubs, one image of protest still stands out clearly in my mind: a newspaper photo of a University of California, San Diego, student who had set himself on fire.

I resumed my studies with a vengeance. Among my remaining assignments for Bruce, I had to write two book reviews, which indicated a lot about the course my life was taking. For the first I chose *The Lonely Sea and the Sky,* Francis Chichester's stirring description of his solo sail around the world in 1966. Chichester was a hero, setting a new record and receiving a knighthood for his achievement. His account of his nine-month adventure in *Gypsy Moth IV* revealed how he managed to overcome illness, injury, and danger, nearly capsizing in rolling seas where he was beyond the reach of any outside help. It was an amazing achievement for anyone, let alone a man of sixty-five.

My other choice was the famously indiscreet account of one of the great discoveries in molecular biology, *The Double Helix,* by the American Nobel Prize winner James Watson. The book was originally titled *Honest Jim,*[1] partly in reference to how he blundered his way to triumph (as in Kingsley Amis's *Lucky Jim*), and partly a disarmingly frank reference to how some believed he had used another's data for his greatest discovery. (He had even toyed with doing an article entitled "Annals of a Crime.")

It is worth a short diversion here to describe Watson's tale because of the relevance that the man, his science, and his life would come to have in my own story—something I never could have imagined at the time I was reading his account of how he and Briton Francis Crick had found the structure of DNA. Watson described how he and Crick had provided a much needed dose of "genetic fresh air" to the stuffy world of British biology. After the two men worked out the chemical structure of the DNA molecule, Crick had bragged in the Eagle pub that they had found "the secret of life," a claim that, made partly in jest over a beer, was typical of this pair. As Watson put it, "We broadcast it as fast as possible, knowing that if we would wait, someone else would inevitably think out the right answer and we would have to share the credit."[2] They celebrated the mistakes of their buttoned-up rivals and refused to play by the rules. Whether they used sound methods or flashy ones did not matter—all they wanted was the answer as quickly as possible. Crick himself admitted they shared "a certain youthful arrogance, a ruthlessness . . . an impatience with sloppy thinking."[3] Given the staid standards of 1953, Watson and Crick were the original bad boys of molecular biology.

Behind them and their achievement was Maurice Wilkins, who had first excited Watson with his pioneering X-ray studies of DNA and who worked alongside another key figure in the story of the double helix, Rosalind Franklin of King's College, London. Watson depicted "Rosy" as an aloof and irritable intellectual: She squirreled away data she couldn't understand, made Wilkins's life wretched, and treated men like silly schoolboys. Wilkins, in turn, grew deeply frustrated with her reluctance to acknowledge the viability of the double helix. A pivotal scene in Watson's book features him striding into Franklin's lab early in 1953 to inform her that she did not comprehend what her findings really meant.

Franklin was so angry that when she leaped from behind her lab bench, Watson recoiled out of fear that she might hit him. Retreating, he bumped into Wilkins, who then showed him the best of Franklin's X-ray photographs of DNA. Numbered 51 and taken in May 1952,[4] it revealed a black cross of reflections, and would prove the key to unlocking the double helical structure. "My mouth fell open, and my pulse began to race,"[5] Watson recalled. He was sure he was looking at a helix and that he and Crick were on the right track. Wilkins had been telling him as much since 1951, and, coincidentally, Crick had just developed a theory of how a helix would look under X-ray diffraction. The DNA structure came as an epiphany, "far more beautiful than we ever anticipated,"[6] said Watson, because the complementary nature of the base pairs of DNA (the letter A always pairs with T, and C with G) revealed how genes were copied when cells divide.

The earliest written record of this mechanism can be found in a remarkable letter written by Crick to his son Michael on March 17, 1953:[7] "You can now see how Nature makes copies of the genes. Because if the two chains unwind into two separate chains, and if each chain then makes another chain come together on it, then because A always goes with T, and G with C, we shall get two copies where we had one before. In other words, we think we have found the basic copying mechanism by which life comes from life. . . . You can understand that we are excited."

The discovery of the double helix established a recurrent theme in the subsequent history of science: the right of access to data. Half a century later, in his book *A Passion for DNA*, Watson admitted: "There were those who thought Francis and I had no right to think about other people's data and had in fact stolen the double helix from Maurice Wilkins and Rosalind Franklin."[8] But he later explained that the reason King's College did not claim more credit was simple: King's did not ask the most basic question of all when it came to the race to find DNA's structure. How are we going to win?

My hard work at school paid off. I received all A's for that semester and the two that followed, as did Barbara. We partied all night when we found out that we had been accepted as advanced students at the University of California, San Diego (UCSD), in La Jolla. Money was a still a problem, however. Tuition fees of $900 per quarter may seem small by today's standards, but they were beyond what was covered by my California State scholarship. I had to find a way to support Barbara and myself while studying full time. As married students, Barbara and I were eligible for student loans, but my father agreed to provide one at no interest for our tuition—but only if I would sign a promissory note. His apparent lack of faith was troublesome, but I was grateful for his help.

My Genome and My Brother

While swimming in the waters off San Diego with my younger brother, I saw a large fin less than twenty yards away. Although we could easily escape the shark, I began to panic: Keith, who was born with nerve deafness, had taken out his hearing aids to swim and could not hear my shouts or those of people in a nearby boat. I had no choice but to swim after him, get his attention, and point to the shark. We must have looked like cartoon characters when we eventually got into the boat, almost leaping out of the water.

In the 1950s and 1960s society could be cruel to anyone with impediments, and Keith had a hard time when he followed me into school. He was the baby of the family, and we were all very protective. We look after our genes, after all, just as ants, bees, and other social insects do. Are there any hints in my own genome of why Keith became deaf? After all, we share half of our genes. A number of studies have linked deafness with variations in genes. One, called DIAPHI, was linked with hearing loss in studies of a large Costa Rican family; another, TMIE, was linked to deafness in mice and in several Indian and Pakistani families. I do not have either of these mutations.

Another candidate is CDH23, which seems to play a role in the hair cells deep in the ear. Named after the hairlike projections on their surfaces, hair cells form a ribbon of vibration sensors along the length of the cochlea, the spiral, seashell-like structure in the inner ear that detects sound. I likewise checked out this gene to see if anything was amiss, but again, no known problems surfaced. Ultimately, I will have to analyze Keith's genome if we are to discover if genes were responsible for his compromised hearing.

We moved to Fifteenth Street in Del Mar, into a small apartment at the back of a house with an ocean view. I kept my nineteen-foot sailboat in nearby Mission Bay, on a mooring I had made out of an old engine block and a stainless steel beer keg. The ocean was still a big presence in my life. It was a short walk to a good surf beach in Del Mar and not far from Black's Beach, a long stretch of sand with intermittent reefs where swimsuits were optional. My brother Keith came to study nearby, at San Diego State University, and we often went swimming together, training for the La Jolla Mile, an open ocean competition.

When not on the beach or sailing, I was in class. Among my teachers was Gordon Sato, a diminutive man of Japanese descent who had been assigned to a relocation camp during World War II, become an American soldier afterward, and then returned to California to study biochemistry. Sato always seemed to have beautiful women around his lab to "teach [him] new languages" but usually showed benign indifference toward his students. However, I excelled in his class, and he seemed to want to encourage me. I had become fascinated by the cell culture methods he had developed in which tissues could be dissolved with enzymes to yield single living cells that could then be grown in plastic dishes. Sato seemed to recognize that I might have more potential than just practicing medicine; as we sat chatting one day in the sunshine, he asked me if I was interested in basic research. In fact, I had an idea that I wanted to follow up on that was based on an experiment I had done on cells isolated from the heart of a chicken embryo.

A few days later Sato told me that the distinguished biochemist Nathan O. Kaplan, a leading enzymologist, wanted to see me, not least because he was impressed with my unusual background. Although I was still only an undergraduate, Kaplan urged me to come up with an idea for a research project that would stimulate his curiosity. It did not take me long to find one: I wanted to study the "fight or flight" response caused by adrenaline. Then came the moment when I made an important leap in my personal evolution from medic to scientist and asked how the adrenaline made the cells beat faster. I assumed that someone already knew the answer to that question, but surprisingly no one did, even though this mechanism is crucial to our survival. For a few days I read the scientific literature. I began to learn about receptors, the proteins in cells that drugs and hormones interact with as the first step in their actions. One theory, favored in Britain, was that adrenaline worked inside the cell. The prevailing American view was that it worked somehow on the cell surface. I told Kaplan we could end this dispute by using the sheets of coordinated contracting heart cells to study the actions of adrenaline. He liked the idea, and not only would he give me a chance to do my project, but I would have a small lab of my own. By then Kaplan had more than forty scientists working for him, crammed into several laboratories, and those who coveted the idea of having their own space were not amused to discover that the one spare lab had gone to an undergraduate with no research experience.

The experiment I designed involved opening a twelve-day-old fertilized chicken egg. I used forceps to make a hole in the top of the shell and then to lift the contents into a petri dish. Each embryo was translucent with large eyes. Its red, beating heart was visible through its skin, and I removed it with surgical scissors, minced it, and then used an enzyme to digest the collagen that glued the heart cells together. After a day in which the cells had been incubated at body temperature with a growth medium containing sugar, amino acids, and vitamins, I examined them under a microscope and witnessed what seemed like a miracle. The tiny cells I had liberated from the chicken had attached to the plastic dish surface and flattened. Every one of them was contracting, like thousands of tiny hearts. I watched them for hours and then witnessed a second miracle: Over the course of days, as the heart cells divided and began to touch one another, their beating became synchronized until eventually the entire plate of cells, one cell layer thick, would contract as one.

Kaplan and the other scientists were as excited watching the heart cells toil in the culture dish as I was. When I squirted in a little adrenaline, the effects were magical: The heart cells immediately began beating faster and harder. Flush out the adrenaline, and they would slow to their normal rate. Add more adrenaline, and they would take off again. On discussing my findings with Kaplan, we came up with a novel way to unlock adrenaline's secrets. I had no idea that answering the basic question of where it acted in cells was going to occupy me for the next decade.

Back east, in the NIH laboratory of the Nobel Prize winner for chemistry, Christian B. Anfinsen, a young scientist named Pedro Cuatrecasas had attached insulin to tiny beads made from sugar molecules (sepharose) and found that due to the size of the beads, the insulin could not get inside fat cells. However, the insulin could still deliver its hormonal action by stimulating the fat cells to take in glucose and convert it to triglycerides. This was a simple and elegant way to prove that insulin acted on a receptor found on the surface of fat cells.

I could do something similar to find out where adrenaline acted, drawing on expertise in Kaplan's own lab. There, Jack Dixon was studying enzyme activity by attaching these large protein molecules to sand-grain-size glass beads. Nate suggested that I get together with Jack to see if there was a way to chemically attach the adrenaline molecule to the beads and still retain its biological activity on heart cells.

This required some effort. We came up with a long "molecular arm" that could chemically attach to the glass at one end while holding the adrenaline molecule far enough away from the bead so it could still reach the hypothetical adrenaline receptors on the surface of a cell. We made the first batch of "adrenaline glass beads," and after extensive washing to get rid of any free adrenaline, they were ready for testing.

Using a micromanipulator, a device to move objects over tiny distances, I placed some glass beads near the heart cells. Nothing happened, which was a good sign: The adrenaline was not leaching off the beads. Slightly twisting the knobs of the micromanipulator, I gradually moved the beads to kiss the heart cells, which immediately jumped to a new pace. In elation, and due to the same mechanism, my own heart jumped, too. I moved the beads away, and the cells resumed their normal rhythm. I repeated the experiment with untreated glass beads: nothing. Kaplan greeted my results with childlike enthusiasm. He would grab colleagues, students, and friends—in fact, anyone in the vicinity—and urge them to take a look at the small television screen attached to the microscope as I moved the beads on and off the heart cells.

Kaplan suggested that I descend two floors to seek the opinion of Steven Mayer, head of pharmacology, on the potential of our findings for publication. This was my first introduction to the politics of science. Mayer agreed to see me but was cool at first and seemed reluctant to acknowledge that a discovery in his field could be made in an enzymologist's lab. But his curiosity got the better of him, and he ended up suggesting some important control experiments to support our findings and to rule out other possible causes or artifacts. (Little did I realize that I would spend three years carrying out these experiments.) To do this, Steve suggested that we could use drugs that specifically block the action of adrenaline on the various hypothesized receptors, such as propranolol, a "beta-blocker."

The experiment was simple in theory, as most are. It was easy to observe that the heart cells beat harder and faster in the presence of the beads and that the effect was inhibited by the beta-blocker, but it was hard to quantify this effect. Adrenaline stimulates two responses in heart cells: It increases the rate of beating and the force of the contraction. To work out how to measure the change in the force of contraction, I consulted John Ross Jr., head of cardiology, who linked me up with Peter Maroko, a likable and knowledgeable cardiologist who was using dogs to study heart attacks. We decided to place the glass beads on different parts of the surface of the dog hearts to see if they had any effect. My Navy experience of reviving hearts made me a hit with the surgeons on the team, and I was accepted immediately.

The results could not have been more dramatic. When the adrenaline glass beads were placed on most areas of a dog's heart, nothing happened; however, when we touched the pacemaker region, the sino-atrial node, the heart immediately took off with a rapid beat. The second the beads were removed, the pace slowed to normal. In place of my crude stopwatch measurements we now had reams of paper from an electrocardiogram and specific force transducers recording the effect of adrenaline in all its detail: Add adrenaline, and the distance between the blips on the trace fell as the dog's heart beat faster while the readout from the force gauge widened significantly.

Ross called Kaplan to tell him he was impressed with the results. Kaplan had just been honored with a membership in the National Academy of Sciences and decided that this was just the kind of work he would like to use to invoke one of his new privileges: With me, Jack Dixon, and Peter Maroko he could submit a paper on the discovery to the academy's prestigious journal, the *Proceedings of the National Academy of Sciences* (PNAS). I was elated. Three years after Vietnam I was publishing my first paper on a discovery that had been made mostly as a result of my own curiosity.[9] And to top it all, I was still an undergraduate. I felt a profound sense of satisfaction that had much to do with escaping the limitations of my early education; I was now working with the scientific elite and succeeding. Although the practical applications of the research were yet to be determined, the gratification was greater than winning at swimming, even than what I had felt when treating the children at the orphanage.

By then I had been doing everything I could that would help me get into medical school. Once a week, sometimes on weekends, I would drive my VW Beetle to a Tijuana clinic for the poor to help treat people with major genetic deformities, which ranged from removing extra digits (polydactyly) to removing a basketball-size benign tumor from the belly of a young girl whose family was convinced she was pregnant. But my gut instincts were beginning to tell me that my true calling was in research. Sato argued that I could affect far more lives with scientific breakthroughs than I could by working with one patient at a time, and I found myself repeating those arguments to my brother Keith. Still, attending medical school did not preclude a scientific career.

Decision time came when I went for an interview at the University of Southern California on a boiling day with the air thick and yellow. After a two-hour-long session in a dingy office in the medical school, the interviewer concluded that given my interests in research, I probably would not be happy in a clinical program. Given how much I disliked the claustrophobic surroundings, I found myself in agreement. (It seemed appropriate that my interviewer was a proctologist.) By that evening, as I swam in the sea off La Jolla to wash off the sweat and grime of that disgusting day, I had made up my mind: I wanted to continue my research with Kaplan, whom I liked very much.

When I told him of my decision the next morning, Kaplan seemed very pleased and called Palmer Taylor, the husband of one of his postdoctoral fellows, Susan Taylor, who had a lab across the way from mine. Palmer said they would be pleased to have me the following fall, despite my late application. All I had to do now was make certain I graduated from the University of California. I had all the credits that I needed from classes plus the independent study credits for my research with Kaplan. However, there was one last hurdle that gave me nightmares.

The founding class of John Muir College, University of California, San Diego, had a spoken language requirement. I had by then switched to Spanish from French to aid in my work with patients in Mexico, but I was not sufficiently competent in either language to pass an oral exam. I proposed an alternative that would be useful and also fulfill the spirit of a language requirement: I would translate a scientific paper on cultured cardiac cells that had recently been published in a French scientific journal. The dean agreed to my proposal and gave me one week to turn in my translation. The work was much harder than I expected; I was surprised by the amount of slang and vernacular used that was beyond the scope of my dictionaries. The examiner was impressed with my work, however; I passed and obtained my bachelor's degree in biochemistry with honors in June 1972, a little more than three years from my very tentative start at the College of San Mateo. I was hired by Kaplan to do a summer research project that involved purifying several enzymes (proteins) and producing kilogram quantities of a key—and expensive—vitamin used in his lab.

University life in the early seventies was a far cry from today. Drug use was common and considered relatively safe. Indeed, it was one of the pharmacology faculty who suggested I try cocaine, telling me that I would be impressed with the effects. Across the hall from my lab a student with a chemistry background found a creative way to pay for his tuition at medical school by working late into the night making batches of LSD.

In these days before the AIDS pandemic, attitudes to sex were equally relaxed. Some professors on the medical school admissions committee always seemed to have an unusual number of attractive undergraduates hanging around their offices. The undergraduate dishwasher for my lab did not wear a bra, in common with many women of the day, and because she was fond of wearing transparent tops, too, an unusual number of people—all male—found one reason or another to pay a visit to my lab.

My graduate career began in September 1972. I found it very satisfying that my adrenaline studies were by then being taught as proving that adrenaline worked on cell surface receptors. Meanwhile, I worked several hours each day in the lab attempting to find out more about the hormone's effects on the heart. John Ross and another cardiologist, Jim Covell, suggested that I examine cat papillary muscles—cylindrical muscles about a millimeter across and half a centimeter long that ensure heart valves close at the right time during a heartbeat. Many scientists have found them a useful tool for studies of the mechanical properties of cardiac muscle because the component cells are aligned in a linear fashion. The challenge was to remove the heart and extract the muscles before the organ deteriorated, which happened rapidly. Like all other scientists I dislike using animals in research, although in this case I knew I was providing a more humane death (using an overdose of Nembutal) than the usual fate the animals suffered in the public shelters that supplied them. By tying a piece of suture to one end, the muscle could be anchored in a bath of salt solution that had oxygen bubbled through it. The free end of the muscle was clipped to a tiny strain gauge that could measure accurately the force of contraction. The responses to adrenaline stimulation were dramatic, much more than with the heart cells—as little as one tiny heated glass bead caused clear-cut effects. This setup enabled the very accurate testing of a variety of drugs and hormones on the heart muscle to help sort out adrenaline's mechanisms of action.

Among the experiments that Steve Mayer had suggested was to study what effect cocaine had on the glass bead–adrenaline responses. One of the many ways in which cocaine was thought to work was by blocking the movement of adrenaline into nerve endings. Mayer had a supply of cocaine (along with other narcotics) that he kept in a safe in his office. We found that cocaine boosted the effect of the adrenaline on glass beads—which explains why the drug can cause chest pain in those who abuse it—and we concluded that cocaine must have additional sites of action on the membrane of cardiac cells. Even though I had a full class load, I now had to write a second paper for *PNAS*, which was published early in 1973. My final exams, as a result, fell short of the medical school class mean by 0.5 percent. My career had hardly gotten started when it was under threat once again.

Although other students had been kicked out of graduate school for falling below the class mean, I had published more papers in quality journals by the end of my first year than most doctoral students managed in five, and so was given a second chance, on the condition that I take an oral exam from a panel of senior professors. I did well, and they suggested changing my grade from an F to an A, although they ultimately settled on a B, which allowed me to stay on. Over the next two years I completed eleven more major papers in addition to my medical school teaching assistant duties, which included carrying out open-chest surgery on dogs.

I began to think of new uses for immobilized drugs and enzymes. One idea came to me while listening to a lecture by J. Edwin Seegmiller on gout, known as the "disease of kings" due to its association with rich foods and alcohol consumption. Gout causes a buildup of uric acid in the bloodstream and, ultimately, uric acid crystals in joints and other tissues. The result is intense pain, arthritis, and even death. The condition is linked with the metabolism of purines, which are nitrogen-containing compounds that are components of DNA. All mammals except humans possess an enzyme called uricase that breaks purines down into a soluble product. Gout patients are treated by injecting uricase derived from pigs, but they often develop an immune reaction in which the body attacks the porcine enzyme as foreign, thus limiting the therapy.

I had a wild idea: Why not pass blood over immobilized uricase, housed in an extracorporeal shunt, which would enable a bypass of the body's blood circulation and prevent an immune reaction from forming? Kaplan was interested, while Seegmiller was cautious but thought it worth trying. I faced dozens of technical questions and hurdles. How should I link the uricase to the glass beads? How could I test the idea, since we could not use patients for a treatment this experimental? What were the risks of using a shunt? Kaplan encouraged me, as he always had, by saying that most scientists tend to talk themselves out of doing an experiment by thinking of all the ways it might fail, and "If you believe in the experiment, then just try it."

With the help of Jack Dixon we attached the enzyme to beads and kept it active. To everyone's surprise and my delight, not only was it active but it seemed even more active than the native enzyme. I had developed a unique way to pass blood over the uricase beads by using a cardiotomy reservoir, a soccer ball–sized plastic bubble with nipples on both sides to attach, via tubing, to an artery and to a vein. The blood enters a central chamber that is surrounded by a fine mesh to remove blood clots before the blood passes back to the patient. I added the enzyme beads to that same chamber, using the filter to keep them in place.

The challenge was now how to test my "enzyme reactor" and on what species. Seegmiller pointed out that Dalmatians had a defect in their metabolism that made their uric acid levels rise when on a high-purine diet—which is not hard to induce since meat is rich in purine. A Dalmatian was placed on an all-meat diet, and its uric acid levels soared. Within four hours of perfusing its blood through the immobilized enzyme reactor the uric acid levels fell back to normal and the dog recovered. But my critics remained unconvinced: Perhaps the uric acid levels in dogs were nowhere near high enough to mimic those found in gout. To complicate matters, dogs also have some uricase.

Seegmiller suggested using a bird as a trial subject instead. Birds do not have any uricase—hence the white color (which is uric acid) of bird droppings. Given the size of my enzyme reactor, the only bird that could cope with having it plumbed into its circulatory system would be a very large one. In fact, according to the chief veterinarian, I needed a bird weighing between 60 and 80 pounds. He knew of a nearby farm that could help supply me with a turkey of that size but warned me that it would be relatively old and probably quite temperamental. This turned out to be an understatement.

When the 75-pound turkey arrived in a truck, it created quite a stir with its difficult demeanor and six-foot wingspan. Only the surgical suite in the cardiovascular lab was big enough for the task at hand. After we put two operating tables together and set up our equipment a rope was tied around the turkey's neck and it was coaxed out of its cage. Our experimental subject was by now looking bigger and meaner than ever, but once the bird had been restrained—a job that took four technicians—I managed to get a blood sample to test for uric acid levels.

The vet was unsure of what to recommend as an anesthetic and suggested we try pentobarbital, a barbiturate, using the same dose per pound of body weight as we would a dog. I infused the appropriate dose into the bird, which was still being held down. Then the vet told me he had remembered something about turkey physiology: We might need more pentobarbital. On cue, the bird seemed to agree, turning his head and looking menacingly at me. The veterinarian urged patience, but after a few minutes with no change I repeated the dose. The bird relaxed a little but was not sleepy—not that I had the slightest idea what a sleepy 75-pound turkey really looked like. We decided to double the dose. Perhaps the big bird was now looking a little more dazed. Then again, perhaps not. I injected three more big doses. At last, Big Bird was out cold.

We all lifted him onto the two operating tables where we had the enzyme reactor and blood pump. I was about to cut an artery to install the shunt when the bird suddenly blinked, then moments later chaos followed as the bird flapped its wings, bringing the stainless steel tables up and down with a crash. The technicians scrambled to hold the bird while a very large dose of pentobarbital was again injected into its wing vein. The second time around I got quite far in getting the shunt established and the experiment going. Each time the bird blinked, more pentobarbital was injected. Finally the bird lay motionless on its back before me, everything under control.

Just as I began to relax again, the turkey woke up, and tables, enzyme reactor, IV bottles, and people went flying as it tried to flap. The lead technician had reached breaking point and injected the whole bottle of the drug into the thrashing turkey, stopping it immediately. He asked if he could terminate, and I assented; the experiment was clearly a failure. Even if the uric acid level had dropped dramatically, no one, including myself, wanted to confirm the fact.

Before the bird arrived, there had been much discussion about what to do with it after the experiment. The consensus was that we should roast it at a beach party for graduate and medical students. The dead 75-pound bird that was to be the guest of honor now presented new problems. We had not thought of how to pluck it, and a turkey that size could take half a day to cook. The veterinarian stepped forward with a confession and a plan. It turned out that he and his fellow students survived veterinary school by learning how to cook large animals in the autoclave: A period of time steaming in this glorified pressure cooker would make the feathers easy to remove, and the turkey could then be rushed to the beach barbecue and browned over the fire. I was quite surprised how precisely he knew to set the autoclave for the turkey, and with steam under pressure it cooked in record time. But we were not done yet.

My biggest worry was the large doses of pentobarbital that had been injected into our dinner. Would the entire beach party be put to sleep? Everyone argued that the heat from the autoclave would break down the drug, and further aid would come from the final roasting. I agreed, and the prize bird was packed off to the beach. While I was the hero of the hour as more than one hundred feasted on the turkey and beer, I could not bring myself to eat my failed experiment. I watched my fellow students carefully for signs of drowsiness, but I seemed to be the only one who was fatigued. I went home early for some badly needed sleep.

Despite my turkey troubles, Kaplan was impressed with the data that I had accumulated on the immobilized uricase and suggested that I write it up for publication. I started preparing the paper whenever I could find time around my regular research on adrenaline and my course work. I began to be struck more and more by the high caliber of the scientists with whom I was working, although I did not fully appreciate this until later on in my career. Kaplan himself was considered one of the world's top enzymologists, having demonstrated that there were multiple forms of enzymes (isoenzymes) with similar but not the same properties: In the case of lactate dehydrogenase (LDH), an enzyme that metabolizes lactic acid, it is possible to tell if someone has had a heart attack by measuring the ratio of types of LDH that leak out of damaged cells into the blood.

One day when Kaplan was feeling particularly proud of my progress while working with him, he started a conversation about my scientific pedigree, noting that it stretched back over several generations of biochemists, mine being the fourth. Kaplan, who was third in line, had spent his early career working under Fritz Lipmann, with whom he had discovered a key biochemical intermediate in our metabolism, coenzyme A. In 1953, Lipmann had received the Nobel Prize in recognition of this work. In his book *The Wanderings of a Biochemist,* Lipman described how the results of one piece of research had often been a stepping-stone to the next along an unpredictable route, as I myself would discover. Kaplan became philosophical and slightly emotional as he described how he considered Lipmann his father in science and Otto Meyerhof, the German-born discoverer of basic metabolism—notably the role of an energy molecule called adenosine triphosphate (ATP)—and a 1922 Nobel laureate, his grandfather. Kaplan stopped short of saying I was, in effect, his son but made it clear that that was how he felt. I willingly accepted him in a scientific paternal role.

Whenever Kaplan's famous friends and colleagues, including Lipmann, would visit the university, he would host a big party at his home in their honor. He did not usually invite anyone from the laboratory but would ask me along to be bartender, which allowed me to meet great names in science, such as Carl and Gerty Cori, who shared the Nobel Prize in 1947 (Gerty being the first American woman to win the prize) and Ephraim Katchalski, a biochemist who was working on immobilized enzymes. (Katchalski changed his last name to the Hebrew name Katzir in 1973, after being elected president of Israel by the Knesset.) Another frequent guest was William McElroy, the university chancellor, who was best known for working out the biochemistry of the firefly's glow. He was appalled by my less-than liberal use of a shot glass and taught me to "pour drinks like a biochemist." He would turn the scotch bottle upside down into a large glass and slowly count to three, a ritual he could repeat up to four times in a party.

Despite my success in San Diego, my life was still overshadowed to some extent by Vietnam. After rumors began to circulate that a massive protest was going to be held near where I lived, to halt a train carrying napalm to ships docked in the harbor, strange things began to happen. First, my home telephone suddenly began to sound different, with more background noise. Conveniently enough, a repairman seemed to be in constant attendance, sitting in a small booth on the pole immediately outside my second-story living room window. One day he showed up at our door as one of three FBI agents who insisted on interviewing us. At the end of a long discussion, they indicated that they had checks allegedly written by Barbara that had been linked to international money laundering. As well as needing a sample of her handwriting and fingerprints, they would, by the way, need mine as well. Before leaving, they warned us that we had better be prudent or they would deport Barbara. They said they would be back in two days, which just happened to coincide with the date of the protest.

On that day, hundreds of police in riot gear lined up in front of the railroad track, which runs on a cliff above the ocean. By the afternoon a group of around five hundred had assembled on a grassy area between the main street and the tracks. As soon as I saw the police moving in to surround the chanting crowd, I grabbed Barbara's arm so we could escape quickly. Just as we reached the street outside the police line, the first plumes of tear gas appeared. As helicopters swept overhead, people started running in every direction, while the police arrested anyone they could. We made it back to the apartment and watched events unfold from our front window until nightfall, when the last groups of protesters being hunted under helicopter searchlights were arrested. By Monday the telephone repair booth was gone. The FBI never returned, and Barbara and I resumed classes.

I had been wanting to test an idea that was inspired by one of Kaplan's earlier breakthroughs. Could my beating heart cells in tissue culture be used to study biochemical changes associated with a heart attack? Perhaps I could simulate what happens when a blocked artery starves the heart of blood—and thus oxygen—by measuring the release of marker enzymes, LDH and creatine kinase, from my heart cells when the oxygen levels fell. From the first experiment onward it was clear that I had a real hit. The enzyme released in the single cells mimicked what happened in actual heart attacks. One level of enzyme release was associated with the cells' ability to recover, whereas higher levels were an indication of cell death. This could potentially be a wonderful tool to screen drugs that could protect heart cells or promote recovery. Kaplan was so excited by the implications that he asked me to contribute a section to a major cardiovascular center grant application. I was pleased at the request, even though it involved a lot of extra work and took weeks of discussions and presentations. The grant was finally finished and circulated to the large team associated with the cardiac center. Then the unthinkable happened: Kaplan himself suffered a heart attack.

He was perceptive enough to notice the symptoms of sweating and chest pain, and immediately went to the university hospital, where his blood was evaluated using his own method. Enzyme levels showed that he had had a relatively minor heart attack and was expected to recover fully. I visited him often. The practice at the time was to sedate the patient heavily for several days, in theory to keep stress levels low, but it also had other effects on Kaplan, including causing him to be somewhat delirious at times. He once asked the head of cardiac surgery to open his chest and apply my adrenaline beads to make his heart work better.

While authors fret about their words being stolen, scientists worry about the theft of their ideas without attribution. My first encounter with this form of intellectual appropriation came as my mentor lay ill in the hospital, and Jane (not her real name) decided to use this opportunity to take over my project—by removing my name and Kaplan's from the cardiac center grant and substituting her own. The cardiovascular team assumed it had been done with our permission, because Kaplan was obviously too ill to handle this burden.

We discovered what happened a month or so later when the program head, John Ross, sent me a copy of the cardiac center proposal as a courtesy. When I found Jane's name instead of my own, my brain almost exploded. Had Kaplan betrayed me for the benefit of another scientist? I rushed down the hall to his office, threw the grant on his desk, and shouted, "What the hell is this?" Kaplan was unaware of the switch and was equally outraged, and when he calmed down, he argued that an important official's career could be jeopardized by the affair, an outcome Kaplan did not want for various political reasons. I was young, naïve, and not the least bit satisfied or placated by his logic.

Then Nate told me how he, too, had been a victim of intellectual theft early in his career. After he had written up his discovery of coenzyme A with Lipmann for publication in the *Journal of Biological Chemistry,* Lipmann sent a draft to a senior colleague to review. They heard nothing in response, but in the meantime the journal sent Lipmann a manuscript to review: It was Kaplan's and Lipmann's own paper on coenzyme A, with their names removed and replaced by the name of Lipmann's colleague. Lipmann called the editors of *JBC* and got the authorship restored. The paper became famous, contributing to Lipmann's being awarded the Nobel Prize.

Although Kaplan argued that the "truth always comes out in the wash," I still believe that fraud should not be swept under the carpet for the good of the scientific community. Something much greater is at stake than individual reputations—the credibility of science itself. I certainly felt that the bad guys had won when Kaplan told me a few days later that my grant was going to be funded via Jane: I should have felt good about my ideas being validated, but I found it hard to do so when they had been appropriated by someone else.

Throughout this time my research on immobilized adrenaline was moving forward. One unresolved issue was whether the means that we used to tether the adrenaline to the beads interfered with its action. Like many simple questions, this required a lot of work to answer. There was only one report of a chemical bond to the same region of the adrenaline molecule, so we had to take a closer look. Lyle Arnold, who had been a postdoctoral fellow with Kaplan, helped me use the method of nuclear magnetic resonance, NMR, to reveal where the chemical bond on adrenaline was inserted. Thanks to Chip's NMR skills, we showed that it was located on the ring in a position compatible with biological activity.

The final hurdle was to reveal whether the adrenaline was still fully active when we attached a very large, bulky molecule instead of a glass bead. For this task I sought the help of the head of the chemistry department and member of my thesis committee, Murray Goodman. A polymer chemist, Goodman had several large polymers to which we could bond adrenaline; if the combination of polymer and hormone was still biologically active, it would prove that substituting glass beads for the polymer could also be active.

Goodman set me up with a postdoc, Michael Verlander, who had been making polymers of two modified amino acids, hydroxypropylglutamine and para-amino-phenylalanine. We made polymers in two sizes—one so big that it would take time to diffuse into tissues, the other small—and checked them to ensure they did not break down to release adrenaline. Test results on cardiac tissue were clear from the very first injection: They worked and were almost identical in activity to the native hormone. I was ecstatic because I now had all the answers for my critics and was looking forward to the publication of the results, once again in the prestigious *PNAS*.

I wrote the paper after several revisions, and within a few weeks it was ready to be sent in with Venter, Verlander, Goodman, and Kaplan as authors. Then Kaplan called me into his office to tell me that Goodman wanted a change in the order of authorship, with Verlander in the coveted spot of the first author. I reminded him that it was my study, based on our ideas, and that Verlander had simply been the chemist who made the polymer. Kaplan agreed but argued that I was already becoming famous for a graduate student, and it did not matter where my name was as long as it was on the paper. Verlander, on the other hand, needed a career boost. Verlander was a colleague and helpful scientist, and while I relented, I still felt my contribution was the key one and that I deserved to be first author. In the end, Kaplan's instincts were correct in that he and I got almost full credit for the work, but the concession still stung.

By that time Kaplan had discovered from a colleague that he had been nominated for the Nobel Prize for his work in enzymology. He was accordingly looking for a dramatic and newsworthy discovery to persuade the Nobel committee that he was a deserving recipient, as had Lipmann for work that Kaplan had himself contributed to. To that end, Kaplan's right-hand man indicated that I could have a blank check to try some of my ideas on adrenaline receptors and their underlying enzymes. One obvious extension of my research was to go fishing for receptors, the sites in the body where the adrenaline hormones bind. I wanted to use the bead-bound drug to pull the receptor out of a complex mixture of cellular proteins as a means to purify and study it.

Here again I turned to my old friend, the turkey. Earl W. Sutherland Jr. of Vanderbilt University had used turkey erythrocytes (red blood cells) for his work on deciphering the mechanism of the action of hormones on cells, work for which he was awarded the Nobel Prize in 1971. According to his research, the hormone attaches to a receptor on the surface of the cell, activating the enzyme adenylate cyclase to form a molecule called cyclic AMP, which acts within a cell. As early as 1960, Sutherland had suggested that this mechanism underpinned the effects of many hormones. His peers, however, rejected his suggestion that just one chemical, cyclic AMP, could be responsible for the numerous effects that are known to be caused by different hormones. Today we accept that cyclic AMP is one of the "second messengers" that help hormones achieve their effects. My intent was to use the adrenaline on the glass beads to capture the adrenergic receptor in blood and presumably the adenylate cyclase that was linked with it at the same time.

To accomplish this I needed a lot of turkey blood, and I called the vet with whom I had worked on my Big Bird–surgery fiasco. He made arrangements for me to visit a turkey farm about an hour's drive from the medical school. All too aware of how much help I had needed the previous time, I persuaded Jack Dixon to come and assist me in drawing blood from the turkeys before they could do the same to us. We must have looked quite a sight when we arrived in our white lab coats and goggles. "Are you the boys here to get the turkey blood?" the farmer asked. He grabbed a turkey, flipped it over, and confidently held it down while I quickly removed 50cc of blood from a wing vein. The first experiment with the blood was successful: The immobilized adrenaline did seem to pull the receptor out of the blood to enrich it. But it soon became clear that to isolate pure receptor was going to be a major undertaking, and though I did not know it at the time, I would be drawing turkey blood for years to come.

By now, after only three years in graduate school, I decided that I had accomplished enough and that I was going to write up my thesis and attempt to graduate. Typically, a Ph.D. candidate would have one or two submitted or published papers after the conclusion of his thesis work, which averaged five to six years. But I was already prepared to go to the next stage since I had submitted or published twelve papers, half of them in the *Proceedings of the National Academy of Sciences* and others in serious journals, including *Science.*

Barbara and I had recently moved to a two-bedroom town house nearer to the campus. I did not want to be distracted by anything, so I moved out of our bedroom into the spare one, which I had converted into an office for writing. After my experience in Vietnam, I liked having the day free to swim, sail, and surf, and would start writing in the evening, with my greatest productivity often occurring between midnight and 3:00 A.M.

While I could easily have pasted together my published papers and submitted a "Frankenthesis," I wanted to do more to put the research findings in perspective for myself and to go into detail on some theoretical issues that my discoveries might raise for the new field. I was particularly fascinated by how my data were at odds with prevailing theories on how drugs and hormones worked on receptors. Existing theories assumed that drugs and hormones reached a similar concentration essentially simultaneously in all cells of a tissue and that the response was directly proportional to the number of receptors occupied by the drug or hormone. In other words, if half of its receptors contained a given hormone, then a muscle would respond at one-half of its maximal level. My data suggested that the responses were not in any way directly proportional to the percentage of the receptors occupied in the tissue. I wanted to argue that given the short time that it took the cardiac muscle to respond to a hormone, the hormone could not possibly have reached every one of its cells. For the whole muscle to act in a coordinated manner to make a heart beat as a unit, there must be a propagation of the hormone signal via a few cells to the whole muscle.

Because this went against prevailing thinking, it was a challenge to make the arguments convincing enough for others to consider. To back up my suspicions I needed to understand the rates that molecules diffuse or move through liquids or solids and the mysteries of boundary layers where, for example, an indolent layer of liquid lies adjacent to a vessel wall even in a well-stirred medium. In other words, I had to move into unfamiliar scientific territory, which was all the more daunting because I lacked the appropriate mathematical background. But I eventually prevailed.

I was drawn to the sea during this intense period of thinking and writing. As a respite from my monastic existence, I decided to go on a sailing trip a hundred miles or so to Mexico. I used my nineteen-foot lapstrake (a traditional form of construction) wooden boat that was almost as old as I, having been built in Denmark in 1949. In retrospect, this trip seems a little foolhardy to me now, but at the time I felt I had enough experience from sailing my little boat up the coast and back to Catalina Island, navigating the ocean by taking bearings from the stars and using a simple home-built Heathkit radio direction finder. It was the most fantastic feeling to be on a small open boat on what seemed like huge ocean swells, in the warm sunshine and with no land in sight. At night I had only the stars and radio signals to guide me.

Two postdocs in Kaplan's lab, Chip and Ron Eichner, thought that it would be a great adventure to join Barbara and me. There was confusion over who was bringing supplies, and we ended up with very little food to feed the four of us for two days. Our food crisis was made worse by my having packed the cooler with dry ice from the lab, which carbonated our meager provisions.

The first day was glorious. We sailed downwind in the sun to the Coronado Islands, almost directly off Tijuana, only to be chased off by the Mexican police. I was not sure what to do, because the wind and seas had intensified to 18 knots, which prevented us from easily turning back. After a few more hours we found refuge in a small bay and anchored by tying to a huge piece of kelp. Although we were all exhausted and hungry, the four of us did not get much sleep in the tiny boat. The next day, with calmer seas, we sailed back to Mission Bay in San Diego, and when we hit the dock, we rushed to get some food.

In the days before computers and word processors, getting a thesis typed was a big deal—all the more so when your wildly ambitious 365 pages of hard work would cost 50 cents per page for the drafts and $1.25 per page for the final version. After three months of intense effort I had ten copies of my telephone book–size magnum opus ready to distribute to my thesis committee, which included Kaplan as chair, Gordon Sato, John Ross Jr., Steven Mayer, and Murray Goodman. They seemed surprised at the size and scope of what I was handing them, Kaplan most of all. He joked that he was not ready for me to graduate yet because he had no money to pay me as a postdoctoral fellow. Once they had read and approved the written thesis, I had to prepare to defend it before an audience.

My anxiety rose because it seemed to take forever to find a convenient time for my thesis defense, given the packed calendars of the committee members. To add to the stress, the defense was scheduled to be held in the main auditorium of the medical school, which could hold a large audience, as if to underline the attention that my work had drawn. As I walked down the stairs to the auditorium with Kaplan he gave me only one piece of advice: "Know more than anyone else in the room on your subject, and you will be okay." I found his counsel strangely comforting because I felt I really did know more about my subject. But then I remembered that I was about to face a senior committee, three of whom were department chairs.

The auditorium was packed, but I tried not to let that unnerve me. I now know that if I am well prepared, I can speak publicly with ease and the speech just flows. I have a strange ability to feel as though there is nothing on my mind, as if I can preview what I am about to say and edit it as I go along. That day I spoke for ninety minutes nonstop, without notes, and then answered questions for another ninety. I was stunned to see that after three hours no one had left the auditorium. The most tense moment came when the committee got up to form a huddle. After a few minutes, they walked over to me, led by Kaplan, and said, "Congratulations, Dr. Venter." It was December 1975, a little more than seven years and five months from the moment I had stepped off the plane on my return from Vietnam.

I overcame my early lack of an academic background to earn a Ph.D. from the same university that had provided the backdrop for so many key times in my life. I used to drive by UCSD as a child when my parents went to and from San Diego each summer and while I was in the Navy, on my way to and from Newport Beach. The university had always left me slightly envious of those who had the ability and privilege to attend it. Now I would be the first in my family to earn a doctorate degree. I was proud, happy, relieved, and tired.

But even as my peers, friends, and family congratulated me, I was looking ahead to what would come next. Just as my IQ test had paved the way for me in the Navy, so, too, did my doctorate create opportunities. The path of least resistance would be to spend up to five years in a postdoctoral position, akin to intern and resident positions in medicine, or serving as an apprentice. Kaplan wanted me to pursue this route, either remaining in his lab or doing a postdoc with Sydney Udenfriend, a distinguished biochemist who was then the head of the Roche Institute of Molecular Biology in New Jersey.

A year earlier, however, I had received an unusual offer that would short-circuit a traditional career path. Neil Moran, chairman of pharmacology at the Emory School of Medicine in Atlanta, had visited to review the big cardiovascular grant application that I had contributed to with John Ross. Moran believed that I could skip the postdoctoral phase and immediately assume a medical school faculty position at Emory, a choice Steven Mayer urged as well. In short, I had three job offers when others in Kaplan's lab had trouble getting even a single interview after dozens of applications, so depressed was the job market at the time. But I did not want to follow in the shoes of Mayer and Moran, who did traditional pharmacology. And I noted that all the stars of UCSD had cut their academic teeth in other institutions, often much less desirable ones. It seemed to me that if I wanted to end up somewhere prestigious, I had to prove myself elsewhere first.

To Kaplan's annoyance I looked further afield. Although there had been rumors of a new department of molecular pharmacology opening at Stanford, led by a well-known scientist, Avrum Goldstein, Kaplan warned me I could not wait for it to open and needed to make a decision soon—after all, I had job offers on the table. I made arrangements to visit Emory but found myself intrigued by an offer of a faculty position in the School of Medicine at the State University of New York in Buffalo. A group of leading researchers there was focusing on the molecular mechanisms of neurotransmitter receptors. The group included Eric Barnard, who studied the nicotinic acetylcholine receptors that cause muscle contraction; David Triggle, who was working on the protein pumps that move calcium into cells; and Sir John Eccles, who had won the Nobel Prize for his work on how charged atoms (ions) acted on nerve cells. I organized a visit immediately after the trip to Emory.

The interview in Atlanta went well, and Moran struck me as a likable and decent man. But the lab itself seemed dark and claustrophobic, and I felt that in some unfathomable way it represented old science. I was also a little alarmed that several faculty members asked me why, with my success, I wanted to work there. When I met a member of the Emory genetics department, he joked that the people portrayed in *Deliverance*—a film Barbara and I had recently seen—were not actors but his research subjects. With some pride he said that he could tell me the real story behind the movie. I boarded the flight to Buffalo knowing that Atlanta was not for me.

Although I was very skeptical about moving to Buffalo, I received a warm reception from the team there, which made an effort to impress me and did. The combination of the university, the medical school, and Roswell Park Cancer Institute offered academic science of a depth that surprised me. To complement the institutional expertise on receptors was an interdisciplinary graduate program on biological membranes founded by Demetri Papadopoulos that included Daryl Doyle, the chair of cancer biology at Roswell Park, and George Poste.

I was made a generous offer: substantial laboratory space, an allowance that would permit me to start my research immediately, and a salary of $21,000 a year—$9,000 more than I could earn in La Jolla. Barbara was part of the deal, too, being offered a postdoctoral research position at Roswell in a breast cancer laboratory with a good reputation. I accepted and arranged to start in July, to allow Barbara time to finish her thesis. We sold our house for $25,000, after buying it a few years earlier for $14,000 (it sold again a few years later for more than $100,000—so much for my supposed money-making instincts). We rented a furnished town house in Del Mar. At last, Barbara defended her thesis and together we went through the graduation ceremony at the university—my last notable career milestone in La Jolla. My parents came to stay at our place and we ended up renting another for the weekend. We celebrated there by making love by the fireplace.

We were about to launch the next phase of our careers in Buffalo, where I could leapfrog being a postdoc to take up a junior faculty position. My former English teacher, Bruce Cameron, and his wife, Pat, presented us with a going-away gift of tickets to see *A Chorus Line* in San Francisco. One of the songs we heard that evening would come to haunt me. After driving day and night we arrived in a city that, since it was Independence Day in 1976, seemed uninhabited, almost dead. A single line from the musical played over and over in my mind: "Committing suicide in Buffalo is redundant."

4. STARTING OVER IN BUFFALO

It is on record that when a young aspirant asked Faraday the secret of his success as a scientific investigator, he replied, "The secret is comprised in three words—Work, Finish, Publish."

—J. H. Gladstone, *Michael Faraday*

Soon after we arrived in Buffalo, Barbara discovered that she was pregnant. The timing could not have been any worse. As far as our marriage was concerned, we had now jumped out of the frying pan into the fire.

When we had lived together in California, Barbara and I had been equals, engaged in the same adventure to get our Ph.D.s. We had studied together in a friendly competition and lived together in harmony. Now that I was a medical school professor with two laboratory technicians, two graduate students, and a postdoctoral fellow, I could function independently. But Barbara, who was a postdoctoral fellow working for someone else, felt as if she were standing in my shadow and was also very unhappy in her laboratory. We were in substantial debt because of the cost of our education, and far away from the blue skies and warm surf of La Jolla. Now that she was pregnant, life was even harder for her, and our relationship came under great pressure.

While cracks began to appear in my private life, the first day of my professional life in Buffalo could not have been much worse. The morning I arrived, I was invited to the thesis defense of a graduate student of Peter Gessner, a senior professor in the department, who I can only presume wanted to show off his protégé to the new guy from the West Coast. The problem was that I had just arrived from a university with an intellectually tough, critical, and unforgiving climate, where fools were never suffered gladly, and you quickly learned not to take scientific criticism personally.

As I tore into the student for her trivial thesis, ignorance of the basics of her field, and poor science, it didn't occur to me that I was also implicitly attacking her thesis advisor and one of the senior figures in the department. Because I had been a self-starter in La Jolla, it had never registered with me that a graduate student might not be doing her own work but that it was an extension of her advisors'. Nor was I aware that the culture in a university could be one in which academics seemed to have implicit deals: "I'll pass your student if you pass mine." With my long hair, ponytail, and scraggly beard, enhanced by the polyester and bell-bottomed excesses of my seventies wardrobe, I must have made quite an impression on my new peer group. No wonder that for the rest of my time in Buffalo, fellow distinguished professors would balk at the idea of consulting this parvenu scientist for advice on their own students' quality.

But there were some great times in Buffalo, too. My son, Christopher Emrys Rae Venter, was born March 8, in the middle of a snowstorm and just a few weeks after the notorious Blizzard of 1977, when Buffalo was effectively cut off, wind gusts reached 70 mph, snow drifts topped twenty-five feet, looting broke out, and people died in their automobiles. I had taken part in deliveries while in the Navy, but nothing came close to watching the arrival of my own son. It was love at first sight. Having to pay off our student loans, we could not afford any domestic help, and neither of us wanted to or could take time off from work. Having the luxury of an office, unlike Barbara, I volunteered to take Chris with me to work and did so for several months.

At first Chris slept in a drawer in my office file cabinet, but as he grew, I finally bought a crib. Although I was happy to spend what time I could with him, some of the faculty in nearby offices would complain when he cried. I was trying to be a full-time father, a full-time scientist building a research program, and a medical and dental school professor, a situation that created a great deal of stress for me and for our marriage. I began to take refuge in a well-established pastime: I would purchase scotch by the half gallon and have a drink or two almost every night.

By a year or so after my arrival, I felt I had developed a panoramic view of the local academic landscape. Buffalo probably peaked in the late 1970s or early 1980s, after which its lifeblood of top scientists began to trickle away. Sir John Eccles was leaving just as I arrived, a fact no one had mentioned when I was being recruited; the team that moved me into my apartment had just packed his belongings. The biochemist Eric Barnard departed within a few years to head up a big program in London, Demetri Papadopoulos was lured to the University of California, San Francisco, and George Poste joined SmithKlein. Witnessing this exodus of talent, I realized that I needed to establish my scientific reputation quickly to avoid being stuck in Buffalo forever. Fortunately, I was immediately successful in applying for my first National Institutes of Health (NIH) grant, although the funds came with the telling critique that my proposal was too ambitious.

My work in La Jolla had been on adrenaline receptors, locating them in the fatty membrane that surrounds nerve cells in the autonomic nervous system, which is responsible for processes that are not consciously controlled, from ejaculation to pupil dilation to heartbeat. The logical extension of this research was to isolate and purify the receptor protein in order to study its structure at the molecular level, which was the key to understanding how it worked. No proteins of the rarity of the adrenaline receptor had ever been successfully isolated and enriched, let alone ones embedded in the cell membrane. The first step was to develop a way to measure the concentration of protein isolated from the membrane. We spent close to a year perfecting a probe to detect traces of receptor protein, a radioactive iodine atom linked to a beta-blocker drug that we knew would bind to the receptor. Once we could radioactively label the receptor, we could try endless methods to remove it from the lipid membrane, gauging our success by the level of radioactivity left on the glass filters that we used for the job. The next step was to try to purify the receptor from the vast assortment of other cellular proteins that had come along for the ride from the membrane. Dozens of different approaches were attempted, hundreds of times in all.

Nothing in my Ph.D. training had prepared me for directing this kind of effort, a long, complex journey and trial by fire in which I learned mostly from my mistakes. And, of course, I made many. I began to understand how to manage, stimulate, guide, encourage, redirect, and train students, thesis students, postdoctoral fellows, and technicians, and to manage the complex relationships among them. I learned how to fire and, most important, how to hire the good candidates.

University research is driven by graduate students and postdoctoral fellows, so the competition among labs for top students is fierce. It helped that I was working in a hot field, which meant I had some great applicants. One, Claire Fraser, had just graduated with good grades from Rensselaer Polytechnic Institute, a top science and engineering school in upstate New York. She had already been accepted by Yale but still turned up to be interviewed: She was engaged to a banker in Toronto, and Buffalo was the closest American city with a reasonable university. I was impressed by her but figured that she was Ivy League material and that I would not see her again.

Months later I learned that she was coming to my department. After she had completed her required trial period in several labs, I was delighted when she chose mine for her graduate work. In 1979 we published a paper on how to dissolve adrenaline receptors using a detergent, a useful trick for our studies.

Claire seemed to have a quick grasp of the new approaches that I had brought into the laboratory. Our strengths seemed to be complementary: I was an ambitious scientist trying to move on several research fronts, using various technical approaches, to gain visibility on an international level. Claire, a buttoned-up New England girl, the daughter of a high school principal, was a logical worker whose interest in details of a specific subset of the problem was compatible with my somewhat chaotic, enthusiastic, high-energy-driven, goal-oriented broader approach.

I wanted to give her a challenging problem for her thesis, and one idea was to use a monoclonal antibody specific for the receptor to pull the receptor out of complex protein mixtures so it could be purified. At that time science was buzzing with the news that a process to make monoclonal antibodies had been invented in the Laboratory of Molecular Biology, Cambridge, England, by Georges Köhler and César Milstein, an advance that would earn them the Nobel Prize. Each polyclonal antibody in our blood derives from single white blood cells that expand into millions of copies of themselves (clonal expansion). Köhler and Milstein had developed a method to isolate individual white blood cells and single (monoclonal) antibodies.

Claire agreed to take on a receptor antibody project, and we quickly developed a method to see if we had antibodies that could compete for the radioactive drug binding to the receptor. This ambitious approach soon began to pay off. Our first success came with asthma, which has long been linked to adrenaline receptors. It was also a subject in which I had an abiding personal interest because I suffer from allergic asthma.

Some of the most common asthma drugs, so called beta-receptor agonists, work by stimulating the adrenaline receptor, which in turn causes the airway muscles to relax. Many theories had been put forward as to why adrenaline receptors might be linked to asthma. For example, there was evidence that people with asthma have fewer adrenaline receptors and therefore the airway smooth muscle constricts in response to other stronger influences. In turn, perhaps there were fewer free receptors because asthma sufferers make their own antibodies that block the receptor. We now had a way to test these ideas. Working with Len Harrison of the National Institutes of Health in Bethesda and an asthma specialist at the Children's Hospital in Buffalo, we obtained blood samples from patients. We were surprised when we found that some of the most severely affected had what appeared to be antibodies that affected the adrenaline receptor. This was a significant finding. Len, Claire, and I submitted our findings to the journal *Science*, the premier scientific journal. We celebrated when we received word that our paper would be published.

My Asthma and My Genes

Like many other people I reach for my inhaler in smoggy conditions. Genetics contributes to susceptibility to asthma, and researchers have focused on the glutathione S-transferase (GST) family of enzymes, which helps detoxify compounds such as carcinogens, drugs, and toxins. The thinking goes that the better the body can use these antioxidants to defend itself, the better it can protect itself from airborne pollutants, detoxify harmful particles, and limit the corresponding allergic reaction.

Two of these enzymes are called glutathione S-transferase M1, or GSTM1, and glutathione S-Transferase P1, or GSTP1. GSTM1, found on chromosome 1, occurs in two common forms in the population: present and null. People born with two null forms cannot produce the GSTM1 protective enzyme at all. About 50 percent of the population falls into this category. Meanwhile, a common variant of the GSTP1 gene on chromosome 11 is called ile105, and people born with two copies produce a less effective form of GSTP1.

In a California study led by Frank Gilliland, variants in this family of antioxidant-related genes were linked to reactions to diesel exhaust particles. Diesel particles caused volunteers who lacked the antioxidant-producing form of the GSTM1 gene to have significantly greater allergic responses compared with the other participants. The group that both lacked GSTM1 and had the ile105 variant, which represents approximately 15 to 20 percent of the American population, experienced even greater allergic reactions. Analysis of my genome reveals that I am among this particular group since one of my GSTM1 copies is deleted and I carry ile105. As a depressing bonus, given its detoxifying role, GSTM1 deficiency may make me more susceptible to particular chemical carcinogens, and there is also an association with lung and colorectal cancers. The good news is that asthma is also linked with the absence of another gene from this family, glutathione S-transferase theta 1 (GSTT1) on chromosome 22, and in this case both copies of this gene are present and correct in my genome.

Thinking about the drugs that I had taken to treat the condition, I had an idea. One major treatment for asthma uses steroids (glucocorticoids), which reduce inflammation and trigger protein synthesis in cells. Could it be that they worked by boosting the synthesis of adrenaline receptors and therefore made the cells more responsive to adrenaline? We came up with a simple plan: We would measure the number of receptors on the cell surface using the radioactive drug, then add the steroid hormone and see if the radioactivity—and hence the level of receptors—rose with time. We were delighted when we discovered that the receptor density more than doubled in twelve hours. This paper has been frequently cited ever since as showing the mechanism of action of glucocorticoids in asthma.

Nineteen eighty saw my team publish a flurry of other scientific papers. One was on a second type of cell membrane receptor where adrenaline acted, called the alpha-adrenergic receptor. We found a way to reconstitute adrenaline receptors isolated from the membranes of red blood cells of a turkey (involving lots more turkey bleeding) into other cell membranes, a step critical for the long-term study of the receptor protein. We showed that the receptor density (the number of receptor protein molecules) changed substantially during the course of the cell cycle (that is, when a cell divided). The crowning achievement of the year came with Claire's monoclonal antibody work when she succeeded in making the first monoclonal antibodies, which bound to a receptor that handled messenger chemicals (neurotransmitters). To top it all, I was able to use those antibodies to obtain exciting new information on how structurally similar the family of adrenaline receptors on nerve cells are to one another. The Nobel laureate Julius Axelrod sent the study paper off to the *Proceedings of the National Academy of Sciences* for publication.

While 1980 was a transitional year for my science, it also signaled the end of my marriage to Barbara. Claire had broken off her engagement soon after moving to Buffalo and was now living with a new boyfriend, but because of all the time she and I spent together in the laboratory, Barbara assumed that we were having an affair, although our relationship was only professional. Nonetheless, Barbara herself had begun a relationship with a faculty member from Galveston and soon she got a teaching position in Texas, leaving Christopher and me in order to live with her friend. Although this was a difficult period, I was relieved that she did not try to take custody of Chris. Being a single father was one of my greatest challenges and at times the most rewarding of my life.

Claire and I now began to see each other, and the student who dated her professor soon became an open scandal at the medical school. The situation was complicated by the fact that Claire continued living with her boyfriend, and I was still legally married, though I was separated from my wife. Although divorce proceedings were under way, they had become acrimonious, and Barbara sued for custody of Christopher. The New York courts were unsympathetic to single fathers and ruled in her favor. I still find myself wondering if our troubled young marriage would have survived had counselors been as readily available then as they are now.

Given how much time I had spent with him, not having Christopher as part of my daily life was very painful, and I became deeply depressed. As time went by, I could take some solace in the thought that Barbara had begun to really care about him. The courts granted me frequent visitation rights, and at a young age Chris became a frequent flyer. Barbara would go on to three more marriages and became a very successful patent lawyer. After a twenty-year hiatus, I'm happy to say that we are now friends once again. Journalists have often asked me that tired old question: If I could change something in my life, what would it be? Easy: I would have loved to have been able to raise my son.

That year saw another major professional clash, one that would teach me an important lesson. Although the pharmacology department had a seven-year up-or-out promotion/tenure program, I believed that I deserved an early promotion. After all, I had obtained almost half of the grant money coming into the department (many full professors had none), I had six top students and postdocs, and my publication record was the envy of my peers. I had been approached by other universities and had strong letters of recommendation, so I issued an ultimatum: Give me early tenure, or I will leave.

Peter Gessner, whose doctoral student I had so undiplomatically torn into on my arrival in the department, was in charge of the tenure review panel. After a drawn-out three-month process, the committee ruled that I would have to wait a few more years to join their ranks. As a matter of honor I felt that I had no choice but to leave as soon as I could, but my departure took place sooner than I expected, thanks to an immediate offer from the biochemistry department. There I was given a lab that was three times bigger than my existing one, a teaching commitment that was a fraction of that in pharmacology, a promotion to associate professor, and a pay raise from $23,000 to $32,000. My life improved immediately.

Although I still did not have much money, I began to indulge myself a little. I figured that I could afford to slide a bit further into debt and bought a $3,000, eighteen-foot Hobie Cat, an extremely fast racing catamaran, which enabled me to compete in the weekly races that took place in the strong winds that gusted off Sherkston Beach, near Crystal Beach on Lake Erie, in Canada. Each weekend I would make a pilgrimage to the races with my boat in tow behind my blue diesel Mercedes.

As well as its speed, I loved the technique used to sail a catamaran. Connected to a trapeze (a waist harness attached to a wire strung to the top of the mast), I had to use my body as ballast to stop the sails from tipping over the light hull. It was exhilarating perching on the edge of the windward hull as it stood high out of the water to stop the boat from flipping over. In big waves great skill was required to keep the boat from pitchpoling, when the bows were buried in a wave and the boat catapulted end over end.

It took some persuading for Claire to join me on my sailing adventures. In our first outing I managed to slip on the hull while I was out on the trapeze, and we were moving at 25 mph. When Claire looked back to talk to me, I was gone—now at the bow of the boat, swinging on the wire. She started screaming and was about to jump in and swim to shore when I swung back as though I had never left. "Don't ever do that again," she screamed hysterically.

Sailing became my means of escape and a source of some great adrenaline rushes. One weekend, sailing in big waves on Lake Ontario with six people on board, the Hobie Cat buried itself in a wave and continued to sail forward, with only our heads, the mast, and sails visible. We rose slowly, like a submarine surfacing, as we sailed on while everyone screamed with excitement. Witnesses were so amazed by the sight that they tracked us down to buy us all drinks.

I was fearless, having at that point experienced no injuries to teach me to respect the waves, and I was soon winning more races. But I was about to be given a tough lesson during a single-handed race, a hard-fought battle against good competitors in which I was far enough ahead that it was clear I would win. I was so confident that I began "hot dogging," showing off by flying with one hull standing almost vertical. To prevent us from capsizing I sat on the upper hull, shifting my weight back and forth. Then the tiller snapped. I fell backward as the boat began to flip over in what seemed like slow motion, and

I was certain that I would end up in the water. However, I landed on the lower hull instead, breaking my right clavicle, which runs from the shoulder to the chest bone, and the bone on the top of my shoulder. I hit my head and passed out, rolling into the waves without a life jacket, but luckily the ice cold water and the pain from my shoulder brought me around, though witnesses said that I had floated facedown for about a minute. By now the pain was intense, and I could not get out of the water without the help of the rescue boat. I was on serious painkillers for weeks. But as soon as I could, I was back sailing.

By now Claire was focusing on finishing her doctorate. Her thesis defense went well, and after the celebrations had died down, I asked her to marry me and continue to work in the lab. She joked that the answer was yes as long as I agreed not to use that as a recruitment technique again. (The truth is, it took a while for her to make up her mind about me. When asked by *People* magazine years later whether it had been love at first sight, she said, "It was for him—but not for me.")

At least I knew that Claire was not marrying me for my money. She had seen my financial situation go from bad to worse as I attempted to deal with the wreck of my first marriage: the huge credit card bills, the child care expenses, and the legal fees. I had to sell my beloved Mercedes to help pay my debts and could not even afford to buy an engagement ring; Claire had to take out a loan for me to get one.

News of our pending marriage spread quickly through the medical school, but by now the scandal had run its course, and there was little more fun to be had by gossiping about the "affair" between the student and the professor.

We decided to get married on Cape Cod. Although Claire came from a Catholic family who wanted us to marry in a Catholic church, my divorce made that impossible. Instead we found an historic church in Centerville, Massachusetts, that was surrounded by the graves of former sea captains and not far from Claire's parents' home in West Hyannisport. We set the date for October 10, 1981, and the wedding and reception were joyous events. Gary was my best man, and for this one occasion, my family felt like one. My father showed obvious pride and delight in how I had transformed myself from a surfer into a medical school professor and seemed to be serious about settling down with my new bride. A wedding photo of my beaming father with three of the bridesmaids, all hanging on his every word, reveals that there was a warm person inside, one whom I was grateful to know, if only briefly. After the wedding I got into the habit of talking with my parents almost every week, usually my father.

We did not get much respite from science during our honeymoon, which consisted of a single night at an old inn on Nantucket. The very next day we flew to Paris, where I was scheduled to give a lecture, then on to London to write an application for a major National Institutes of Health grant with Claire (how romantic!) and to take part in a Ciba Foundation workshop on receptors. There we met a man who would become both friend and mentor, Martin Rodbell, who was attending with his wife, Barbara. Marty would win a Nobel Prize in 1994 for the discovery of G proteins—short for guanine nucleotide binding proteins—molecular "switches" that are coupled to receptors, such as the adrenaline receptors that I had studied, and allow or inhibit biochemical reactions inside the cell. Marty was a kindred spirit, being one of the few people at this level of research who had experienced combat—in his case as a radioman in World War II. He had been to the brig more than I, such was his rebellious spirit.

At one point in the Ciba meeting, I flew to Belgium to participate in a special workshop at the Princess Lillian Cardiology Institute. There I got on famously with another Nobel-to-be, James Black, the British discoverer of beta-blockers. After we had a heavy lunch, washed down with two bottles of wine, we both thought we were waxing very philosophical at the meeting (at least, that is, until we received the transcript to review for publication). Then, after a formal dinner with the Belgian king and princess, it was back to London, Claire, and the Ciba conference, where Marty Rodbell told me of a cunning way to figure out the rough size of a receptor protein: Simply bombard a cell with radiation until the protein stops working. The more radiation you need to turn it off, the bigger the molecule. With the help of the Korean scientist C. Y. Jung, I would apply this approach to gauge the size of the receptor proteins we were attempting to isolate.

Claire and I started married life in a two-bedroom apartment in Buffalo. She got an appointment in the department of biochemistry, and we continued to work together. I had by then cleared my debts and had even sent my father an early Father's Day card with a check that paid back the loan he had made at the height of my financial problems. By then our relationship had changed profoundly since its lowest ebb on my return from Da Nang. When I went to San Francisco that year to give a talk, he picked me up at the airport, looking gray and unusually run-down, but I did not think much about it.

My Father's Genetic Bequest

The most common cause of heart disease is atherosclerosis, in which calcium, along with fats and cholesterol, collects in the blood vessels to form buildups called plaques, which can trigger a heart attack or stroke. A protein called apolipoprotein E (Apo E) is responsible for regulating levels of certain fats in the bloodstream, variants of which have been linked with heart disease and also to Alzheimer's, the progressive and devastating neurodegenerative disease.

I can summon my genome on a computer screen and examine chromosome 19, where the Apo E gene is found. Some nine hundred letters long, it comes in three common forms, known as E2, E3, and E4, which differ in terms of two "letters" of code. E3 is the most predominant spelling in European/Caucasian populations and in health terms is the "best." By contrast, the 7 percent of people who have two copies of E4 are at higher risk of early heart disease, as are the 4 percent of people with two copies of E2, who are more vulnerable to high-cholesterol, high-fat diets. Carriers of E4—which differs by just one letter from the "good" E3—have also been associated with increased risk of Alzheimer's disease. I have one copy of E3 and, unfortunately, one for E4 too, placing me at risk. Although there is no evidence from either side of my family of Alzheimer's, it is possible that carriers of this gene, like my father, died too soon for this awful degenerative disease to manifest itself.

By reading my own book of life, I have been given a chance to address these potential conditions, because they involve a biochemical imbalance that can be treated. Diet and exercise offer one way. I am also taking a statin, a fat-lowering drug, to counteract its effects. The same statin also shows some degree of prevention of the symptoms of Alzheimer's disease.

Many more genes are linked with forms of coronary disease, from heart attacks to high blood pressure due to narrowing of blood vessels (stenosis). My genome carries lower risk versions of some of these genes, notably TNFSF4, CYBA, CD36, LPL, NOS3. But hundreds more are involved, and it will take years for us to understand the complicated way they interact with one another.

Back in Buffalo, I had planned to call him on Father's Day, but early on the morning of June 10, 1981, my mother rang to tell me that my father had died in his sleep. He was only fifty-nine; his doctor diagnosed sudden cardiac death. Claire and I left on the next flight to comfort my mother and make the arrangements for the funeral and his cremation. We interred his remains in a military cemetery at the Presidio in San Francisco, and because my father felt that religion did more harm than good, my mother refused to have any religious symbol on his tombstone, leaving it to stand out among the crosses and the stars. I wept, for his passing was a wrenching transition for me and my family. I was glad that my father had at least lived long enough to see some of my success and for me to establish a positive relationship with him. While Vietnam had taught me that life is finite, nothing underscored that fact more than losing my father while he was relatively young.

The last of my burdensome teaching commitments disappeared in 1982 when I was recruited to become deputy director of the molecular immunology department at Roswell Park Cancer Institute by Heinz Kohler, its new director. I had beautiful new labs, was made a full professor, and my salary doubled. I also got Claire a pay increase and a staff scientist position. We moved out of our apartment into a four-year-old ranch-style house, and our outlook on everything began to improve. By 1984 my team was publishing a paper every two or three weeks.

Thanks to Kohler I felt as if I had it all under control; however, it was Kohler who almost brought my new life to a watery end one windy day off Sherkston Beach on Lake Erie. Heinz had been keen to join me on the Hobie Cat, and we had put together a large crew to keep the boat stable. We were enjoying an exhilarating sail, flying down the waves and becoming airborne a few times, when Kohler asked if he could steer, assuring me that he knew what he was doing. Moments later, with Claire and me helplessly looking on from the trapeze, he turned into a big wave and we began to pitchpole. The catamaran catapulted Kohler into the air, and he hit the mast halfway up as the boat flipped end over end.

When the boat turned turtle, Claire was submerged, tangled in wires and lines. After I had freed her, she was so mad that she wanted to push Kohler under water and drown him. Kohler himself was by then bleeding a fair amount and looked as if he were going to go into shock. Even though we were more than a mile from shore and in a howling gale, Claire started swimming toward the beach in a panic. I restrained her while trying to help Kohler, and as the waves broke over us, we managed to right the boat and return safely. A few weeks after Claire vowed that she would never board the Hobie Cat again, I placed an order for a Cape Dory 25D, a single-hull cruising sailboat.

In 1983, I received an offer that would change my life and the course of my science, from the powerhouse of medical research in America, the National Institutes of Health in Bethesda. There, thanks to the "intramural" programs, funding was almost automatic, as long as the director had faith in what you were doing, which meant there was no need to apply for grants. It was especially flattering to be recruited by them, for at that time the NIH preferred to grow its own research teams rather than import them.

Ouch!

That sums up how I felt when my team told me what my future had in store when it comes to a gene called SORL1, responsible for a protein called Sortilin-related receptor. Variants of this gene are implicated in late-onset Alzheimer's disease, the most common cause of dementia. Reported in 2007 by an international team, the link marked only the second genetic variant linked with this type of Alzheimer's, with Apo E4 being the first.

Variants of SORL1 were found to be more common in people in four ethnic groups with late-onset Alzheimer's than in healthy people of the same age, and it is thought that the risky variants of the gene aid the formation of protein deposits in the brain, called plaques, believed to be responsible for the relentless mental decline caused by the disease. Or, to put it another way, when SORL1 works properly, it has a protective function, perhaps by recycling the protein that forms the deposits. Thus, the reduction of SORL1 in the brain, which has been seen in postmortem brain tissue of patients, increases the likelihood of developing Alzheimer's.

Although there is a much less established link between SORL1 and the devastating disease than with Apo E4, it turns out that my genome carries all the risky variants in one part of the gene and some of them in another. Ouch, indeed. Given that I already have the Apo E4 variant, my team has taken a closer look. We do understand how the part of the gene where I have some risky variants alters the way this gene is turned into a protein by cells, and thus the link with dementia. However, it is unclear how the variants in the second sensitive region (the so-called 5′ risk allele cluster) raise my risk.

The NIH seemed like the ideal setting in which to pursue and advance my research. In the decade I had spent characterizing adrenaline receptors, it had become increasingly clear that traditional methods to purify the extremely low amounts of receptor in each cell were never going to provide sufficient protein to obtain its amino acid sequence, and thus to enable us to figure out its molecular shape, the key step toward understanding how it works. I wanted to use the new approaches of molecular biology to sidestep this problem in order to provide unprecedented understanding of the fight-or-flight response.

Kohler and the Roswell administration made an attractive counteroffer, and in some respects I was tempted to stay. I had many good memories of Buffalo, which had helped me build my scientific career from scratch, and where I had made the transition from a graduate student to a full professor. I had also witnessed my son being born there and met Claire, my wife and collaborator. But I had also had to deal with the death of my father and a painful divorce, and there was also the fact that I was in a weak academic culture. Above all else, I was still driven by my experiences in Da Nang, and I wanted to accomplish so much more. Around a dozen members of my team were prepared to move to the NIH with me, and we began the chore of filling out the paperwork to become government employees. Despite his impending loss of so many staff members Kohler cheered considerably when he learned that I would leave my grant money behind in Buffalo. I was now ready to begin a new phase of my life, as a civil servant.

5. SCIENTIFIC HEAVEN, BUREAUCRATIC HELL

The large molecules were broken down, usually by suitable enzymes, to give smaller products which were then separated from each other and their sequence determined. When sufficient results had been obtained they were fitted together by a process of deduction to give the complete sequence. This approach was necessarily rather slow and tedious, often involving successive digestions and fractionations, and it was not easy to apply it to the larger DNA molecules. . . . [It] seemed that to be able to sequence genetic material a new approach was desirable.

—Frederick Sanger, Nobel lecture, December 8, 1980

I found myself walking down a suite of windowless rooms in a dark corridor on the second floor of what was simply referred to as Building 36. This was to be my new scientific home in the National Institute of Neurological Disorders and Stroke, one of the buildings in the sprawling complex of the National Institutes of Health in Bethesda, Maryland. To set up and equip my laboratory, I had been given hundreds of thousands of dollars, and to launch my new career in molecular biology, I now had an annual budget of over one million dollars. Having brought most of my team from Buffalo along with me, I could make a rapid start.

All around us on the NIH campus worked hundreds of the nation's finest researchers, with whom we could collaborate. Directly beneath us was the lab of Marshall Nirenberg, the scientist who shared the Nobel Prize for deciphering the genetic code to show how each of three "letters" of DNA in a gene spell out an amino acid, the building blocks of proteins. He and his NIH peers could teach us new things and help spark our own novel ideas. The techniques and interests I picked up in Bethesda had a profound influence on the rest of my life, laying the foundations for my future interest in reading genomes. I was in scientific heaven.

But I have learned again and again that pleasure always comes with a price. Every beautiful week of sailing is matched by a punishing ocean passage through gales or storm-tossed seas. To enjoy the top science to which I was exposed at the NIH, I had to deal with lumbering government bureaucracy. I was supposed to join the NIH at the highest civil service level: Grade 15, step 10. The problem was that the personnel officer assigned to my lab had apparently never made an appointment at that level. Typical of when an apparatchik runs into an unexpected problem, he took the path of least resistance: He dropped my folder in a bottom drawer and forgot all about it. I would call to ask about my salary and was assured that something *had* to be happening because they could not find my file. Fortunately, Roswell Park was continuing to pay my salary as the principal investigator on various grants, or I would have had no income for several months. When finally confronted, the personnel officer confessed that he had been so afraid of screwing up the procedure that he decided to do nothing at all.

That was not the only source of friction. A tenured position had been promised to Claire, as well as positions for key people such as Doreen Robinson and Martin Shreeve to enable them to follow me from Buffalo to the NIH. But shortly after arriving I was summoned to the office of the scientific director, Irv Kopin, and told that Claire would indeed have tenure but only after a year or so: They felt that her scientific credentials fell short of their standard.

The hunt for a new home likewise proved harder than we had expected, due to the relatively high cost of living near Washington. A smaller and older version of our former $80,000 house in Buffalo was going to cost hundreds of thousands of dollars. Our real estate agent, Barbara Rodbell, was the wife of Marty Rodbell, who had been instrumental in bringing us to the NIH. Barbara was new to selling houses, and we were her first (and only) client.

We bought a place in Silver Spring, Maryland, that we could barely afford, a two-bedroom $105,000 town house, and moved in with Cesar, our six-month-old keeshond puppy, named for his royal command of our living space. I also found a mooring for my 25-foot Cape Dory sailboat—an indulgence that hardly helped our financial problems—at Hartges Yacht Yard in Galesville, Maryland. The West River opened there into the Chesapeake Bay, offering me thousands of miles of shoreline and great anchorages. The local economy was driven by tobacco and crabbing, which lent the place a wonderful old-fashioned feeling, helped by the presence of an ancient general store that was complete with a woodburning fireplace and a checkers table.

My professional and private enthusiasms were to come together one weekend when I planned to explore the bay by sail. Being addicted to speed, I always kept a keen eye on the rearview mirror for unmarked police cars as I drove to Galesville. I would look for anything unusual, such as the brown Ford Fairlane with two noncasual men in the front seat. Whenever I changed lanes or turned on the complex route to Galesville, this particular brown Ford was behind me. On arriving at Hartges I soon forgot about the mysterious tail. But after my day of sailing, it was there again as I returned to Silver Spring. I began to fret about the Ford and wondered if my antiwar protest days had come back to haunt me now that I was a senior government employee. I might not have been so paranoid about being followed if I had known what an important role the vehicle's occupants were to play in the future direction of my research.

The following Monday morning my fears seemed to be realized when I found two men in dark suits and pencil ties waiting in my cramped NIH office. As I entered, they stood and presented me their ID cards, which indicated that both were with the U.S. Department of Defense. They explained that they wanted to talk to me about using my research to help detect nerve poisons and associated biological warfare agents in the field. This request made sense, because the receptor proteins I was studying were the selfsame receptors that are the targets of nerve toxins. Despite the grim subject matter, I was relieved to hear that they were interested in my science, not me, and invited them to sit down.

The basic idea they wanted me to pursue was simple: Was it possible to detect the presence of a nerve agent by exploiting the same proteins that the agent binds to in the body to cause harm? Could these proteins be used as a hook to fish out even minute traces of an agent in the air and then, with the help of clever chemistry, announce their catch with a little flash of light to help warn anyone around that they are under attack?

Although the particular adrenaline receptor that I was working on could not be obtained in enough quantity for this purpose, the nicotinic acetylcholine receptor, where the messenger chemical acetylcholine acts, was available in sufficient amounts and, I thought, would make a good candidate for study. That is probably exactly what they wanted to hear. Acetylcholine transmits nerve signals to a variety of muscles, including the diaphragm, which controls breathing. Nerve agents such as Tabun (GA), Soman (GD), and Sarin (GB) kill by blocking the action of a crucial enzyme called acetylcholinesterase, which breaks down and removes acetylcholine. The agents act within minutes by overloading the body's transmission of nerve impulses: They short-circuit the nervous system, ultimately paralyzing the diaphragm so that the victim suffocates.

Because acetylcholine also acts at a number of sites in the brain, the receptor was of interest to civilian scientists, too. Nicotine triggers one of these receptor sites on nerve terminals, increasing the use in the brain of dopamine, another signaling molecule, which in turn plays a role in what neuroscientists call the "reward pathway" to cause the craving experienced by many smokers.

The acetylcholine receptor had been successfully purified by John Lindstrom, who was then at the Salk Institute in San Diego. John was pleased to provide me with the protein for a hefty fee, which I thought reasonable given the Herculean effort it then took to isolate it. In this way he would receive defense money to help support his basic research, and I could figure out how to help the government to detect nerve agents.

But the bureaucratic brain of the government is not a unified and coordinated entity. NIH bureaucracy made it harder than it should have been for me to take money from the Department of Defense. Scientists did not usually bring government money *into* the NIH; they were there to spend it. Eventually an inter-agency transfer of $250,000 was arranged and deposited in a special NIH account in my name, though I was informed that I was perhaps too entrepreneurial and that there really was no need for me to generate extra money.

Once my lab was set up, we were ready to isolate or to clone the adrenaline receptor gene from the human brain, which would provide us with its molecular structure and new hints on its mysterious workings. In this context, cloning means copying the gene, usually done by putting it in the laboratory version of an E. coli microbe so that, as the bacteria multiply, so do copies of the gene that we want to study. For all intents and purposes, cloning a gene also means finding it in the cells and/or genome so that one can reveal the DNA recipe the gene uses to make the protein—in this case the receptor in the brain that responds to an adrenaline rush. To do this we had to isolate the receptor protein, work out its sequence of amino acids, and then deduce the possible DNA codes that could spell out these sequences.

That description makes the process sound easy. In practice, back then, it took ten years of grinding toil. Thanks in some part to my own efforts, the same job now takes a fraction of the time, even a matter of days. But let's return to the 1980s and the traditional slog necessary to isolate and study rare proteins made in the human body.

To enrich the minute trace of receptor protein to the point where we could begin the gene hunt, I wanted to use a recently developed technique called high-performance liquid chromatography (HPLC). In this method, components of the membrane of (in this case) human cells are dissolved in detergent to break up fats and lipids and then separated by passing them through columns of materials. The progress of each protein depends on its size or charge, so smaller, nimbler molecules pass through more quickly than the larger ones.

To use this new process I needed experienced people. I hired Anthony Kerlavage, who had done HPLC protein purification work with Susan Taylor at the University of California, San Diego. To develop strength in the molecular biology techniques required to study the protein I built a fledgling team comprising Fu-Zon Chung, a postdoctoral fellow from North Carolina, and two laboratory technicians from Buffalo, Jeannine Gocayne and Michael G. FitzGerald.

As my team closed in on making purified receptors, we turned out thirty scientific papers on different aspects of receptor structure and function. One of our more important findings was how parsimonious Nature is when it comes to the design of receptors, using the same structure over and over again with just minor variation. Our analysis of the structure of muscarinic acetylcholine receptors (a subtype of acetylcholine receptors) and alpha-adrenergic receptors (a type of adrenaline receptor) revealed that they were very similar in basic structure even though they recognized different neurotransmitters in the body. This was a surprise to the scientific community, where the different receptors were regarded as essentially different fields of study. As a result many dismissed our findings, until a few years later when the receptors' genes were sequenced and revealed to be very similar.

Within two years our effort was making substantial progress, though our morale was to suffer a serious setback just as our goal was in sight and we were obtaining our first amino acid sequences from small amounts of purified receptors. We were beaten to the prize by Robert Lefkowitz's group at Duke University, which had teamed up with the pharmaceutical company Merck in a major effort to purify and clone the adrenaline receptor from the red blood cells of the turkey. His triumph was an exciting moment for everyone in the field, including me. I called my disappointed team together and reminded them that these were early days in our field and most discoveries had yet to be made. We should forge ahead and take full advantage of Lefkowitz's breakthrough to clone the rare adrenaline receptor from the human brain.

To make progress we could take advantage of how complementary base pairs in the genetic code—each half of the double helix—join together. In their insightful 1953 breakthrough, Watson and Crick had figured out from the way the base pairs paired up how a single strand of DNA could easily be copied into a complementary strand. This in turn revealed how cells replicated the DNA in their chromosomes when they divided. Of the four base pairs in the genetic alphabet, they found that A always paired with T, and C with G. Separate the double helix into two daughter helices, and this simple rule helps to grow a complementary strand. Likewise, if you create a single strand of base pairs, the same rule ensures that it will stick only to a complementary strand—giving you, in effect, a DNA probe that would attach to a single gene in the huge terrain of the human genome.

We could exploit this complementary base pairing in two approaches to hunting down the genetic code for the human adrenaline receptor. First, we could use the small section of human receptor protein sequence we had obtained to deduce the corresponding DNA sequence and use it as a probe to search for the gene in the human genome. Second, we could exploit the legacy of evolution: The turkey adrenaline receptor gene was likely to have a very similar genetic code to the human receptor. In other words, we could make DNA probes from the turkey receptor gene itself: If they bound to complementary human DNA, they would reveal the equivalent human gene. By placing a radioactive label on either kind of probe we could gauge our success by locating the radioactive hot spots where the probes had attached.

But, as ever, there was a practical problem to overcome. We first needed to get the human genome in a form suitable to conduct these experiments: Even packaged in the form of chromosomes, as it is in cells, the code is too big to handle. Cutting up the human genetic code into manageable pieces would prove crucial for this gene hunt. If you take the full complement of DNA in a human cell and break it up into pieces, you end up with what scientists call a library. There are two basic types of DNA library: genomic and complementary.

A genomic DNA library is created by using special enzymes, called restriction enzymes, to chop up human chromosomes into small pieces consisting of approximately fifteen thousand to twenty thousand base pairs of DNA each. To be able to handle these pieces of human DNA, we also need a way to copy and store them, just as a book needs to be printed and bound. In the copying process, each fragment of human DNA is attached to the DNA of a phage, a virus that infects bacteria and multiplies within them. When the treated phage is used to infect the bacterium *Escherichia coli* (E. coli), it carries the human DNA with it. Paint the surface of a petri dish with E. coli that have been infected this way, and clear spots will develop on the bacterial lawn where the viruses have killed the E. coli. These spots are called plaques and contain millions of virus particles—and therefore millions of copies of the original human DNA fragment.

The complementary DNA library begins with another type of genetic material found in cells, called messenger RNA, which is used to carry out the genome's orders to make proteins. Only 3 percent of our genetic code is responsible for coding proteins, and so by focusing on the RNA used to make them, we end up with a much more concise "working copy" of the human DNA code. (A typical gene can be spread over as many as one million base pairs; in contrast, the edited RNA might be only one thousand base pairs in length.) In other words, this library exploits the way that Nature's copy editor turns the whole genetic code into a much smaller stretch of messenger RNAs, which represents only the subset of genes required for a specific cell or tissue type.

Messenger RNA is by its nature transitory and unstable; otherwise we would simply isolate it from a cell and read it directly. However, by using an enzyme called "reverse transcriptase," the RNA can be copied into a stable form of DNA. This is called complementary DNA, or cDNA. By isolating human RNA to make cDNA, you end up with the condensed version of the human genome containing the protein-coding genes in an easy-to-read form.

As with the genomic library, each volume of the complementary library has to be put in a form in which it can be handled and copied. Here again, Nature provides the solution. Complementary DNA libraries are made by isolating the messenger RNAs at work in a tissue, converting them into complementary DNA fragments, and inserting those fragments into a plasmid, a small ring of

DNA that carries the instructions for a bacterium. Once again E. coli bacteria provide the "printing press" for the books in this library when they are infected with the plasmids. Because each bacterium will contain a different segment of human cDNA when the bacterium replicates (divides into daughter cells), both the mother and the daughter cells will contain the same human gene fragment.

Even though these DNA fragments are invisible to the naked eye, it is possible to see them with the tools of science. By spreading out the E. coli containing the human cDNA thinly, so that in all probability no two bacteria are touching each other, separate colonies will begin to grow. Eventually, when visible as spots, each colony will contain millions of identical bacteria cells (clones), all of which have exactly the same piece of human cDNA. On a small petri dish it is possible to have tens to hundreds of thousands of such isolated colonies, and thus a vast library of human DNA.

Using either type of library, the hunt for the receptor can now begin, exploiting the way complementary DNA sticks together. Using a filter paper it is possible to remove DNA from the petri dishes growing E. coli with either genomic or cDNA libraries. The filter paper is then soaked overnight in a solution of the DNA probe for the receptor. To determine whether it has bound to complementary DNA in the libraries, the probe is labeled—made radioactive—by swapping some of the phosphorus atoms in the DNA with a radioactive isotope known as P-32. The filters are then washed to remove any of the radioactive-probe DNA that has not attached to any of the DNA fragments, dried, and placed into an X-ray film cassette for a few days. Positive colonies and plaques—areas where the radioactive probe has stuck on to target DNA—are revealed on the developed X-ray film as black spots. By aligning the film with the petri dishes, the positive colonies or plaques can be identified and their DNA isolated and amplified.

It is not always easy to find the unstable messenger RNA that codes for rare proteins. In the case of an elusive membrane protein, such as the adrenaline receptor, there are only a few thousand protein molecules in every cell. As a consequence, the RNA message that codes for the receptor protein is rare as well. Using the genetic material from human brains that had been donated to medical science, we had to survey more than one million cDNA colonies to find just one that contained the message to make an adrenaline receptor. Once we had grown a colony to produce a sufficient quantity of this DNA, we could then read it by a process called DNA sequencing—working out the order of the four nucleotides (CGAT) that form the "rungs" in DNA's outer backbone of sugar and phosphate. Reading the order of these base pairs in DNA could be accomplished with two basic sequencing methods. One was developed in the Medical Research Council's Laboratory of Molecular Biology in Cambridge by Frederick Sanger, a dedicated researcher (and fellow lover of boats) who once said he was "all right at the thinking, but not much good at the talking." The second was the work of Wally Gilbert at Harvard, described as a "power broker, a man of grandiose aims."[1] Although Sanger and Gilbert shared the Nobel Prize in 1980 for their achievement, most sequencing performed in the past decades is a direct extension of the method devised by Sanger, a scientist with an astonishing capacity to crack some of the toughest problems in biology. Sanger awed an audience with his first partial DNA sequence in May 1975 and went on to deliver the first complete sequence of a viral genome: the 5,375 base pairs in the genetic code of a bacterial virus (phage) called phi-X174. Sanger then sequenced the 17,000 or so base pairs of DNA in the human mitochondrion (the energy factory found in our cells), marking the first human genome project. Given his remarkable record, perhaps it is not surprising that Sanger is a double Nobel Prize winner (the first, in 1958, for his work on the structure of proteins, when he unraveled the fifty or so amino acids that make up the insulin molecule). Despite his huge influence, and how he is revered by his peers, Sanger is a quiet, modest, and unassuming man: "I am not academically brilliant."

The DNA reading method that Sanger pioneered in Cambridge with Alan Coulson involves making numerous copies of the DNA molecule by using an enzyme called DNA polymerase. For the polymerase to replicate DNA, it must be placed in a soup of DNA building blocks, the nucleotides. The enzyme reads from each end of the original DNA strand using the nucleotides to make new copies of it. Sanger's contribution was to add another ingredient to this soup: "terminator nucleotides," each radioactively labeled with P-32, which are so named because they randomly stop the action of the polymerase when they are incorporated in the growing copy, marking the end of the growing chain with a radioactive period. Because this can happen at any stage in the process of making vast numbers of copies of DNA molecules in the test tube, the result is a mixture of DNA fragments of varying lengths, each finishing with a radioactively labeled C, G, A, or T, depending on which base contained P-32.

These fragments are then driven by an electric field through a slab of gel that can separate the DNA molecules according to their size. Now one can read the sequence, because the largest pieces of DNA will take longer to go through the gel. Since the label on all four nucleotides—C, G, A, T—is the same and produces the same black mark on an X-ray film, one has to run four separate experiments, one for each letter of the code. Once the DNA polymerase has been used with each of the four different terminator nucleotides (so that in one batch all the Cs are labeled, another all Gs are labeled, and so on), they are put on four adjacent tracks on the same gel. When the fragments separate, one track shows the DNA fragments that end with a C, one those that end with a G, and so on.

The gel is then dried and placed for a few days against an X-ray film, leaving four parallel tracks of black bands. Then someone has to study the film, starting with the first band from the four letter tracks and then moving to where the next closest band appeared. If the first, smallest piece of DNA is in the C track, for example, then C is the first letter. If the next black mark is in the A track, then an A follows, and so on. In this painstaking way, recording each letter in order a few hundred times for each sample, the sequence emerges. This made for tedious work, where many things could and often did go wrong. If any one of the four reactions on a gel failed, the whole experiment was worthless; many times the lanes would not run parallel to one another, so the farther one went down the gel, the less able one was to compare the black marks and gaps on each track to read the sequence. The gels could and would often crack on drying. Reagents could and would go bad as we waited days and days to get an answer; weeks could pass without producing any usable data.

I found the room for interpretation particularly frustrating because I had such high hopes for molecular biology. Too many times I had seen science driven less by data and more by the force of a particular personality or the story on which a professor had built his career. I wanted the *real,* empirical facts of life, not those filtered through the eyes of anyone else. You either have the sequence or you don't. It is either accurate, within the limits of the method, or it is not, usually because of sloppy work. I noticed that many sequencing labs often made a guess when a base was unclear, rather than leave a question mark. After weeks and weeks of struggle we found that we were able to obtain a cDNA clone for the human brain adrenaline receptor. The excitement grew when we realized that its sequence was significantly different from that of the turkey receptor. We were on the brink of our first big success in Bethesda.

However, as we determined the final pieces of the sequence, it became clear, or at least highly likely, that we had not finished the job: We were missing the starting end of the gene, the point at which DNA usually has the genetic equivalent of a punctuation mark. As mentioned earlier, only a small percentage of our DNA corresponds to genes that code for proteins. To help the molecular machinery of cells distinguish them, there are the genetic equivalent of capital letters and periods.

Just as a capital letter starts a sentence, most genes mark the beginning of the protein coding region of DNA with what is called the start codon, ATG (which codes for the amino acid methionine). Just as capital letters often occur within sentences, methionine is not uncommon within protein sequences, so that extra information is required to verify that an ATG marks the true start of a gene. One can, for example, look for a nearby stop codon, one of the molecular "periods" at the ends of DNA sentences that tell our molecular machinery when to cease building a protein.

Because we knew that we had only one segment of DNA from the brain library that corresponded to the adrenaline receptor, we needed another library to find the rest of the gene. Our only hope was to make a radioactive probe from the DNA near the missing piece and use it to rummage through a second library of human genetic material to find a complementary piece with the missing end of the gene.

Once again we were looking for the equivalent of a needle in a haystack, though the amount of DNA to survey was smaller this time. The DNA fragments in the second genome library were, on average, 18,000 base pairs in length, so that each of these pieces (clones) represented only 0.0006 percent of the three-billion-letter genome. Rather than look for one clone in a million, we were looking for one out of approximately 167,000 clones. Within a couple of weeks we had several promising leads to test. One 18,000-letter-long clone appeared to have the end of the gene in it, and so we set out to sequence it.

The plan worked. At last we had the entire gene sequence assembled in the computer and wrote up the first molecular biology paper of my career.[2] We submitted it to *FEBS Letters* because I knew the editor, Giorgio Semenza, and he promised a quick turnaround. We soon came out with the sequence of the first human brain adrenaline neurotransmitter receptor gene. Given the steep mountain we had had to climb from the purification of the receptor proteins to working out how to read the code, we all felt as though we had accomplished something significant. It was a wonderful moment for me to see the DNA sequence that until then had only been part of my imagination—like walking out of a pitch-black cave into the sunlight. Even today it still seems remarkable that one can use human-scale machines and technologies to visualize molecular codes. The associated paper marked the turning point of my career. I had left a safe field in which I had become comfortably established and taught myself and my team a new discipline, molecular biology. We were ready to rocket forward.

But it was very clear to me even then that obtaining the sequence of a gene or protein represented only one step toward understanding the way adrenaline works. The end of this particular journey marked the beginning of many new paths. For example, by inserting the human receptor gene in mouse cells and growing them we could mass-produce the receptor for use in all kinds of experiments. We needed to do more sequencing to follow up on our earlier discovery that a range of neurotransmitter receptors reacted with the same antibody. To confirm our belief that they had a common evolutionary heritage, we needed to compare the sequences of a large number of receptors. Our effort to read the DNA of several different receptor genes marked a first—unknowing—move into a field that at that time did not exist: genomics.

This major change in direction of my science was also spurred by the arrival of the latest issue of the journal *Nature.* An article by Lee Hood's group at the California Institute of Technology described a remarkable possible advance in DNA sequencing technology. By using four different fluorescent dyes instead of a single radioactive probe, the four different Sanger sequencing reactions could be combined into a single lane on a sequencing gel. As a DNA fragment ran toward the bottom of the gel, it passed in front of a laser beam that activated the dye. The glowing dyes were easily detected by a photomultiplier tube and the data recorded by a computer. The four colors, representing the different nucleotides, provided a direct readout of the genetic code, transforming the analog world of biology into the digital world of the microchip.

I had already collaborated with Lee Hood and his postdoctoral fellow Michael Hunkapiller on receptor proteins, and now I wanted to work with them again. Having just spent close to one year sequencing a little more than one thousand base pairs of genetic code using the traditional radioactive methods, a miserable experience, I immediately saw the value of the Caltech automated approach. I contacted them and learned that Hunkapiller was going to head the development of this method into a commercial DNA sequencer. He had joined Applied Biosystems (ABI), a biotech company that until then had been making DNA synthesizers. I talked to Hunkapiller and my local ABI sales representative and told them that I wanted to buy one of their first machines. That suited ABI, since an NIH purchase would lend their technology prestige and visibility. After many discussions, everything was set in motion for my lab to be the first test site for the new sequencing method. Now all I needed was $110,000 to pay for it. Ernst Freese, the basic science director of my institute, was against buying new and unproven technology, however, and offered me $250,000 to buy a protein sequencer instead.

The problem was that the receptor proteins I wanted to study were too low in abundance to be analyzed effectively with such a sequencer. Freese and I argued about the relative merits of protein versus DNA sequencing and I lost. After a few days of feeling despondent, however, I remembered the special account with $250,000 for my biological warfare detection work for the Department of Defense. I told Freese that I felt so strongly about trying the new technology that I would use this money and take responsibility for the risk. He could still have fought me, but I think by now he had become impressed by my determination, and the order went to ABI.

February 1987 saw a remarkable delivery to Building 36 on the NIH campus: my new DNA sequencing machine. This was my future in a crate. I was so excited by my baby that I decided to deal with the lack of lab space by putting the machine in my own office and to take personal charge of helping to establish the new method of DNA sequencing. Alongside me was the technician Jeannine Gocayne, who had just received a master's degree from Buffalo. I thought she could easily have gone on and finished her doctorate, and although she seemed to lack the confidence to do so, I had enough faith in her skill to ask her to help me churn out DNA sequences with the new method.

We quickly went to work and looked over every part of the machine. The business end was the "electrophoresis box," which contained a vertical sequencing gel about the size of a legal pad. The gel had sixteen lanes so that sixteen samples could be run simultaneously. (Because we had to run four standards to ensure that the machine was working properly, in reality it could only cope with twelve samples.) At the bottom of the gel was a scanner that went back and forth reading any signals from fluorescent dyes into the computer. A single run took sixteen hours and produced data that would have taken a week to obtain using the old method.

After several weeks of stamping out the bugs, we were soon seeing very pretty data, with up to two hundred base pairs of genetic code per DNA sample. The problem was that the machine's software was primitive and unreliable, an obstacle we encountered when Jeannine and I took turns reading the color peaks while the other recorded the genetic code. We both could recognize specific sequence patterns, such as the ATG codons that mark the start of genes and the restriction enzyme binding sites. Later, our software engineers spent a lot of time getting the computer to accomplish what Jeannine and I could do easily with a trained eye.

A key stage in the process of DNA sequencing involved the use of DNA polymerase, the enzyme that copies DNA, which works with the help of a small piece of DNA called a sequencing primer. You can picture how the polymerase and primer work together by imagining the process involved in repairing a damaged train track where for a stretch one rail has been removed. The track represents the DNA double helix, and the railroad crew—the DNA polymerase—begins to put down new track at the last point where there were two intact rails. DNA polymerase can be fooled into starting at a specific point on DNA by a short piece of synthetic DNA (the primer) that binds to the specific bases of the DNA sequence to create a short stretch of double-stranded DNA.

When I trained as a biochemist with Nate Kaplan, I learned to measure everything accurately, to check the purity of my reagents, and not to rely on commercial providers' claims of purity. Every time we ran the machine, I would measure the quantity of DNA and sequencing primer to produce the correct ratio between reactants and products in the chemical reaction. This attention to detail turned out to be of critical importance: ABI told us that initially no one else could match the results we were getting from the sequencer; in fact, few were getting any results at all. Most of their customers were so disappointed that they returned their machines. We felt we had made sufficient progress on our own machine to use it for a project to sequence two new receptor genes that we had isolated from a rat heart: the beta-adrenergic receptor, which modifies cardiac pumping activity in response to adrenaline, and the muscarinic receptor, which slows the heart rate from vagus nerve action. We sequenced both genes quickly, and as a control did some manual sequencing by the Sanger method for comparison's sake. When we published our results in the *Proceedings of the National Academy of Sciences* in the fall of 1987, they marked the first data to be obtained by an automated DNA sequencing method, one that I had read about in *Nature* only a year before.[3] My science career would never be the same again.

Now that we had cloned, sequenced, and made (expressed) the adrenaline receptor, we could use molecular biology tools to ask questions about the receptor structure and function. What made it recognize adrenaline? Once it did bind to adrenaline, what happened next? What did the receptor molecule actually do? What controlled its synthesis and degradation? What was its molecular structure in the membrane of our cells?

Underlying these questions was a determination of the picture of the receptor as a three-dimensional protein within the cell membrane. The complicated shape of a protein cannot easily be inferred from its DNA sequence, and solving this problem remains one of the grand challenges of biology. This knowledge is all-important, for it is when one of the vast numbers of molecules milling around in our cells is of precisely the right shape and decorated with the right charges to bind with a receptor that it can produce a vital reaction such as making a heart beat harder or regulating cell growth.

Everyone working on the structure of the adrenaline receptor molecules had noticed a key structural feature: seven stretches of amino acids that, according to computer predictions, would form a corkscrew shape, an alpha helix. These helices tend to span the lipid membranes of cells. Recall that receptor molecules are the key communicators from the outside of the cell to its inner workings, as my glass bead experiments had helped prove years before. The adrenaline receptor spanned the membrane seven times; presumably the seven "fingers" formed a pocket that gripped adrenaline and in this way somehow changed the rest of the receptor molecule to announce the arrival of the messenger chemical. Because the environment outside of the greasy cell membrane contains water, we figured that the amino acids that docked with the adrenaline would be hydrophilic (water loving), and that they would be negatively charged, because adrenaline bears a positive charge; we did find a few amino acids with these qualities. Other amino acids in the receptor sequence—for example, proline—usually play a role in protein structure by creating odd angles or kinks.

By that time we could even see what would happen if we redesigned the receptor protein. A molecular biology technique called site-directed mutagenesis, or "protein engineering," allowed us to do some clever experiments. By making changes in the genetic code of the receptor gene, we could change the sequence of amino acids and thus the protein structure. In this way we could begin to dissect the inner workings of this once elusive molecule by studying how well the altered receptor protein worked—for instance, if it still bound to adrenaline and whether other drugs liked to bind to it, too. And if so, did the receptor behave as if it had encountered adrenaline?

Here I have to confess to being an old-fashioned biochemist: I like to think not only about the mutations that change the shape of proteins, but about how those changes are reflected biologically. Many geneticists are content with finding an association between a piece of DNA and a trait and leaving it at that. To me that is a bit like getting starstruck by meeting someone who knows somebody famous: "I have this friend who knows Madonna!" But I wanted more than this; I wanted more intimate knowledge, and not just of Madonna. I wanted to understand the receptor biology that animated Madonna, and everybody else, for that matter.

We eventually made dozens and dozens of amino acid changes to the receptor protein. In 1988 we published two key papers that described how we found amino acids that affected the way the adrenaline molecule bound and activated the receptor but surprisingly had no impact on the beta-blocker drugs that also bound to the receptor, such as propranolol. The only conclusion from that finding was that the spots on the receptor protein that bound to receptor activators such as adrenaline (so-called agonists) were different from the spots on the receptor protein that attached to the receptor blockers like propranolol (called antagonists). Our simple picture of how receptors worked now had to be revised. Hormones had always been thought to work like a key in a lock—the lock being the receptor—while antagonists were just faulty keys that could not turn all the cylinders in the lock. Now it seemed as if they could act on a different part of the lock to prevent it from working.

All this speculation would have been much easier if we had had a model to show what the adrenaline receptor really looked like. I remembered how, when I first joined Kaplan's lab, the three-dimensional structure of the enzyme lactate dehydrogenase was being finished by Susan Taylor using X-ray crystallography. Her data led to a four-foot-by-four-foot-by-four-foot model of the protein, helping to reveal how in a wide variety of organisms, including plants and animals, the enzyme catalyzes the interconversion of the biochemicals pyruvate and lactate in basic metabolism. I wanted to do the same for the adrenaline receptor. But to take a "picture" of the receptor would require an X-ray study of the protein in crystal form, which in turn would require gram quantities of purified receptor protein—about a million times more than we had available at that time. I scanned the literature and found that there had been some substantial successes with using yeast to mass-produce proteins, so I hired an expert, Dick McCombie, to attempt to make enough for the X-ray study.

It was around this time that I also had the first discussions about a project that would one day propel my research into the limelight. While I spent most working hours fine-tuning my automated DNA sequencer, I began to follow some of the early discussions about the then wild-eyed notion of sequencing the entire human genome. (Readers are referred to *The Gene Wars,* by a Duke University genome commentator, Robert Cook-Deegan, for a detailed history.) One of the first had been held in May 1985 at a workshop organized at the University of California, Santa Cruz, by Robert Sinsheimer, who was driven by the thought that a big biology project of this kind could put his university on the map. By the time I became involved, Renato Dulbecco, a Nobel laureate at the Salk Institute, La Jolla, had advocated in the journal *Science* sequencing the genome to help win the war on cancer, and Sydney Brenner of the United Kingdom's Medical Research Council had urged the European Union to undertake a concerted program. Human genome discussions were also being driven by Charles DeLisi, a mathematical biologist at the Department of Energy. The involvement of the DOE might seem odd, but in fact it had been charged with understanding the effects of radiation, notably on the genetic code of the *hibakusha* ("those affected by the bomb") who survived the atomic weapons dropped on Hiroshima and Nagasaki.

The idea of sequencing the human genome was considered everything from impossible to wrong by most of those who commented at the time, and was opposed by the National Institutes of Health. (James Wyngaarden, its director, quipped that the DOE plan seemed "like the National Bureau of Standards proposing to build the B-2 bomber.")[4] Even Brenner had joked that the task was so epic and the technology so limited that sequencing should be meted out as punishment to convicts—say, 12 million bases for a typical crime. But I was immediately drawn to the thought of having a database of the sequence of every human gene. I had just spent the better part of a decade attempting to decode just one of the (then) estimated 100,000 human genes. If we could determine the structure of every human gene within fifteen to twenty years by undertaking a massive effort, then I was all in favor. My field would certainly benefit, as would many others. As far as I could see, the only genuine controversy involved was whether the project was feasible. It would certainly be impractical to use the original Sanger method, with its radioactive markers, cracked and diverging gels, and endless frustrations; making an automated attempt on the genome was an entirely different matter.

I started to think more and more about how I might participate in such an effort. I would have to expand my laboratory, but real estate at the NIH was limited. After discussions with Ernst Freese (by then a convert to the cause of sequencing, thanks to my success) and Irwin Kopin, intramural director of the National Institute of Neurological Disorders and Stroke (NINDS), I was offered space in the Parklawn Building in Rockville, Maryland, across the street from the Food and Drug Administration. The move there would allow me a four- to fivefold expansion of my program and a doubling of my team to more than twenty scientists.

Still, the decision was not straightforward. I was uneasy about moving off campus—space there was so highly prized that some of my peers had fought legal battles over which floor they occupied in Building 36. But two incentives were included to encourage me to go to Parklawn: I would have minimal administrative duties, and I would be on the planning committee for a new building (Building 49) on the main campus, into which I would move my group on its completion. I agreed and we moved to Parklawn in August 1987.

At that time most had scoffed at the thought of the ABI machine being used to tackle a project of the scope of the human genome. The Japanese had already offered one alternative, making headlines in 1987 by announcing they wanted to build a new machine that was capable of sequencing a million base pairs per day. (The goal was eventually scaled back to 10,000 bases a day.) But to me the real solution was a no-brainer, as the history of manufacturing had shown. If you had one sewing machine and wanted to double your output, you added a second one. In order to double our DNA sequence output, we needed a second DNA sequencer. I knew I could match the goal of the Japanese by adding a few more ABI machines to my lab. Parallel processing was the simple answer.

I began to negotiate with Freese to add $750,000 to my annual budget so that I could buy more DNA sequencers. Although fearful that other lab chiefs would complain about my disproportionately large budget, he sidestepped the problem by rebranding part of my lab the NINDS DNA sequencing facility. That tactic enabled me to get my machines as long as I would help fellow investigators sequence pieces of DNA involved in their research. I agreed because I knew that very little molecular biology was going on in the neurology institute, and there would therefore not be much demand. One exception was Carleton Gajdusek, who had won the Nobel Prize for his work on kuru, an unusual brain disease related to BSE (mad cow disease). I purchased three additional sequencers, and instantly my lab became the largest DNA sequencing center in the world.

While I was eager to begin sequencing immediately, it quickly became clear that we did not have all the methods in place to proceed in a cost-effective manner. We needed a strategy that made the best use of the automated sequencing machines and took account of the fact that they could handle only a few hundred base pairs of DNA sequence at a time.

One strategy was to produce the longest sequence possible from a single run, reading a few hundred base pairs at a time. Then, using DNA code from the end of this sequence, we would make a new primer that would mark the starting point for the machine to read the next few hundred base pairs of the adjacent DNA sequence. This process would be repeated again and again until the end of the gene or piece of DNA in question was reached. This technique, called "primer walking," took days, because a new primer had to be customized for every step of the way. Primer walking was made even more time consuming and expensive with the ABI sequencer because the primers were not simple stretches of DNA but had fluorescent dyes chemically attached. It was immediately clear to me that the primer walking method would not be a practical way to sequence tens of thousands of bases, let alone the billions in the human genome. (While clear to me, it was not so clear to others. This was debated for several years.)

The principal alternative to primer walking was small-clone shotgun sequencing: smash up the 18- to 35-kilobase clone into pieces small enough to be sequenced and then figure out how to put the short sequences back together again. There were various versions of this method, depending on how the DNA was broken into smaller sections. While he did not adopt a true random (shotgun) approach, Fred Sanger, in his pioneering 1982 work on decoding the forty-eight thousand base pairs of the bacteriophage lambda, used a novel way to break up, or fractionate, the lambda genome. He took advantage of research by another Nobel laureate, Hamilton Smith, who had discovered restriction enzymes, the molecular scissors that are able to cut DNA at precise, unique sequence sites. For example, the restriction enzyme ECORI will cut the sequence GAATTC but leave it untouched if there is a single letter change (GATTTC). After cutting the DNA with a variety of restriction enzymes into fragments small enough to be sequenced, Sanger used a special map of the DNA under study to reconstruct the lambda genome. This map showed the sites on the DNA pieces where the restriction enzymes were known to act, which could be used as landmarks by Sanger to align one piece with another.

Imagine cutting up a copy of *The New York Times* according to a rule (equivalent to the restriction enzyme) that directs you to snip out a section in front to the word *today* when the word *with* occurs on the page. Now repeat the process but using the words *and* and *that*. Even if one cannot read, knowing the identity of these words (the restriction sites) on every piece of newspaper to result from the cutting experiments would help put the paper back together again.[5] For viruses, Sanger's restriction enzyme method was the only available technique, but since it was manual, slow, and tedious, it was not an approach that was going to provide the sequence of a bacterial genome, let alone the 3 billion base pairs of the human genome.

While the DNA sequencing effort was advancing, our receptor research was also booming. We looked for the equivalent of the adrenaline receptor in one of the most intensively studied organisms in biology, the fruit fly (*Drosophila melanogaster*). We isolated and sequenced the fly gene, called the octopamine receptor, which was probably an evolutionary precursor of our own adrenaline receptor. In insects the chemical messenger octopamine plays a similar fight-or-flight role as adrenaline does in humans. I was making great progress and was increasingly in demand as a speaker at international conferences, as well as doing some consulting work for the biotech and pharmaceutical industries. But this research path, one that I followed since I had walked into Kaplan's lab almost two decades earlier, was about to come to an end.

I was lucky to be in an environment where we were encouraged to go in new directions. Because my lab nestled in the NIH intramural program, my research budget gave me the security to take huge risks. Most of my peers preferred caution even though we were *supposed* to be adventurous, because we did not have to write grant applications and were thus free of the conservatism of peer review where, in the endless competition for limited funds, committees prefer to give grants to the projects and people they perceive as safe bets. Human genomics offered sufficient allure to encourage me to make a leap, although it was into the unknown, and the dividends for our undertaking were still uncertain.

I could have moved into genomics while also maintaining my receptor research as a kind of insurance against failure. But after five years at the NIH, Claire had still not received tenure and did not seem likely to. Because she was one of many in my lab, it was impossible for the committee to distinguish her unique contribution to what was a highly successful program. Fortunately, the National Institute of Alcohol and Drug Abuse wanted to establish a receptor research lab and offered her the opportunity to form her own group. Instead of trying to split the receptor research program or have her start from scratch, I decided that I would attempt to build a new career in the emerging field of genomics and take with me only those who were interested in that field while letting her take over my receptor work.

This decision seemed logical, but it was a very hard one to make. As well as breaking up professionally with someone with whom I enjoyed working, I was walking away from nineteen years of effort, both failure and success, in the receptor field. What if my new efforts in genomics were a failure? What impact would this new arrangement have on our marriage and daily relationship? I referred to the move as our lab divorce: Claire would look after my receptor baby, and I would have genomics. On the other hand, I hoped that the move would strengthen our relationship because Claire would now be able to receive more objective feedback about her work and feel a greater sense of accomplishment for striking out on her own.

Entering the field of human genomics meant getting up to speed not only with the science of the community but also with its politics. I was fortunate in that when I made my move, I encountered Rachel Levinson, whose husband, Randy, was working in Claire's new institute. Rachel was the executive secretary of a new working group appointed by the NIH director Jim Wyngaarden to explore what the NIH could contribute to a human genome effort. I enjoyed talking to Rachel, who was personable and knowledgeable. Rachel suggested that I speak to Ruth Kirschstein of the National Institute of General Medical Sciences, who wanted to ensure that any NIH genome effort would be carried out within her institute and therefore come under her control and direction. Rachel set up a meeting so that I could tell Ruth about what was already under way with DNA sequencing within the NIH intramural program. I was invited to participate in a key meeting in Reston, Virginia, that Rachel was organizing between February 29 and March 1, 1988, on behalf of Wyngaarden to help position the NIH at the heart of the genome effort.

The Ad Hoc Advisory Committee on Complex Genomes was memorable on several counts. This was my first encounter with most of the major players in human genetics, such as the Nobel laureates David Baltimore, Wally Gilbert, and Jim Watson. The meeting also marked the real beginning of the genomics field and saw a surprise announcement by Wyngaarden: None other than Jim Watson was coming to the NIH as associate director of the newly formed Office of Human Genome Research, lending the project much needed scientific credibility.

Watson's appointment generated many undercurrents and was my introduction to the intense politics, lobbying, and maneuvering that would come to dominate the genome effort in years to come. Although Watson himself admitted to feeling uneasy about the challenges that lay ahead, one informed commentator believed that the position held for him "the attraction of power, but not so much the power over people as power over the future of science."[6]

A few things that I heard in Reston about the direction of the program shocked me. Watson argued that our goal was to work out the sequence and let future generations of scientists worry about understanding it. I had always believed that interpretation was crucial to making the sequencing both efficient and worthwhile. Many of those present likewise contended that once we had the genome sequence we would simply have to run it through a personal computer to find where the actual genes were located. But real life is never that simple. Another surprise came when Lee Hood described the ABI sequencer his team had developed as the equivalent of Henry Ford's model A. He wanted to build a Ferrari before starting sequencing in earnest, which would probably take years.

I just wanted to get on with it. After the meeting I went to see Ernst Freese and told him that I was going to make a commitment to advances in genomics and I wanted his backing because my budget depended upon it. He was generally supportive but added the caveat that because our institute had a mandate to work on neurological diseases, to get his endorsement I would have to target regions of the genome relevant to neurological disorders and the brain. This seemed like a reasonable compromise. Before I left, Ernst mentioned that he had been a postdoctoral fellow with Watson years before. When Watson became the head of the NIH genome center, Ernst promised to bring us all together so I could show Watson my DNA sequencing data.

There were several regions of the genome that I thought would be fruitful places to start my sequencing—for example, the tip of the short arm of the X chromosome, where many disease-related genes, including fragile X, a type of mental retardation, had been mapped. Another promising site was the tip of the short arm of chromosome 4, where we expected to find the cause of Huntington's disease, a devastating form of neurodegeneration. To keep my hand in the receptor field I would stay abreast of data on the mapping of receptor genes. After much reading and many discussions I developed a preliminary plan to sequence the X chromosome as a way to jump-start the human genome project. I now know that it is impossible to jump-start anything in a field that is so bogged down in laboratory politics that science and data come in a poor second to personality, pride, and ego.

Shortly after Watson arrived on the NIH campus, Ernst Freese set up our meeting, and he and I walked together to Building 1, where the NIH director had his office. After brief introductions, I showed Watson the readouts from the sequencers in my lab and told him I thought that they worked sufficiently well to get on with a human chromosome. He asked if I had one in mind; and I presented my plan for the X chromosome. Jim was fascinated with my data and its quality. He told me that just the day before he had visited Lee Hood's lab at Caltech and that Hood was not even attempting to use the automated DNA sequencer that he himself had invented, preferring to stick to the old radioactive method. Why had I been successful when everyone else had abandoned the ABI sequencer? I told him about the effort I had put into the sequencing process, notably getting the primer chemistry right. He asked about my scientific pedigree, and I explained how I had trained with the biochemist Nate Kaplan. "That explains everything," said Jim. "You are a biochemist."

Even today I can laugh at how I misread this remark. I had taken it at face value as a compliment since I myself was proud of my training, which stretched over four generations of biochemists. Only later did I learn that Watson held biochemists in particularly low esteem. Years later, at an NIH symposium celebrating our sequencing of the human genome, I joked that I looked back fondly to the good old days when the worst thing Watson ever called me was a "biochemist."

How much money did I need to get started? Jim asked. A few million dollars, I replied, but Watson insisted that that was not sufficient and that it would cost more like $5 million to start the sequencing of the X chromosome. He asked me to write a few pages to outline the program for the $5 million budget. He was to appear before Congress the next day, and he said he would ask that they add that amount on to the budget so we could get moving.

During his testimony, Watson was as good as his word. He said that I had the best sequencing lab in the world and that he needed $5 million to start sequencing the X chromosome. Around that time, at a lawn party at the Cold Spring Harbor Laboratory, Long Island (about thirty-five miles from New York City) where Watson was director, he had boasted: "The Human Genome program is going to succeed. I've got this guy who can get automated sequencers to work."[7] I was floating on air, confident that my effort to sequence the X chromosome would actually get the green light. I would lead the first funded human genome project.

I quickly put together the draft proposal that Jim had requested. But the problem was that, however well intentioned, Watson was a federal bureaucrat, and like any other federal bureaucrat, he was afraid to make decisions on his own. A week later there was much hand-wringing at an NIH meeting over whether it was premature to actually start sequencing; could the money be better spent on gene hunts or finding landmarks on the genome (mapping)? Perhaps it was no accident that such doubts were being expressed by university-based researchers who were not taken with the thought that a large chunk of the money that they coveted could end up with an intramural NIH researcher and not them. Jim was combative, however, and pointed to me as the guiding example of how to get the job going, explaining that he wanted to fund my team to do a demonstration project on sequencing the X chromosome.

Robert Cook-Deegan recalled how the politics of sequencing "were intense, as there was considerable opposition to large-scale sequencing among molecular biologists, even for model organisms such as yeast, nematodes, and fruit flies. The opposition was even stiffer when it came to sequencing human DNA. . . . There was also considerable disagreement about the ideal sequencing method and the best strategy. . . . The fierce opposition forced Watson to retrench on his initial commitment."[8]

Watson told me afterward that he wanted a much longer proposal from me, of around twenty pages, to create the impression that he was following some type of peer review process and not just pushing my plan through. I soon realized this ostensibly reasonable request was going to cause me problems, because by acceding to it, I would be setting a terrible precedent. Here I was, an intramural scientist who did not need grants, applying to an extramural program for funds. Several senior scientists, including Maxine Sanger, who had worked with Wally Gilbert on his sequencing method, made it clear at a public meeting what a mistake I was making. Scientists at NIH had always fretted about bringing outside peer review into the intramural program and absolutely did not want to write proposals for funding their research. I could feel the $5 million slipping out of my fingers because of the one-size-fits-all mentality that plagues any bureaucracy. Jim assured me that that was not the case and urged me to get him the paperwork.

After working diligently with my team for weeks on drafting a strong research proposal, I decided to hand the document to Watson in person at the annual spring meeting in the laboratory that he directed, Cold Spring Harbor. His response did not come quickly, and when it did, it was a longer version of what he had said before: The scientific community was simply not ready for genome sequencing, and he would alienate his supporters, allies, and advisors if they thought he was showing favoritism to me as an intramural scientist in the NIH.

Now those advisors wanted me to write a full-blown NIH extramural program project grant application, which would involve the formation of a study section not only to pore over what I had to offer but to consider rival grant applications from the likes of Lee Hood and Wally Gilbert. My view of Watson, his leadership, and his ability to deliver on his promises fell another notch.

Perhaps it was a premonition of the clouds gathering over my research, but I decided it was time to test my sailing skills by experiencing a storm at sea. I had by then traded my much-loved Cape Dory 25D for a larger Cape Dory 33, named *Sirius,* after the Dog Star in the constellation Orion, the brightest star in the sky. After making the thousand-mile round-trip from Chesapeake Bay, "down east" in the coastal Atlantic Ocean to Cape Cod and back, over two summers I had experienced some squalls and rough weather but now felt that *Sirius* and I were ready for a bigger challenge. Storms were romantic and frightening, and both were good reasons to endure one.

For East Coast sailors there was one obvious way to earn a "blue water" merit badge: sailing the seven hundred miles to Bermuda. By doing so I would, of course, encounter the Bermuda or Devil's Triangle between the island itself, Miami, and San Juan, which is infamous because of its unexplained disappearances, from the five TBM Avengers that vanished shortly after takeoff to the traceless sinking of USS *Cyclops.* The Devil's Triangle is one of the few places on earth where a magnetic compass does point to true north. It also feels the full power of the Gulf Stream, a giant river of warm water within the Atlantic Ocean that forms in the Gulf of Mexico and flows from Florida to Cape Hatteras, up to the Nantucket Bank, and then to Europe, where it helps keep Britain out of the ice age. The Gulf Stream generates its own weather, particularly when the wind is blowing against the direction of the main flow, making the waves high and steep. But these dangerous waters also nurture a great variety of marine life, and dolphins and fish congregate there in great numbers.

While I was daring, I was not foolhardy, and I decided to sail in early May to avoid hurricanes. Even though I was tempted to try a single-handed voyage, I thought (wrongly) that having a crew would be much safer. One person who wanted to make the crossing was Daryl Doyle, then a professor and chairman of biology at the State University of New York at Buffalo, who had sailed on Lake Erie. (Daryl died in 2006.) He brought along his friend Herb. We loaded the boat with fuel and cans of tuna and Dinty Moore beef stew, and on May 14, 1989, I waved good-bye to Claire in Galesville, telling her that I would see her five days later in sunny Bermuda. I deliberately did not check the weather forecast, wanting to deal with whatever blew our way without any advance warning.

The winds were light and our progress slow. As we drifted through the glassy water with increasing frustration, Daryl did the unthinkable and tempted the sea gods by shouting, "I would rather have a full gale than no wind!" Somebody was listening. By 9:00 A.M. on the morning of the seventeenth the waves were more than twelve feet high and increasing. By 3:00 P.M. we were certain that we had maneuvered out of the Gulf Stream and expected the waves to decline; in fact the opposite happened. A radio weather report described a storm over Bermuda. My log entry stated "vertical blue water," and by 6:00 P.M. the wind was so strong we had to sail under bare poles. Because of the danger that the boat could turn its side to a wave and be rolled over, I deployed the Galerider, a half-sphere of heavy webbing that dragged behind us on a heavy line to keep us stern to the waves.

Sitting in the cockpit of such a small boat during a storm of this magnitude was exhilarating and frightening. In the troughs between them, the peaks of the waves stood as tall as our 47-foot mast. As they started to break and push our stern sideways, the Galerider pulled us square, like a giant hand saving us from certain tragedy. By midnight on the eighteenth the wind was down to

25 or 35 knots, but the crew was exhausted. While I attempted to nap, Herb, who was now certain that he would never survive the Bermuda Triangle, had changed our course by 90 degrees to avoid it, putting us all in danger.

When I discovered what he had done, I was angry and sent my two crewmates below for the night, latching the cabin hatch. By now I had been awake for more than thirty hours, and Daryl had deprived me of one legal way to keep focused by spilling all our coffee beans into the bilge. I had a prescription for amphetamine just for such emergencies. Dressed in my foul-weather gear, inflatable life jacket, and safety harness, and with Roy Orbison and Elton John to listen to on my Walkman, I sailed elatedly at 9 knots up and down giant waves in the moonlight, hundreds of miles from land, through the most incredible night of sailing in my life.

The next morning, with the storm easing, I opened the cabin and asked Daryl and Herb to put their harnesses on and come steer while I got a position fix. I later learned—from the state of his mattress—that Herb had been so frightened at the height of the storm that he lay in his bunk pissing himself. Now he panicked at the sight of the huge waves and managed to turn the boat so that one crashed down upon us, almost rolling us over. Water poured down the hatch, and I rushed on deck to find that Herb was dragging in the water behind the boat, saved by his safety harness. There was no sign of Daryl. The next wave washed Herb back on board with a thud. As I turned the boat back on course, I spotted Daryl draped over the boom like a wet rag. Dangling next to him—but not on him—was his safety harness.

We finally arrived in Bermuda at 2:00 A.M. on May 22, eight days and 947 miles after starting out. After my two crew members bolted for the airport, I called Claire, who by then was certain I had perished. She had been checking insurance policies even as I was feeling the most incredible high and sense of accomplishment and survival. (I named my next boat *Bermuda High* to honor the feeling.) Claire joined me the next day, but I was not much company, sleeping for two days straight. I had no idea then that Bermuda would mark the beginning of a much longer, tougher, and ultimately more rewarding battle for survival in which my science, my marriage, and my reputation were ultimately at stake. When that voyage ended and I had finished sequencing the human genome, I felt the same incredible elation, the same visceral thrill I had experienced sailing into Bermuda eleven years earlier.

Back at the NIH I once again set to work on the X chromosome sequencing proposal, which had now grown to more than sixty pages. As I had feared, I began to run into considerable opposition from within the intramural program and decided to seek allies in the human genetic community. I contacted C. Thomas Caskey, at the time the chairman of the Human Genetics Department at Baylor College of Medicine and an X chromosome expert, who agreed to collaborate with me. He proposed, as I had already done with Freese, that we start with a region on the short arm of the chromosome, called Xq28, to which many disease genes had been mapped, including Fragile X. Tom's lab was already working with a young researcher at Emory University named Stephen Warren, who had assembled a library of cosmid clones—pieces of human DNA around thirty thousand base pairs long—covering the Xq28 region many times over. Warren came on board along with Anthony Carrano, the director of the human genome project at the Lawrence Livermore National Laboratory, a DOE lab. Because of NIH politics I printed our ambitious 1989 proposal on plain paper, not extramural NIH grant forms. Our aim was to sequence the X chromosome over a twelve-year period, but the real substance of the proposal was a plan to sequence the 4.2 million base pairs of Xq28 over three years and, at the same time, reduce the cost of sequencing a base pair from $3.50 to $0.60. (Today, our costs are $0.0009 for every base.)

I eventually received a letter stating that an interview—a "reverse site visit"—had been scheduled for March 29, 1990, at the Crystal City Marriott Hotel in Arlington, Virginia. What was then called the National Center for Human Genome Research had a Special Review Committee comprised of many scientists who wanted to play a starring role in the future of genome sequencing. They included Bart G. Barrell from Cambridge, Ronald Davis from Stanford, Glen Evans from the Salk Institute, Thomas Marr from Cold Spring Harbor, Richard M. Myers from the University of California, San Francisco, Bruce Roe from the University of Oklahoma, and F. William Studier from Brookhaven National Laboratory. Jane Peterson attended to represent Watson's genome center.

Over the years that followed this depressing encounter, almost every member of the committee went on to apply for his own center to sequence the genome. With the benefit of hindsight it is no wonder that, on the day I appeared before them, they seemed to be united by one purpose: to stop anyone else from getting any money before they did. The line of questioning made it clear that the committee believed the technology was not mature enough, the maps were insufficient, and we were not proposing new approaches. As Tom Caskey was about to fly back to Texas, he told me it was the most humiliating experience of his career.

I later learned that two rival proposals had won over the committee. One was from Lee Hood, who planned to sequence the T-cell receptor region, a vital part of the body's immune defense system; the other was from Wally Gilbert, who proposed to sequence the first genome from a living species, the microbe *Mycoplasma capricolium,* which causes pneumonia in sheep and goats. Remarkably, he wanted to use a new method that lacked any preliminary data to suggest it could really work.

Watson was as upset with the grant review outcome as I was and told me that he would set up a new review if I was willing to try once more. Over the next several months we negotiated what the revised proposal would contain and came up with other chromosomes for study. Ewen Kirkness, a postdoctoral fellow in my lab, had been hunting various receptors for the neurotransmitter GABA and had isolated a new one on chromosome 15 in the middle of a region associated with two human genetic conditions, Angelman's and Prader-Willi syndromes. These unusual diseases were hardly understood until the late 1980s, when scientists realized that the origin of a gene can make a difference in the way it is used: "Imprinted" genes from the mother's chromosomes dominated some processes, while genes from the father's dominated others. Prader-Willi syndrome, which causes mental retardation, obesity, and several growth abnormalities, occurs in people who inherit both copies of chromosome 15 from their mother rather than one from each parent. Angelman's, a syndrome that causes mental retardation, characteristic jerky movements, and epilepsy, is also linked to chromosome 15, but its sufferers lack functioning genes from their mother. Mark Lalande of the Children's Hospital, Boston, was working on imprinting and, keen to see this region sequenced, pledged to work with us. I also added regions linked to Huntington's disease. The faulty gene responsible for this devastating neurological disorder had been mapped to the tip of the short arm of chromosome 4 some eight years before, but the actual gene had not yet been found even though a large consortium of scientists—about sixty in six different groups—had been brought together by Nancy Wexler to do just that. Nancy had a personal reason driving her on: She and her sister were "at risk," since their mother, uncles, and maternal grandfather had died from this hereditary, untreatable, and fatal brain disorder. The consortium was using every method available to help her, except for genome sequencing.

Some of the consortium members argued that sequencing was an untested method and would not likely work, while others worried that it would sap energy and resources away from their own efforts. I countered that I had independent funds from the NIH and that they had nothing to lose; not a single one of their researchers was interested in trying the genomic sequencing approach in any case. As I was to learn over and over again, a surprising number of human geneticists are often more concerned that they win the race to discover a disease gene than that the race be finished as quickly as possible. My impression was that the Huntington's gene hunters were no exception. Unless they could take the credit, they would turn their backs on new approaches that might independently lead to isolating the gene more quickly.

The critical and understandable exception was Nancy herself, who, like anyone in the shadow of a terrible illness, was willing to do whatever it took to find the faulty gene that had caused so much suffering in her family. Nancy weighed in and insisted that the group try to work with me. No one could come up with an intelligent counterargument, so it was decided that James Gusella of Massachusetts General Hospital would provide my team, led by Dick McCombie, with three clones that spanned one hundred thousand base pairs from the tip of chromosome 4. As it turned out, the clones were at the fringes of the target region where the gene was thought to be. I accepted them, in any case, because I had a bigger goal of establishing rapid DNA sequencing as a valid approach. I felt that if I established a working relationship with the Huntington's consortium, then, if I succeeded, they might give me more promising clones the next time around.

Another target was the muscle-wasting disorder myotonic dystrophy. I had better luck with the myotonic dystrophy group headed by Tony Carrano, a collaborator from the ill-fated X chromosome application. After signing an agreement that said I would share data and credit with them, we received three clones from chromosome 19 that covered one hundred thousand base pairs from the region most likely to contain the gene. Antonia Martin-Gallardo, a postdoctoral fellow from Spain, headed my chromosome 19 team, which I funded with a diversion from my intramural budget. I believed then, and still do today, that success was the best tactic to impress my critics: Good data will always win the argument.

The sequencing got under way quickly using the shotgun approach. The clones were in the form of cosmids, stretches of DNA about 35,000 base pairs long, packaged in a phage so they could be put into E. coli. First we used sound waves to shatter many copies of the DNA into tiny fragments about fifteen hundred base pairs in length. Statistically, by selecting one thousand random fragments and sequencing 300 to 400 base pairs of genetic code from each, we would in theory cover every base pair of DNA in the cosmid at least ten times ($350 \times 1{,}000 = 350{,}000$).

But that did not mean we would not encounter any problems when we attempted to assemble the DNA sequences. Because the software of the day was designed to handle a few hundred sequences at most and could not cope with assembling one thousand, we had to resort to tedious manual methods. To make any significant progress we would need much more powerful computers than we were using and far better software (despite what my peers thought). To meet those requirements I began to hire scientists with backgrounds in computing.

One of them was Mark Adams, a Macaulay Culkin look-alike from the University of Michigan with big glasses and a can-do attitude, whom I had interviewed toward the end of 1989. I remember being impressed that this thin young man had started a software company on the side while still in graduate school. Mark was excited about genomics and agreed to start the following spring. We purchased powerful Sun computers, and I sought out software engineers to help develop new ways to interpret the genetic code that was already beginning to pile up in our computers. We were also looking at DNA in a new way, thanks to a blind programmer, Mark Dubnick, who used a special keyboard system that talked to him at a very fast rate, faster than the rest of us could understand. We even attempted to use it to read the genetic code and play it back to us in the form of music, to see if that sort of realization helped us detect changes in genetic structures.

But whatever approach we tried, we found that actual interpretation of the human genetic code was almost impossible, even after we had managed to assemble the long stretches of a chromosome sequence. Software that could effectively find and define genes within bacterial sequences simply did not work on the more complicated human genome, where genes are broken into small segments (exons) by meaningless DNA (introns), rather like a TV movie is broken up by meaningless commercials. In this way a gene is often spread in bits and pieces over hundreds of thousands to millions of base pairs of the genetic code. We employed state-of-the-art programs to hunt for them, but the software could not distinguish between real genes and noise generated by the random nature of the four-letter genetic code.

I came up with an approach to validate each predicted gene by looking for its counterpart in messenger RNA. Whenever a real gene existed in the human genome, there would be a corresponding messenger RNA molecule, the abridged version of a gene that contains only the base pairs of genetic code that a cell needs to make the protein. By turning this fleeting RNA into cDNA that could be sequenced, we had a way to confirm our gene predictions.

We began testing cDNA libraries from different human tissues, principally brain and placenta. Our probes were of gene sequences we had predicted from the computational analysis of the chromosome 4 and 19 DNA. If a cDNA clone could be found representing a gene predicted from the genetic code, its existence would be proof that the gene was real and not an artifact. However valid the logic of this method, it required months of hard work to confirm just a few real genes in our sequence. No wonder that, when cDNA approaches were put forward (notably by Sydney Brenner and Paul Berg) in early discussions about the genome project as an alternative to genome sequencing, they were quickly shouted down.

But despite such obstacles I still felt that I was on the right track. My conviction deepened in 1990 when I was invited to take part in a symposium series in Japan sponsored by the DNA sequencer manufacturer Applied Biosystems. There I found that my genome research was viewed as the leading effort of its kind. Many Japanese groups had chosen to concentrate on cDNA clone isolation and sequencing, and I had long conversations with them about my use of cDNA clones to confirm predicted genes in the genome sequence. They were excited by my data because it validated their own methodology. Two Japanese scientists in particular made a strong impression on my thinking. The first was Hiroto Okayama of Osaka University, who with Paul Berg had developed one of the major methods for obtaining cDNA clones and was now working on a way to make sure they covered the entire gene sequence, so-called full-length cDNAs.

This was a reference to a major problem that plagued the cDNA field: the instability of mRNA when isolated from tissues; it had a tendency to break into smaller fragments before it could be copied in its entirety. Other problems involved the enzyme reverse transcriptase, which was used to convert ephemeral mRNA messages into more stable cDNA. The enzyme would fall off the mRNA before it finished the job. I was only too familiar with this phenomenon: When I had used cDNA for my work on the adrenaline receptor, I missed a piece of the gene from one end of the clone.

Another inspiration was Ken-ichi Matsubara, director of the Institute for Molecular and Cellular Biology at Osaka University, the leader of the Japan genome effort and advisor to Monbusho (Japan's Ministry of Education, Science and Culture). Both men were convinced that sequencing full-length cDNA clones was going to be an essential part of, or even an alternative to, the genome sequencing effort even though this approach had been soundly rejected by American and British genome experts. I would come to share another bond with Matsubara: Watson, irked by how wealthy Japan appeared to be freeloading on the genome research funded by other nations, had written to Matsubara and threatened to withhold scientific data, using language that generated headlines such as "The Human Gene War."[9] "If there is going to be a war, I will fight it," Watson had declared. "You never get anywhere in the world by being a wimp."

On my twelve-hour flight home all I could think about was the full-length cDNA sequencing approach under way in Japan and how easy it would be to analyze the human genome if all the cDNA clones from the human body were already isolated and sequenced. I thought about my ten-year-long struggle to find a single gene, just one cDNA clone; about how inefficient it was to use the shotgun method to read the code of one thousand or so genome sequences to locate one gene—usually only a fraction of a gene—and about how hard it was to find a corresponding cDNA clone to prove the existence of that gene.

All of a sudden it came to me at thirty-eight thousand feet over the Pacific Ocean: I was using the right sequencing technique but on the wrong DNA. What if I could combine the rapid random shotgun sequencing method with cDNA clones? What if I just randomly picked a cDNA clone and sequenced it in one go? My sequencing machines were reading four hundred or so base pairs of genetic code at a time, which would be more than enough to find a match in a genomic database. It was like discovering an index to the human gene catalog.

Knowing that a given sequence was obtained from a cDNA clone made from fragile mRNA isolated from the human brain, for example, told us it inherently contained the key information that (a) it was part of a real, expressed gene, and (b) the gene was essential to the function of the brain. Sequences from the genome told us almost nothing by comparison. If I switched to sequencing one thousand randomly selected cDNA clones, I might discover hundreds of genes for every one I identified via traditional genomic sequencing. I was so excited by the idea that I could not wait to get back and try a critical experiment.

As soon as I walked into my lab the next morning I gathered my senior team together, certain they would share my excitement about this new approach, but I was met with a brick wall of skepticism and doubt. The bottom line from McCombie and others was that my eureka! idea had a high likelihood of failure and would probably end up drawing resources away from the genome sequencing effort. They used the same arguments that others had in questioning cDNA approaches. Current dogma said that gene expression in human tissues would involve only a small number of highly expressed genes that would swamp out any signal from the rare, lowly expressed ones. Any messenger RNA we intercepted would likely be skewed toward these dominant genes. But while this argument made sense for some tissues, it was unlikely to apply to the human brain, which depended on vast numbers of genes, some expressed at very low levels, to enable us to think. A study by a Scripps Clinic team suggested that up to one-half of all human genes are used in the brain.

I remembered Nate Kaplan's advice not to talk yourself out of doing experiments: The answers ultimately lie in how the world actually works, not in the dogma of the day. Fortunately, I found someone on my team who was intrigued by my idea. Mark Adams had joined us from the University of Michigan only a week earlier, and we had yet to consider what he wanted to do in the wake of the demise of the X chromosome sequencing project that had initially attracted him to my lab. We discussed using human brain cDNA libraries to try my cDNA random selection and sequencing experiment, and he agreed to start right away. I did not learn until years later how McCombie and others had attacked Mark behind my back for taking this idea seriously, so concerned were they about a potential diversion of resources. In fact, they had no reason to worry, because we had enough funding to push forward on many fronts. If we needed more money, I would get it somehow.

The latest grant attempt for genome funding from Watson, based on sequencing in gene-rich regions rather than sequencing an entire chromosome, was by now taking shape. Watson and I met that October at my Genome

Sequencing Analysis Conference on Hilton Head Island, South Carolina. I outlined my plan to go after the Huntington's region of chromosome 4, the myotonic dystrophy region of 19, and the Prader-Willi region of chromosome 15. He agreed that my scheme was solid, because it would appeal to the disease gene hunters, and he liked the fact that it could help end the long hunt for the gene that caused Huntington's.

Watson also assured me that this time I would receive a favorable outcome, since the review committee was being chosen with a real interest in seeing the genome project move forward. However exasperated I had been by his lack of follow-through on his three previous promises, we were still on good terms with each other, and I felt that he genuinely wanted to see my program get funded. After all, we were both true believers in genomics and sequencing the human genome.

There was one other incident I remember from this conference that would take on future significance. During a one-day workshop that I had organized with Watson, he got into a shouting match with pharmaceutical company representatives over who would have the rights to gene patents from the genome project. He had also called for a mutually agreed-upon policy on the release of sequence data, preferring it to be done as soon as investigators were confident that the data were accurate. In years to come this spat would come to haunt me.

Suddenly I found my scientific life transformed. The results emerging from the sequencers in my lab made it clear that sequencing a random selection of cDNA clones was going to be a very big winner. I was jubilant, but a tremendous amount of work remained to be done in order to ensure that we were not being misled by our data or somehow making an error. With only three hundred to four hundred base pairs of genetic code from the cDNA, the challenge was to confirm that we had sufficient information to identify the corresponding gene. We wanted to show the full value of the approach, including mapping the cDNA sequences back to the genome. We also tried manipulating cDNA libraries to see if there were efficient ways to damp down the effects of the common genes so we could "see" the rarer ones.

As we studied more and more cDNA clones, the lab became feverish with excitement. In 1990 fewer than two thousand human genes had been identified and sequenced, of which only 10 percent—such as the adrenaline receptor—came from the brain. Every day that our sequencing machines ran we were able to discover twenty to sixty new human genes. Every day! I found this figure almost hard to contemplate, since it represented ten times as many human genes as we had decoded in months and months of genomic sequencing, and sixty times more than I had managed with traditional methods in a decade of unbelievable grind to find the adrenaline receptor. We were going to turn biology upside down.

6. BIG BIOLOGY

Progress in science depends on new techniques, new discoveries, and new ideas, probably in that order.

—Sydney Brenner, 2002 Nobel Prize Winner in Physiology or Medicine

In science the credit goes to the man who convinces the world, not to the man to whom the idea first occurs. Not the man who finds a grain of new and precious quality but to him who sows it, reaps it, grinds it and feeds the world on it.

—Sir Francis Darwin, First Galton Lecture before the Eugenics Society (1914)

It all seemed so clear, so logical, and so very simple. I knew how to make real progress in unraveling the secrets of the human genome. I knew how to realize this huge ambition by focusing our efforts on the few percent that made proteins and not being distracted, for the moment at least, by the other 97 percent that contained regulatory regions, DNA fossils, the rusting hulks of old genes, repetitious sequences, parasitic DNA, viruses, and mysterious stretches of who-knows-what. While it would ultimately be important to understand the genome in all its head-reeling complexity, I could learn the critical aspects of what it was saying about life itself simply by looking at what it was telling the rest of the cell to do. Little did I realize that by focusing on the much smaller quantities of genetic material carrying out the genome's instructions, not the actual genome itself, my express way to genes would be seen through the warped lens of politics as a threat to the very survival of the genome project.

At the end of 1990 I wanted to tell the world how well we were doing with the cDNA approach and called *Science* magazine to discuss a paper. The editor was interested and urged me to put together a major article on our new method and gene discoveries. I was well aware that the first paper on the cDNA sequencing method had to be as thorough as possible. I met daily with Mark Adams to discuss the progress of the project. By now the rest of the lab had been fully convinced we were on the correct path forward. While Mark worked on novel ways to make cDNA collections for us to study, I contacted a fellow NIH lab chief, Carl Merill, for help in mapping the new genes back to the genome. Others were developing new software to identify the genes.

I also began to discuss our findings with other scientists and gave some public lectures on my team's efforts. I soon discovered that Sydney Brenner, a molecular biologist working in Cambridge, England, was also thinking of sequencing randomly selected cDNA clones. The South African–born son of an illiterate Jewish cobbler, Sydney is garrulous, creative, and probably the most brilliant living molecular biologist. (He went on to win the Nobel Prize in 2002.) Because I had tremendous respect for him I called to tell him that I had heard he was doing some similar work to ours. I explained that we were very near to submitting our paper for publication in *Science* and asked if he would consider publishing back-to-back papers if the editors were interested. We could even exchange data to see if we had found some of the same genes. Sydney explained that his effort had not gotten that far yet but agreed that what I said made sense. We planned to talk soon.

After *Science* said it would consider publishing a paper by Sydney Brenner at the same time as mine, I passed on the news to Sydney, repeating my original offer to exchange data and assuring him that I would hold off from submitting my manuscript to give him a chance to get his own ready. He explained that he had a complicated funding arrangement with pharmaceutical companies and the British government and that he was therefore not free to exchange data just then, but he suggested that our bioinformatics people at least touch base. Tony Kerlavage, who was now working on a new database for our sequences, accordingly followed up with one of Brenner's people, but after a few weeks it became clear that there would be no exchange of genome data and no manuscript forthcoming from Brenner. Finally, I called Sydney and told him that we wanted to go ahead and submit our paper. Because his was still not ready and would not be soon, he agreed.

We had one more issue to address before finishing our manuscript: We needed a name for our technique. We were dealing with genes used in tissues, but in most cases we did not have the entire gene, just a part of the gene sequence. Tony Kerlavage, acknowledging an existing mapping tool called "sequence tag sites" or STSs, came up with the moniker "Expressed Sequence Tags" or ESTs, a description I really liked, as did the rest of the lab. We entitled our paper "Complementary DNA Sequencing: 'Expressed Sequence Tags' and the Human Genome Project" and sent it off to *Science* in early 1991.

All the while I had been thinking about my fourth attempt to convince Jim Watson that my lab deserved funds for the genome project. Although the proposal was already under review, it had now become even clearer to me that my EST approach was of incredible value for gene discovery and understanding the genome. I sent a preprint of the *Science* paper to Watson to see what he thought, as well as a query to him via Jane Peterson (the mid-level government bureaucrat who had delivered all the objections to our earlier applications in a depressing monotone) asking whether he would be willing to allow me to alter some of the aims of the pending grant to include the EST approach. *No* was the resounding response. Although Sydney Brenner and Paul Berg of Stanford University, who had shared a Nobel Prize with Fred Sanger for his work on the genetic code, had already argued for a systematic cDNA sequencing effort, the majority of the human geneticists advising Watson were soundly against it for many of the same reasons that my own team had initially opposed it.

In the spring of 1991, the Watson review group met to review my fourth proposal. Once again Jane Peterson telephoned me with the outcome; once again, she informed me that the committee had found the technology too new. No one really knew if it would work, and so on and so forth. It was the X chromosome grant debacle all over again, and to say I was frustrated was a colossal understatement. I was angry with Watson for having reneged on his original promise and for having made me go through a review process that had dragged on for years. I was angry with the review process, too. I felt that science was not its top agenda but, rather, money and control. I knew that I had made a breakthrough that could change genomic science, and I was wasting my time, energy, and emotion on battling with a group that had no serious interest in letting an outsider analyze the human genome.

I began to draft a scathing letter to Watson. For the past two years the bureaucracy that he had helped generate had become a pointless, annoying, and frustrating distraction from science. I had expended considerable effort in writing four proposals for Watson instead of moving genomics forward, all for nothing. I accused him of trying to please all his critics while being afraid to actually lead the project. I told him that I was withdrawing my application so that I could concentrate on my EST/cDNA approach, which did not require his funding. After friends and colleagues had toned the letter down so that it was less of a personal attack on Watson, I sent it off on April 23, 1991, more than a year after I had sent him the preprint of the EST *Science* paper, to which he had never responded.

Jane Peterson telephoned me soon after my letter arrived in Watson's office. She was puzzled, an emotion I had not detected in her before, and asked if I knew that my grant was likely to be funded. I have sometimes been accused of being insensitive, and with justification, but I told her quite honestly that from the response I had received there was no way I could possibly have discerned any scintilla of positive feedback or interest in my proposal. Still angry with the endless foot-dragging and the way I had been treated, I told her in a fit of pique that I was going in a different scientific direction from that in the proposal and that if I could not use the funds in part for the EST effort, then I would do without them.

I have since wondered if I made a major error. Although I will never know where the track would have taken me had I not withdrawn my grant application, Ham Smith, one of my closest scientific collaborators, has argued that I might have accomplished much *less* had I accepted the money from Watson. The affair illustrates one of the great and simple lessons of life: When you come to a fork in the road, you can travel only one path.

My EST paper was by then receiving rave feedback from the scientific reviewers, and it was clear that it would be accepted and published. A minor setback occurred when one of the reviewers suggested that we include a reference to an article that we had not come across and that had been published almost a decade earlier. The paper appeared at first glance to preempt our own work because it described the partial sequencing of 150 cDNA clones that had been randomly selected from the library of rabbit skeletal muscle cDNAs. If this paper established anything, however, it seemed to confirm the view that a cDNA approach was *not* viable. Only a few already known, highly expressed genes had been found by this team, which, of course, is one of the main arguments that had been made against using any cDNA method. Nevertheless, I was curious to see if the paper had had any impact or if it had in any indirect way influenced my own thinking. As a rule of thumb, the more significant a paper, the more it is cited by subsequent work and scientists. We surveyed the subsequent literature and discovered that no one had cited this work other than as a means to obtain copies of rabbit muscle cDNAs. My paper would be published in June, and I was only too happy to include this reference.

I could trace my own EST brainwave to a flight back from a trip to Japan. But, of course, great ideas are often simultaneously conceived by several people who have responded in a similar way to the climate of thinking, and it can be hard to pin down when and precisely how a flash of inspiration is born. Kaplan had taught me that good ideas are a dime a dozen for a smart person, and the only thing that alternately distinguishes good from great is in how an idea is executed—how it becomes reality.

Scientific history is littered with stories of one person having an idea but not following through on it only to see another have a similar inspiration and then prove it to be valid. For example Darwin was not the first to conceive of evolution or even to write about it, but he did undertake a lifetime of research and writing to support the idea as a viable one. The same was true of ESTs. Brenner never did publish his data even though I believe he was thinking in the same way I was. A proposal that cDNA clones from a fruit fly library should be randomly selected and sequenced had also been included in a 1990 grant application to the NIH Genome Center by Steve Henikoff of the Fred Hutchinson Cancer Research Center in Seattle. While I was fortunate to work within the freedom of the NIH intramural program, Steve had to go through the lengthy process of writing a grant application and waiting nine months for a review. For his trouble he received a five-page critique citing all the reasons the method would never work and denying him funding. Steve sent me a copy after my paper was published in *Science* and told me that he posted a copy of the dismal review of his proposal on his lab wall, next to my picture and the *Science* article.

The complex provenance of ideas means their origin is often open to interpretation; just consider the different claims that have been forwarded in different countries for who invented the television, who turned on the first lightbulb, and so on. Even with a clear publication trail to my EST paper in *Science,* my critics have often been confused, or at least seemed to be suffering from selective vision, when it came to determining how the idea came to me. For example, in his recent book on the history of molecular biology and genomics, released to mark the fiftieth anniversary of the discovery of the double helix, Jim Watson described how I had visited Sydney Brenner's lab and had been impressed by the cDNA strategy. "He could hardly wait to return to his NIH lab outside Washington, D.C., where he would apply the technique himself to produce a trove of new genes."[1]

Of course, I had not visited Sydney, and more than a decade before Watson published this account, Tony Vickers, Brenner's boss and the head of the U.K. Medical Research Council's genome effort, revealed what had actually happened in Sydney's lab: "The MRC group was not ready to publish when Venter first broached the issue—its sequencing system had been running for just one month."[2] Watson's book does, however, contain one revealing insight that explains why Brenner never published his data: "Keen to reap the commercial benefits of the sequences, the MRC prevented Brenner from publishing them until British pharmaceutical firms had a chance to position themselves to profit from them."[3]

The commercial ramifications of my own sequencing work became apparent in May 1991. As I was wandering around in an NIH administrative building in search of a meeting I realized I was lost and asked a helpful-looking person in the hallway where the conference room was. He happened to be Reid Adler, the head of the NIH technology transfer office, which determined the agency's patent policies. "Aren't you Craig Venter?" he asked and then explained that he had been meaning to track me down after receiving a letter from a Max Hensley, a patent attorney for the biotech giant Genentech, asking what NIH planned to do with my fountain of gene discoveries. Adler wanted to visit my lab to discuss intellectual property issues. I told him that I did not welcome the idea.

I did not know much about patents and patent applications, and the little I did put me off. When I was at Buffalo, the State of New York wanted to file patent applications on the antireceptor antibodies that Claire and I had made. The major impact, as far as we were concerned, was to hold up our scientific publication while a patent application was prepared and filed. This was all a consequence of the Bayh-Dole Act, passed in 1980 and named after its sponsors, senators Birch Bayh and Robert Dole. The act encouraged the utilization of inventions produced under federal funding but, inevitably, blurred the distinction between discovery and commerce. As a consequence, scientists, lawyers, and economists continue to disagree whether the research of universities and federally funded bodies has been diverted away from the pursuit of basic knowledge, which is freely passed on to all, toward a pragmatic hunt for results that can be of use to industry. Little did I realize that my discoveries would effectively throw gasoline on the fire of this debate.

In the name of technology transfer the NIH had filed patents on the fruit fly octopamine receptor and on the cell lines that made the receptor protein. But when it came to the ESTs and the EST method that we had discovered and developed, Mark Adams and I made a very conscious decision to ignore the government rules because we did not want our publication to be held up while patents were filed on our discoveries. I accordingly told Reid Adler that our paper had been accepted by *Science* and was due to be published within one month and that I refused to delay it for even one day.

Adler visited my laboratory to press me on my opposition to filing any patents. The reason was simple, I told him. We wanted everyone to be able to use our new method and our newly discovered genes to push forward our research in an unimpeded fashion. But, of course, it was not that simple. I, like many scientists, equated patents with secrecy, largely because the initial efforts of patent attorneys do indeed delay public disclosure of scientific discoveries. But I now understood that a patent is a bargain the state makes with an inventor that is actually designed to do the opposite: to make information publicly available while it also awards commercial development rights to the inventor.

For those seeking to protect inventions or discoveries, the alternative to patents is trade secrecy. The formula for Coca-Cola has been kept confidential within the company ever since the syrup used to make the drink was first developed in Atlanta, Georgia, on May 8, 1886, by Dr. John Stith Pemberton, a local pharmacist. Under the patent system an inventor is given a designated period of time in which to commercially develop his idea; after that period has elapsed, the invention can be commercially used by anyone. Had the formula for Coca-Cola been patented, we would all now know what is in that glass of Coke, and, after the patent expired, competitors could make the same carbonated drink.

Patents play a crucial role in drug commercialization. In the United States there is a highly regulated environment in which companies have to provide vast amounts of data to prove that a new drug is safe and effective. The Food and Drug Administration controls which drugs are released for the American market after a clinical trial process that can cost hundreds of millions of dollars. Fewer than 10 percent of the drugs that enter these expensive trials are approved. Because it is relatively easy for a generic manufacturer to duplicate a drug by reverse engineering, no pharmaceutical company, the argument goes, is going to invest the hundreds of millions if not a billion dollars required to create a new drug if it cannot protect its intellectual property.

When Caffeine Kills

I drink endless cans of Diet Coke but, fortunately, carry the benign version of the gene P450 1A2 (CYP1A2), which helps me cope with my addiction. This stretch of my DNA is worth a mention because it again emphasizes how some genes only become harmful in combination with a certain lifestyle—such as drinking a large amount of coffee, tea, or Coke. The gene, found on chromosome 15, is responsible for an enzyme in the liver that helps metabolize caffeine (one of a big family of "cytochrome" detox enzymes); a mutation in it slows this process down and, as a result, increases an individual's risk of having a heart attack. A study of about four thousand people showed that the risk was raised by 64 percent for four or more cups per day over the previous year compared with patients who drank less than one cup per day. The corresponding risk was less than 1 percent for subjects who had two copies of the rapid metabolizing gene—a category into which I fall. This insight may explain why many studies looking at the association between coffee consumption and heart attack risk have been inconclusive.

By the time Adler had approached me about patents, intellectual property rights had become a major issue in genetics, having arisen in the 1980s when the insulin gene was first isolated, patented, and manufactured by Genentech, the very same company that had expressed concerns about my work. Recombinant (produced by a genetically altered organism) human insulin marked a welcome improvement over the standard treatment: People took pig insulin but eventually developed an immune response against the hormone. They required higher and higher doses of insulin to achieve effective therapy but often died early from kidney failure when antibody-insulin protein complexes blocked the kidneys.

Amgen, another biotech giant, isolated and patented the gene that codes for the protein hormone erythropoietin, which increases the production of red blood cells. Erythropoietin was the first multibillion-dollar biotech drug, and along with insulin it sent the seductive message that control of gene patents equaled billions. Scientists, universities, states, and the federal government began to patent every newly discovered human gene. Many scientists and federal officials who once thought that patents inhibited development began to be swayed by a series of studies revealing that discoveries that had been placed in the public domain by NIH and Harvard rarely resulted in benefits to the public because they were never developed.

Adler explained that when it came to ESTs, the law was unclear. There was a strong possibility that our partial sequences would render the entire gene unpatentable because it was obvious how to use the EST to get the full gene and protein sequence. This would discourage the pharmaceutical companies from taking an interest in the gene—as they wouldn't have exclusive use of the data, any drug they developed could be easily copied—and impede the development of new gene-based treatments and therapies. In other words, we could do more harm than good with our publication. But if the NIH patented the genes first, then academics could use them for free, and commercial ventures would be allowed to license them for a realistic fee. He made it clear that it was not up to me to make such decisions and that he could decide independently to go forward with the patenting process.

Mark and I agreed to cooperate with him on two conditions: The *Science* paper would proceed as scheduled, and Adler would make every effort to make his decision public and bring it to the attention of the likes of Watson, to make sure that this was the right approach. I remembered Watson's arguments at the workshop on Hilton Head on gene patents. But I also remembered how he had sold the genome project to Congress as potentially a tremendous boost to the biotech and pharmaceutical industries.

I have often been told to be careful what I ask for because I might just get it. Adler did make an effort to talk to the NIH Human Genome Center and Watson and his lieutenants before submitting the patent application. But on this important issue Watson remained silent. This seemed strange at the time but not in hindsight. Years later when we asked Watson to support a proposal to sequence the human genome rapidly, he told us that he wanted to pretend he did not know what we were planning so that he could act surprised in public by the announcement of our plan. He now had the same opportunity to raise his eyebrows and look shocked when Adler filed the application on the 347 genes that my team had discovered just prior to our publication of our paper on June 21, 1991, in *Science*.[4]

In an editorial, in *Science*, Daniel E. Koshland Jr. said we had found "a shortcut of immediate practicality and great interest to the understanding of the human genome. By providing expressed sequence tags of complementary DNA, a large number of new genes are being uncovered (particularly in the brain). At the same time, lighthouses are being provided along the chromosomes to guide the way for weary sequencers struggling with foggy restriction maps."

But the journal also contained a harbinger of the storm to come in a news article by Leslie Roberts entitled "Gambling on a Shortcut to Genome Sequencing." I was not, in fact, promoting cDNAs as an alternative to genomic sequencing but rather as a "handy adjunct," and had carefully pointed out that I could not discover 100 percent of human genes with my method. Watson, of course, had felt that this was a major flaw, but I had replied, "I won't feel we have failed if we only get 80%–90% of the human genes." Even today, I wince at a sideswipe made in Roberts's article by the British molecular biologist John Sulston, who had quipped: "I would bet quite a lot that it won't be 80%–90%. I think 8% or 9% is more like it."[5]

Sulston worked on *Caenorhabditis elegans,* a little soil worm that grazed on bacteria, in that British birthplace of all things genetic, the MRC Laboratory of Molecular Biology, Cambridge. It had been Sydney Brenner's idea to find out how an entire animal functioned (a vision that Jim Watson once thought too wildly ambitious to be worth funding), and he decided he had better start with something simple: the worm. Sulston had first encountered the nematode in 1969, when he started work with Brenner's group and began the research that would win him, along with Brenner, a Nobel Prize three decades later, not least for a monumental effort to trace the pedigree of all 959 cells in the worm's one-millimeter body. By 1983, Sulston joined forces with Fred Sanger's right-hand man, Alan Coulson, to map and sequence the worm genome.

But his depressingly downbeat estimate of the value of ESTs had shocked me, not least because I had already discussed this work when I met Sulston and his collaborator Robert Waterston, another former Brenner worm researcher who had gone on to set up his own laboratory at Washington University, in St. Louis. At a conference I had also mentioned the desire of Dick McCombie on my team to apply the method to *Caenorhabditis elegans.* Sulston and Waterston did not want to collaborate, parroting the consensus view that it was of no scientific value. That was no surprise, but I was taken aback when they urged me not to publish the EST paper because it might be viewed as undermining the genome project. I have since been told by a senior scientist that Waterston had blocked the publication of a paper to assess errors in DNA sequences, fearing that it could frighten away government funds. And Sulston, in his own "accurate and unvarnished account"[6] of the genome race, published in 2002, admitted that he considered my EST approach a threat to his own work: "If his lab was able to identify a substantial number of worm genes at a time when we had only a few dozen . . . it would probably not help our case for funding."[7] Within a year, though, Sulston and Waterston would mount their own major EST sequencing project for *C. elegans,* and Sulston would state that "ideological stances are best avoided: a good genome project must take data from all available sources and merge it into a complete picture."

Within one month of the EST paper's publication in *Science,* I was invited to participate in a Senate hearing on the genome project being conducted by Pete V. Domenici of New Mexico, who had been a keynote speaker at my Hilton Head meeting. This would be my first encounter with the Senate. To my great relief it was supposed to be a feel-good hearing: Domenici, a Republican, was one of the Senate's earliest and strongest backers of the genome project, thanks in part to the efforts of Charles DeLisi at the Department of Energy, who had once drafted a memo that outlined with great prescience what would come to pass, including interest from the private sector and the need for heavyweight computing. The only other senator present was Al Gore, who was also supportive. Part of the hearing was to discuss intellectual property issues, and even by that time there still had been no public disclosure about NIH's filing a patent application on my team's discoveries.

At the hearing I not only described the EST method and the rapid rate of human gene discovery but also voiced my concerns about NIH's patent efforts, a subject I was glad to get out in the open. The room went quiet as many were startled by this disclosure and then Watson suddenly shouted that it was "sheer lunacy" to file such patents, adding that "virtually any monkey" could use the EST method and that he was "horrified."[8] As Cook-Deegan, a Duke University genome discussant, described the event, "Watson was lying in wait and took aim with heavy artillery."[9] Cook-Deegan, who was Watson's assistant at the time, told me later that Watson had practiced the lines for a week prior to the hearing.

I was understandably shocked by this reaction, given that Watson had known about the patents for months and had many opportunities to weigh in officially as head of the genome center. An onlooker described my startled reaction: "You could see the dagger go in—it killed him."[10] In place of informal discussions that might have defused or even prevented the conflict,

Watson chose to grandstand in front of the press and two senators, and place the blame for the patents on me as a means of discrediting the perceived threat to his budget. Even Watson later admitted he had been too hard on me.

Watson Got It Right. I Am a Primate.

Now that I have analyzed my genome, I forgive Watson for this outburst. Unwittingly, in a convoluted way and without respecting the differences between monkeys and apes, Jim had hinted at a discovery that would come from my own DNA sequence fifteen years later: Parts of my genome are absent from the public genome, from that of the mouse, and even from the reference human genome produced by my effort at Celera,[11] but do appear in the chimpanzee genome. What Jim probably meant to say was something like: "Craig has ape DNA, so perhaps any ape could do it."

My colleague Samuel Levy made the discovery while browsing my genome and comparing it to the public genome held by the NCBI and the genomes of other creatures. He found up to forty-five thousand base pairs of genetic code that are in my genome and in that of the chimp, but do not seem to be common in other people or even in mice and rats. Among my ape commonalities are an additional five hundred base pairs for a DNA region found on chromosome 19 that codes for a so-called zinc finger protein, which binds to DNA to regulate how genes are used.[12] And it seems that my ape version binds to less DNA because it is slightly longer than the apparently "normal" version of this piece of code. (I say apparently since there is so little genome-wide data available, so we are still trying to figure out which flavors are common and which are not.)

One might imagine that this difference would affect the way the genes regulated by this protein are used in my body, but it will take many years to uncover all the details. Who knows? Perhaps, in the race to sequence the human genome, these chimp regions gave me a selective advantage over my rivals in the public program.

Apart from being offended by the politics and intellectual dishonesty of this episode, which struck me as Machiavellian, I was worried that this outburst from an establishment heavyweight would demoralize Mark Adams and the young team that had worked so hard on this breakthrough. Humor is the best way to relieve tension, and here Claire was ahead of me. She showed up the next day in my DNA sequencing lab dressed in a gorilla suit and put on her NIH issue white laboratory coat. The team took turns being photographed as they worked a DNA sequencer with the gorilla, who also sat down to read Watson's textbook. It was juvenile but, as makeshift group therapy goes, effective, and we all got back to work with a vengeance.

The next day I received a call from a *Washington Post* reporter, Larry Thompson, who told me that he wanted to understand the science of what we were doing and asked for an interview. We spent hours going over all the aspects of the new automated DNA sequencer, the computing, and the hundreds of new genes being discovered every day. The following Monday his story appeared and it was firmly about the politics, not the science. This encounter taught me a valuable lesson about the media: The content of a story does not map onto the content of an interview. (Thompson would end up becoming the press person to Watson's replacement, Francis Collins.)

Thompson portrayed the patent controversy as a battle between Watson and Bernadine Healy, the first woman to lead the NIH, a cardiologist who had served as deputy science advisor to President Ronald Reagan. This angle was understandable because Watson had broadened his attack to include Healy for her supposed complicity in what he now liked to call the "Venter Patents." The battle was presented as a clash of titans, the world's best-known molecular biologist against the most powerful administrator of biomedical research. In the scientific establishment these attacks are usually shrugged off. But when it comes to the business community, they can have damaging consequences, as they ultimately did.

The furor continued. In direct contradiction to Adler's initial contention, the EST approach was condemned because my critics said it "would undercut patent protection for those who labor long and hard at the real task of elucidating the function of the proteins encoded by the genes."[13] Eventually the dispute took on an international dimension. "If Craig can do it, so can the U.K.," Watson told *Science.* Sir Walter Bodmer, a highly regarded population geneticist, "warned that if Venter continues with this wholesale patenting, the British may have to follow."[14] More seriously, Tony Vickers, head of the MRC genome effort, had confirmed that "should Venter's U.S. patent application succeed, British researchers would be forced reluctantly to follow suit" and added that the French were also considering patents.[15] For me the most striking feature of all these stories was how Watson had turned Adler's efforts at the NIH into the "Venter patents," effectively turning me into the poster boy for commercialization of research, a scapegoat and a villain, all rolled into one.

The New York Times quoted one researcher as saying of the EST method: "It's a quick and dirty land grab and it's not very innovative" (although he later wrote to me disowning some of the quote attributed to him.)[16] *Nature* warned of a patent gold rush based "on researchers around the world hav[ing] suddenly developed an interest in the working of the Patent Office—as well as in Venter's simple technique." Maynard Olson of Washington University in St. Louis declared the EST approach "a terrible idea" and dismissed it as cream-skimming: "Scientists who advocate such approaches are playing with fire."[17]

But the issue was not, as many seemed to think at the time, ultimately about access to the sequence information. I could at any time make my sequence information public after filing a patent by releasing it to GenBank, since this did not endanger the patent itself. *Nature*'s reporter Christopher Anderson astutely observed what was really bothering Watson and his supporters in the controversy: "Large-scale cDNA sequencing is just not sporting. The genome project was sold to Congress as a 15-year, $3 billion effort to map and sequence the entire human DNA molecule. Now Venter says he can get almost all the genes—the part of the genome most congressmen care about—in a few years, for perhaps $10 million. . . . [C]onvincing Congress that another $2.5 billion still needs to be spent on sequencing the rest of the genome—the 97 to 98 percent that contains no genes—might be difficult."

The "it's not fair" theme surfaced a few weeks later in the December 2, 1991, issue of the *San Francisco Chronicle:* "Imagine that Sir Edmund Hillary, instead of trekking up the slopes of Everest, had taken credit for climbing the world's highest mountain after being flown to the summit by helicopter. The New Zealand climber undoubtedly would have created an international controversy." The NIH patent is "the equivalent of laying claim to entire mountains by flying from peak to genetic peak and planting the NIH flag."[18] But this metaphor is misleading: What was of concern was not which scientists were tough enough to endure the slow monotony of old-fashioned methods but obtaining some key insights from genetics to begin to guide treatments in the clinic as quickly as possible. You might be impressed by that tough ambulance man who had spent a day climbing a mountain to treat your broken leg, but you'd be much more relieved if he had done it in two minutes by arriving in a helicopter.

The first *real* hint of any commercial promise of the genome came when Rick Bourke entered the story. I first heard of this businessman and former squash champion from Norwalk, Connecticut, during a phone conversation with Isadore Edelman, an outstanding physiologist from Columbia University who had rung to say that he was advising Bourke. Not only was he *the* Bourke of Dooney & Bourke, the women's leather accessory company, but he was married to the great-granddaughter of Henry Ford—Eleanor, nicknamed "Nonie." Bourke was interested in genomics and had made overtures to some of the leaders in the field, including Charles Cantor and Lee Hood. Now, Bourke wanted to pay me a visit. Sure, I told Edelman, come by.

The two men flew to Washington in Bourke's private jet. Bourke was clearly excited about our discoveries and wanted to know if I was willing to leave the NIH to become involved in a new genome company. I loved the science that I was able to do at the NIH and felt at the time that I could easily have remained there. And I knew, too, that basic science did not always endure in a commercial environment. As a consultant and scientific board member I have often witnessed what I call the "praying mantis syndrome": The person who starts a biotechnology company, often as a means to fund his science, ends up getting his head bitten off, fired, or demoted when cold-eyed investors eventually demand a switch from basic science to the developments of marketable products. I wanted to keep on doing basic science and decided that the only way I would leave NIH was if I could set up my own research institute.

I told all this to Bourke, who explained that he was still only interested in the commercial opportunities. Nonetheless, he invited me to spend a couple of days in July at his estate in Maine. Bourke's summer house was on Mount Desert Island, home of Acadia National Park, where the mountains meet the sea in one of the most spectacular of all East Coast settings. There I would join Lee Hood, Charles Cantor, Tom Caskey, and three senators: George Mitchell, who was the Senate leader, and the heads of the Budget Committee, James R. Sasser from Tennessee and Paul Sarbanes from Maryland. It was a relaxed gathering, so families were invited, and Cantor brought along his girlfriend, Cassandra. The neighbors showed up, too: David and Peggy Rockefeller.

The house was one of the most remarkable I have ever seen and was my first introduction to a rarefied world of wealth and privilege. I took along my son, and when Chris and I arrived, we were greeted by Bourke's concierge and shown to a guest room in the playhouse, so named because it was separate from the main house and had indoor squash courts, a billiards room, and a serious swimming pool. Outside, a large ramp led down to a dock on the waterfront. During the day Chris and I were given the run of one of the Boston Whalers, and we used it to explore the harbors and ogle the houses and boats, notably the elegant sailing yachts that the Hinckley Company has made since 1928. At the end of the afternoon Chris worked on a wooden puzzle with Nonie. The jigsaw was a work of art worth more than the $2,000 that I had in savings at the time.

The Fat Gene

New hope of treatments to bolster the effects of diet and exercise have emerged in recent years with the discovery of all kinds of genes that influence obesity, from the way that fat cells develop to the brain mechanisms that make us feel pangs of hunger. But a 2007 study in *Science* by Andrew Hattersley from the Peninsula Medical School, Exeter, U.K., and Mark McCarthy from Oxford University was particularly significant because, unlike previous work, it establishes a very common genetic link with mild obesity rather than a rare genetic link with extreme obesity.

The study—conducted over a fifteen-year period and involving 42 scientists—first identified a genetic link to obesity through a study of 2,000 people with type 2 diabetes and 3,000 controls. The researchers then tested a further 37,000 people from Bristol, Dundee, and Exeter as well as a number of other regions in Britain, Italy, and Finland. In every case the same variant in the FTO gene—which is mostly present in the brain and pancreas, among other key tissues—was associated with type 2 diabetes and obesity.

If people carry one copy of a variant in FTO, as does half of the general population in the United Kingdom, it will lead to a gain in weight of 2.6 pounds or put just more than half an inch on their waists and raise their risk of becoming obese by one-third. If people have two copies of this variant in the FTO gene, which is the case in one in six of the population, then they will gain almost 7 pounds more than those who lack the variation and are at a 70 percent higher risk of obesity.

Jiaqi Huang and Samuel Levy scoured my chromosome 16 for this variant of FTO and found that I have two copies of the low-risk version. I do not have this particular genetic propensity to pile on the pounds, with all the associated implications, such as increased risk of diabetes and heart disease. But although this work offers new insights into obesity, at the time of writing FTO is one of many human genes whose actual biological role is mysterious.

That evening we were treated to a picnic unlike any I had ever been to before. The guests, along with a crew, chef, and servers, boarded the *Midnight*, Bourke's power yacht, which took us out to one of the hundreds of rocky islands off the coast of Maine. *Midnight* was anchored, and we lined up to be ferried to the beach in inflatable dinghies. I found myself sitting in one with Peggy Rockefeller, a lovely and very elegant woman. Rocks prevented the inflatable from landing, and I wondered what the drill would be: Would Peggy be carried onto the beach by the servers? She provided the answer by stepping out of the dinghy and wading to the shore, where the crew was preparing a fire and a gourmet feast on the rocks.

As we ate, we talked about the importance of genomics and whether Congress would help fund a genomics company. We discussed whether Jim Watson should be involved. I was surprised when Cantor, Hood, and Caskey urged strongly against it. The serious atmosphere lifted when Cantor's girlfriend asked if there was a restroom. She was directed to the tree line but refused to go in the dark. Bourke's concierge escorted her with a lantern and everyone fell silent. On her return, Tom Caskey asked Cassandra if the noise bothered her. What noise? Why, the whir of the video camera, Tom joked. This was the first of many such conversations over the following months. Bourke, Hood, and Cantor were opposed to the research institute model and preferred a full-blown company. Bourke wanted Caskey to serve as its head, but Tom was not ready to leave Baylor (he would a year or so later leave there for Merck). Hood believed that DNA sequencing was the only way to go forward and wanted me to run sequencing operations. While the group loved to debate these projects, I had my doubts as to whether they were serious about seeing them through. Finally, months later, there came a crunch meeting with Hood at Caltech, where we drew up a science plan and a budget. We discussed it at length with Bourke, and I told them that they needed to make a decision within two weeks or I was going to move on: These conversations were sapping my time and energy.

On my return to the NIH, I was summoned to the office of the director, Bernadine Healy, in Building 1. I was apprehensive and assumed it had something to do with the ongoing patent saga. But she welcomed me into her office, and I was disarmed when she told me that she had heard rumors I was thinking of leaving the NIH. She was worried about a drain of talent from Bethesda. Was there anything she could do to change my mind?

Out poured my frustrations with obtaining genome funding from Watson.

If only there had been an NIH intramural genome effort, I explained, I could get on with my science without having to endure battles with the human genetics community over funding that were driven by politics rather than merit. She asked if I thought the various NIH institutes would back such an intramural program, and I answered her they would—they would all benefit from being able to read the human genetic code. The meeting ended with her giving her support. She had a discretionary director's budget and was willing to use it to aid my program to keep me at NIH. Would I be willing to chair the new Committee on the Intramural Genome Initiative? Of course, I said. Later that day I called Hood and Bourke and told them that my outlook at the NIH had improved immeasurably, and so I had decided to withdraw from their genome company—a decision that would have a number of unexpected consequences.

Since Hood and Bourke now needed someone else to direct their sequencing project, they immediately began to court Sulston and Waterston, who were keen to discuss a commercial venture because it could provide new technologies and vast resources. In Sulston's case it would add up to much more than a proposal on the table from the United Kingdom's Medical Research Council. But Sulston began to realize that his first love, the worm, was not on Bourke's agenda. As I had before them, Sulston and Waterston insisted on continuing their basic research—in their case a ten-year project on sequencing the worm genome. They had also insisted that any discoveries "must remain in the public domain." Bourke had even discussed the matter with Watson, agreeing that genetic information should be made public as long as the funders were given a lead over their competitors in using the information for diagnostic tests and drugs.

By now Watson had started to panic. After me, he believed Sulston and Waterston were the best prospects for serious sequencing, and if they left, he would have no program. To underscore that point and, as he called it, "open the spigot a bit," Waterston had written to Watson to point out that he needed funding from the NIH. "We did not start this project to be outdone by second-rate players with more resources."[19]

Watson flew to London to meet Dai Rees, secretary of the MRC, who had told a newspaper earlier that week that the attempt to lure Sulston was "a flagrant bid to create an IBM of human genetics."[20] Watson, meanwhile, put a U.K. spin on the same story: The nematode project was, he said, "the jewel in the crown of U.K. science. It would be a serious setback for U.K. science to lose it." Then came what for Watson was the real crunch point: "It would also be a major blow to the delicate international collaborations we are trying to set up. The expertise gained from it should be available to all scientists."[21]

Waiting in the wings to help the MRC was a fairy godmother, Bridget Ogilvie, director of the British-based Wellcome Trust, which had been established by the American entrepreneur Henry Wellcome on his death in 1936 and endowed with 100 percent ownership of the drug company that bore his name (then Glaxo Wellcome and later absorbed into Glaxo SmithKline). For decades the trust had used the dividends of its drug holdings in Wellcome PLC to fund projects on the history of medicine, tropical disease, and so on. When the value of those holdings skyrocketed in the 1980s as a result of its anti-HIV drug, AZT, Wellcome's directors persuaded the British authorities to let them break Henry Wellcome's will, which stipulated that the trust retain all its shares in the company barring "unforeseen circumstances."

Two sales of Wellcome stock—in 1986 and in 1992—netted $4 billion and more than doubled the trust's income. Wellcome, now the world's biggest medical research charity, dispensed a flood of funds to cash-starved British scientists.

Aaron Klug, Nobelist and director of the Laboratory of Molecular Biology, regarded Bourke's head hunt as "an extremely distasteful business."[22] Through him and Rees, and with the blessing of Ogilvie, Watson was instrumental in getting the Wellcome Trust to give £50 million over five years to the worm project to keep Sulston on its side. Sulston and Klug together had written a briefing paper for the Wellcome's Genetic Advisory Group that described how to build on the "community spirit" of his worm project to create a human genome project, scaling up to 40 megabases by January 1997, including an effort to look at yeast and cDNAs of the developing rat brain.

To drive the sequencing initiative Ogilvie appointed a senior Wellcome Trust administrator, Michael Morgan, a former university biochemist. Named in honor of the great pioneer in the field, the resulting Sanger Center was set up in a place called Hinxton Park near Cambridge, England, a run-down former engineering lab. ("It had better be good," said Sanger when Sulston asked him for his blessing.)[23] This was a big move by Watson and great news for British science. Ogilvie believes that it had even wider effects, galvanizing genomics worldwide: "We acted like a crystal in a supersaturated solution. We precipitated everyone to get into it."[24]

But this was only the beginning of Watson's efforts to beat off Bourke. As

The Double Helix had shown so vividly, Watson was prone to indiscretion, and after a meeting with Bourke, to whom he took an immediate dislike, Bourke wrote a letter of complaint to Healy dated February 25, 1992, and copied it to senior officials in the Bush Sr. administration. Watson, the letter maintained, was interfering with Bourke's commercial interests, and Bourke added the observation that Watson himself had a conflict of interest, for while serving as head of the NIH genome center, he owned stock in several biotech and pharmaceutical companies. Healy, by now tired of Watson's attacks on her, launched an investigation. As her office told *The Washington Post*, "Dr. Healy does not have the luxury of ignoring ethical questions, even for a Nobel Prize winner."[25]

All this occurred as I began to chair the NIH genome committee, a position that would make Watson grumble to a *Nature* reporter that I now had more influence than he did. Sitting on my committee were the great and the good of the NIH, including the Nobel laureate Marshall Nirenberg; the gene therapist French Anderson from Heart, Lung, and Blood; David Lipman, head of the genome databases at the National Library of Medicine; and the scientific director of Child Health, Art Levine, who, like me, had found his applications for funding rejected by Watson's Office of Human Genome Research, now the National Center for Human Genome Research. Our agenda was to decide how scientists within the intramural NIH could participate in a focused effort to identify disease genes. In December 1991, I had sent a memo to the NIH director listing the committee members and outlining the support for such an intramural genome effort, though I pointed out that there were clear differences of opinion as to what constituted genome research. "The search for human disease genes and the biological characterization of expressed genes were not covered by the mandate of the human genome center," I wrote. I explained that the proposals I had submitted to Watson's center were "a mistake" and gave the reasons: They threatened the ethos of the intramural scientist in the ability to do high-risk research, to set long-term goals, and to change directions and move quickly as the science warranted. The memo concluded that intramural NIH should concentrate on expressed genes, particularly human disease-related genes. On March 10, 1992, the final report went to the NIH director from the committee, and almost every director wrote in its support.

One afternoon soon afterward I received a call from David Lipman. Watson wanted to make a proposal to me, via Lipman and in secret; it had the title

"Gene City." This would be a facility, probably located off the main NIH campus, that would be devoted to gene research as an adjunct to the genome sequencing effort. As a final flourish David told me, "Watson wants you to be the mayor of Gene City."

Given Watson's previous track record I wanted my offer to be "mayor of Gene City" in writing; only then would I circulate it to my committee members. So began a ten-day marathon of discussions and negotiations. Watson's office sent a three-page description of Gene City, dated March 6, 1992, with the dry title: "A Proposal to Establish a Major Facility for Genome Analysis at the NIH."

The memo, to my amazement, sounded much like the proposal I had already circulated to the committee. Even more surprising, all the opposition to ESTs that had been voiced by Watson, Sulston, and others to torpedo my attempts to raise funding seemed to have been forgotten. As Watson wrote,

> *The short DNA sequences, known as ESTs, that are obtained from rapid sequencing of cDNAs have many potential uses in genomic analysis. Initial results have already demonstrated their usefulness in the identification of homology relationships among proteins. The analytical power of the ESTs extends, however, well beyond the information to be obtained from DNA sequence comparisons. Their potential can be realized even more fully in conjunction with genomic maps. A major facility has already been established in France for mapping and for providing the resources developed to the scientific community and to industry. For both scientific and competitive reasons, it is timely to establish a comparable facility in the U.S. NIH is the logical place to do so.*

The French project referred to in the memo was led by Daniel Cohen, of the Centre d'Etude du Polymorphisme Humain (CEPH), who had set up Genethon, a genetics center near Paris that was backed by telethons organized by the French muscular dystrophy association. Cohen was among the first to understand that one of the keys to genome research was the use of automation in a properly designed organization, and he still does not get enough credit for his great achievement.

Using what he liked to call his "monsters," Cohen was producing a map of the entire genome. The monsters in question were the stars of a new method of multiplying and manipulating larger-than-usual pieces of human genetic material, called Mega-YAC (Mega-Yeast-Artificial-Chromosome). While conventional methods handled pieces of genetic code about thirty-five thousand base pairs in length, Cohen claimed he could use YACs to handle ones of about a million base pairs. Cohen was about to unveil a map of chromosome 21 and had managed to map 25 percent of the human genome in only three months.

Although we now know that Mega-YACs had a fundamental flaw of extreme instability, the apparent French success had nevertheless spurred on Watson. He called for:

> *the establishment of a comprehensive research group that can accomplish all aspects of the work and develop the new technology that will be needed to increase the rate of mapping and sequencing. . . . For comparison purposes, the comparable French organization, Genethon, has a staff of 150 people and a 1992 budget of at least $14 million. Taking advantage of the synergism inherent in the several EST-based activities will not only allow the NIH to play a major role in the completion of the Human Genome Project but, in combination with the first-class clinical and basic research programs of the intramural program, to lead the biomedical research world into full realization of the potential that the understanding of human genetics and the genetic basis of human disease offers.*

The final sentence was particularly stunning, given its acknowledgment of the commercial implications of genomics, which, of course, depended on patents: "A facility such as is being proposed can also be envisioned to be the nucleus of a gene-based industrial center ('Gene City') that will include satellite organizations devoted to research and commercialization of gene-based products."

Even today, more than fifteen years later, it makes me both angry and sad to reread this memo. Watson clearly understood the value of ESTs and could have made the same proposal a year earlier, when I had sent him a preprint of my EST paper. His public posturing was presumably driven by a fear that my EST project would draw funding away from the effort to read the genome, letter by letter.

There was, however, one problem with the proposal: One director did not believe it had come from Watson, because the document was written on plain paper. I had been so surprised to get something in writing that I had missed this fact, and indeed, given Watson's public stance on ESTs and patents, and the similarly of the memo to what I had written, I could forgive my colleague for thinking there had been some kind of mix-up. David Lipman obtained the memo once again, this time on Watson's letterhead, and the committee backed it. Now, via David, came a new request from Watson, one that almost floored me: Watson was about to meet Healy for his annual performance review; could I get her on board before the meeting? I would certainly try, I said.

When I outlined Watson's plan to Healy, she was speechless. Was I sure? Yes, I said, and showed her the document on Watson's letterhead. Would I go along with it? Well, I had my concerns, I admitted, but I trusted David, and the rest of the committee supported the concept. Given our assent, Healy indicated that when she met with Watson in two days' time, she would give Gene City her backing. When I told David the news, we were excited and optimistic: Watson was going to make the formal proposal, Healy was going to accept it, and our genome science would be jump-started in an incredible fashion. The plan could not fail.

On the day of the Watson-Healy meeting, David and I waited in my office, chatting about how cool and wonderful it would all be when we were working in Gene City. In truth, we were both anxious; Bernadine had promised to get in touch after the meeting, and the wait seemed to go on forever. When the phone finally rang I picked it up, expecting to hear good news. Instead, she was angry, indicating that she had been set up. I had never heard her this upset and was truly shocked. Did Watson make the proposal? "No," she howled, raging at Watson and raging at me. When Bernadine eventually calmed down, I discovered what had happened.

Watson had turned up for the meeting, which began with some polite small talk. He had had a public spat with Healy once before, in 1985, when he said of the White House Office of Science and Technology Policy: "The person in charge of biology is either a woman or unimportant. They had to put a woman someplace."[26] He was referring to the deputy director for biomedical affairs in the Office of Science and Technology Policy, a certain Bernadine Healy. During the heated encounter that now took place, he began ranting about gene patents, then started to shout at Bernadine. Gene City was never mentioned during this tirade. When Watson's tantrum ended, so did the one chance of détente at the NIH and of us all taking on the challenge of the genome as partners, not competitors. Because of that disastrous meeting and because of Watson's less-than-stellar management qualities, I believe Healy made the decision that Watson had to go. Rick Bourke may have put the gun on the table, but it was Watson, not Healy, who had pulled the trigger.

Watson's days at NIH were now numbered. His exit was handled by Michael J. Astrue, the chief counsel of the Department of Health and Human Services (HHS). In a strange coincidence, the crunch came as Astrue was visiting my lab with Richard Riseberg, the chief counsel for the Public Health Service, so they could understand the science behind the troubling NIH patent application. (Astrue would clash with Healy on the matter, arguing that the EST patents did not satisfy the legal requirements of an invention.) My phone rang; Astrue was wanted on the line. As his visit ended, he told me he had some important calls to make and asked me to step out of my office.

When he finally emerged forty-five minutes or so later, Astrue explained that he had just handled Watson's resignation. Watson resigned on Friday April 10, 1992, amid allegations that his ownership of biotech stocks presented an actual or potential conflict of interest with his government work. But as he was to tell a British reporter later that year, he had been sacked for being outspoken about the "loony" patents and was clearly bitter. Healy was bright, said Watson, but he complained that "she didn't know anything. . . . [I]t is very dangerous to have power and exercise it in the absence of knowledge."[27]

If Watson had known where Astrue's call had come from, he would have been even more embittered and certain that I was somehow behind it. But in reality, I could hardly be counted as a stooge of Astrue, whom I found unimpressive because of his attempts to insert a political agenda into the already overheated patent issue: At one HHS meeting I attended, he said, in effect, we won't be in power forever, and these patents are far too powerful to leave to the Democrats.

As it happened, my own days at the NIH were also numbered as new opportunities beckoned. A senior government official who visited my lab during this period told me: "Son, you are obviously doing extremely well." He did not strike me as being *au fait* with the science, so I asked him why my success was so obvious: "This is Washington, and we judge people by the quality of their enemies, and son, you have some of the best."

By this time one lucrative escape route had presented itself: I had been invited to become the CEO of a biotech company, Genmap. The carrot was a signing bonus of $4 million, an amount that was then beyond my imagination and more than I had made during my entire career. Still, because of my suspicion that pure science would not thrive in a commercial setting, I declined the offer and chose to remain at NIH. That was the first real test to reveal that science was ultimately more important to me than money. But as part of those negotiations, I met Alan Walton of Oxford Partners, a venture capitalist with a doctorate in chemistry. I took a liking to Alan, who understood why I was finally not interested in Genmap, but was still curious to know what, if anything, it would take to get me to leave NIH. Academia was dangerous enough, and I told him what I believed was the unpalatable and unrealistic truth: The only viable incentive would be the opportunity to start my own basic science research institute, a project that would need a funding guarantee of $50 million to $100 million over a decade. I had a deep conviction that my discoveries were of major importance and would only grow more so with time.

Alan did not think my requirements were impossible or unreasonable, or at least he did not say so if he did. He did tell me that only one fund could even consider such a level of commitment: HealthCare Ventures, which was headed by a colorful character, Wallace Steinberg, who was convinced that science was on the verge of unlocking the secret of eternal life—a discovery that he hoped to be the first to benefit from, aided by the invigorating effects of plenty of tennis. After working at Johnson & Johnson, where his reported claim to fame was the Reach toothbrush, the first marketed in the war against plaque, Wally started his venture capital fund and had already launched two Maryland biotech companies, MedImmune and Gene Therapy Inc., with my former NIH colleague French Anderson. Alan promised to give Wally a call, but because it sounded like a long shot, I did not give it another thought and went back to work.

During the previous year, 1991, I had completed my transformation from a neurotransmitter receptor–based biochemist to a genome scientist, and every paper I coauthored that year was a genome-related study. By then I had set a new benchmark for my hunt for brain genes: I wanted to find at least two thousand new ones. This target was not arbitrary. At that time there were fewer than two thousand known human genes in the public databases, so what more dramatic way to underline the power of my methods than to take just one year to double the number of human genes that had previously taken around fifteen years to accumulate? In 1992, the journal *Nature* was pleased to publish our paper entitled "Sequence Identification of 2375 Human Brain Genes." The patent feud rumbled on, however, and the NIH felt obliged to file a second patent application on all 2,375 of those genes. I did manage to get one concession from Reid Adler and the technology transfer office: My method would be put in the public domain (in the jargon it was now a "statutory invention registration") so that any scientists who wanted to use ESTs did not have to obtain a license to do so from the government.

A Gene Fit for Wally

The gene "klotho" was named after the Greek Fate purported to spin the thread of life, because it contributes to longevity. Japanese scientists discovered the gene in mice, noticing that without klotho's protein, the rodents developed atherosclerosis, osteoporosis, emphysema, and other conditions common in elderly humans. Equally, overuse of the gene seemed to extend the life span. It turns out that changes in the spelling of the gene are associated with longevity and common age-related diseases in humans, including coronary artery disease and stroke, strongly suggesting that klotho regulates aging. This seemed to be confirmed when Harry Dietz and colleagues at Johns Hopkins found that having two copies of a less common version of klotho is twice as prevalent in infants as in people over age sixty-five. A variant in one part (exon) of the gene (allele 17 of microsatellite Marker 1) was much more prevalent in the newborn than the old. Perhaps people born with the two copies die sooner than others. I do not know whether Wally had this variant, but I do know that I do not.

The klotho gene product, or klotho protein, is housed in cell membranes. When cleaved by another enzyme, it is secreted into the blood, where it seems to work as an antiaging hormone. The protein seems to boost the cell's ability to detoxify harmful reactive oxygen species and plays a role in how the body handles insulin. The good news is that in my case I also have a blend of klotho gene variants (one from each parent) for the so-called KL_VS allele, which has been linked with a lower risk for coronary artery disease and stroke and an advantage in longevity. But, of course, we all have "good" and "bad" genes, and how their influence pans out when the effects of environment are also taken into account is anyone's guess.

At around this time my work took a new direction, one that would eventually lead me to address the U.S. president and the entire Cabinet in person.

The starting point for this departure was a deadly virus: smallpox, which has killed countless millions throughout history and mutilated many more, damaging its victims' kidneys or scarring their skin with pus-filled blisters. In the eighteenth century alone the brick-shaped virus killed about half a million people annually; when Europeans brought it to the Americas, it wiped out half of the Native American population, which lacked any trace of natural immunity. As late as 1967 the disease was endemic in more than forty countries, with more than 10 million cases.

The last person to be naturally infected with the disease was a Somali cook, who succumbed on October 26, 1977. Two years later the world was declared free of the scourge, and the virus was safeguarded in storage at two authorized sites: the Centers for Disease Control in Atlanta, Georgia, and the Research Institute for Viral Preparations in Moscow, Russia. In an address to the World Health Assembly in May 1990, United States Health and Human Services secretary Louis W. Sullivan stated that technological advances had made it possible to sequence the entire smallpox genome within three years, a key step toward erasing this scourge permanently. Many scientists believed that the preferred first step toward the destruction of the virus was to determine its complete DNA sequence, and in so doing retain the essential scientific information for an entity that would become extinct. Because of my extensive facilities I was asked by the secretary of Health to lead a joint CDC-NIH project to sequence the smallpox virus. I visited the CDC and met Brian Mahy, director of the division of viral and rickettsial diseases, and Joseph J. Esposito, a pox virus scientist. Although at close to 200,000 base pairs smallpox was one of the largest virus genomes, it did not seem to present an overwhelming challenge to sequence. By then my team had been sequencing hundreds of thousands of base pairs of the human genome and thousands of ESTs.

I suggested that by scaling up the shotgun sequencing method we could sequence the virus in one step. No one liked the idea, and for good reason: We would shotgun the genome in a nebulizer device, and there were concerns that a minute amount of live virus might become airborne. An awful earlier mishap underscored how these fears were not misplaced. In July 1978 a photographer in the anatomy department at the University of Birmingham Medical School in England contracted the disease and died a month later. It was believed she became infected when the virus escaped into the air of a laboratory and was carried through the ventilation system to her darkroom. After her condition was diagnosed, the head of the laboratory where the virus was stored committed suicide.

As well as safety concerns, Tony Kerlavage on my team pointed out that we did not have the computer algorithms that would be capable of assembling the smallpox genome from a whole-genome shotgun. At that time the best computer programs in the genomic field had great difficulty dealing with one thousand fragments of DNA, and the shotgun sequencing of smallpox would create three to five times that number.

There was, however, a less straightforward but easier route to sequencing, given these restraints. Esposito had already broken up the smallpox genome by making a series of restriction digests (recall how restriction enzymes recognize and cut DNA at unique and specific sites), thus creating smaller fragments of the genome. My team would perform shotgun sequencing on each of the fragments, one at a time. Once we had settled on a strategy, we were to sequence the most highly virulent Asian major strain, Bangladesh 1975, while the Soviets would sequence a minor strain, all under the oversight of the World Health Organization (WHO).

As required by the agreement, we had an international team involved in the project, including scientists from China, Taiwan, and the Soviet Union, with the main sequencing technician being Terry Utterbach from my lab. All went smoothly with the exception of our Soviet scientist, an affable man with a love of alcohol. I would often arrive in the lab to find Nickolay asleep under a bench with an empty vodka bottle nearby. At first I was told to make the best of it, because I could cause a diplomatic incident by making a fuss. But his drinking reached the point where he had to be admitted to the hospital, where he did not wake up for three days. He was later sent home.

Our effort was billed as mankind's first deliberate attempt to destroy a species (in contrast to the thousands we destroy mindlessly), and the argument went that recording its DNA sequence first would overcome any objection against its forced extermination. At that time I agreed that smallpox deserved the death sentence, since it had likely killed more humans than all other infectious diseases combined (prior to the appearance of AIDS).

This work triggered my first serious discussion about whether genome data should be placed in the public domain at all. I met in Bernadine Healy's office with officials from the defense establishment and other government agencies, who were extremely concerned about the smallpox genome being released publicly. One likened it to publishing a blueprint of an atom bomb, while others talked about putting a barbed wire fence around my lab.

All the while the attempts to lure me away from the NIH into a commercial lab continued. I was contacted by Amgen, which was netting more than $1 billion annually from biotech drugs and was now looking for new ways to invest its profits. A conversation with Lawrence M. Souza, the head of research, and Gordon Binder, CEO, turned to the concept of starting a new not-for-profit research institute in the Washington area. The Amgen team jumped at the idea, though while I envisioned doing basic research in a Venter Institute, they, of course, wanted me to do commercial work in an Amgen Institute. The subject was broached again during a visit to Amgen in Thousand Oaks, California. The company would give me a $70 million, ten-year commitment to establish the Amgen Molecular Biology Institute in Rockville, Maryland. I would serve as the institute president and be appointed a senior vice president of Amgen. The salary would be close to three times my NIH salary and would include stock options in Amgen. I still felt uneasy, but the offer was so generous that I promised them I would discuss it with my wife.

Amazingly Alan Walton's longshot idea of approaching the venture capitalist Wally Steinberg seemed to have paid off as well. HealthCare Ventures was interested in talking to me and sent over its representatives, Deeda Blair and Hal Warner. Deeda is a Washington socialite whose job was to help Steinberg woo scientists away from the NIH into his companies (for a hefty fee). Hal Warner, a Ph.D.-level chemist, was a solid pragmatic worker. Deeda's charms did not work on me, and Hal seemed to be offering me $15 million over three years for a typical biotech startup, exactly the kind of approach I had already rejected. I told them there was nothing further to discuss, but they insisted I at least meet Wally, who was going to be nearby, in Gaithersburg, the following week. I reluctantly agreed to see him.

On the day of the meeting I was made to wait awhile, no doubt to establish Wally's alpha male credentials. Then Steinberg entered with a group that included Deeda Blair; Alan Walton; my NIH colleague French Anderson, founder of Genetic Therapy Inc.; and James H. Cavanaugh, who served on the staffs of presidents Richard Nixon and Gerald Ford and had been a special consultant to President Ronald Reagan. Wally, our MC, made the introductions in a grand, overblown style, heaping particularly effusive praise on French. After his monologue on all the companies that HealthCare Ventures had funded, we discussed what I had in mind: funding for my own basic science research institute, in exchange for which I would offer my discoveries for a limited time—say, six months—to a new biotech company that would develop them into novel therapeutics.

Steinberg interrupted my presentation, frequently asking questions, and then spent another ten minutes repeating the disappointing terms that Hal Warner had already presented. When he finally finished, I thanked him for his time, said it was a pleasure to meet him, and stood up to leave. To make it even clearer what I thought of his offer, I told him I needed to hurry to Dulles Airport for a flight to Los Angeles. When Wally took the bait and asked why, I explained that I had a final meeting to discuss Amgen's offer of $70 million over a ten-year period for a new molecular biology institute that I would head. I detailed the terms including the stock, salary, and position in Amgen, telling him everything—everything except my misgivings about Amgen, of course. Wally sat silently for about thirty seconds and then blurted: "I will give you $70 million over ten years using your model, including equity in the biotech company."

I could tell he was serious, because his associates looked as if they were about to faint. French broke the silence by pointing out that that sum was five times the amount he had been given. My mind was now racing, and Wally watched me intently as I weighed the opportunities. Amgen offered stability, but I would be burdened with several new bosses and a drug company environment that did not appeal to me. Steinberg seemed to be promising the freedom that I was seeking in my research and in my life.

Finally (though in reality only about thirty seconds later), I made my decision and announced, "If you are serious, then I accept your offer." I told them I needed a list of the key terms and time to understand the details of their proposal. I shook hands with Wally and the others and headed for the airport. Overall the meeting had taken little more than fifteen minutes. The subsequent meeting at Amgen also went well, and I told them I would call the CEO, Gordon Binder, with my decision.

Two weeks later I found myself entering a conference room in the Hyatt Regency Hotel in Bethesda, Maryland, accompanied only by Claire, to conclude the deal with Steinberg. Before us sat Hal Warner, surrounded by a dozen lawyers, waiting with the term sheet for my new life as the head of a nonprofit research institute, ready for my signature. I had had no legal representation and felt my trust slipping slowly away. Hal signed the document and then handed me a black Montblanc, the signature pen ("fine writing instrument") of venture capitalists and biotech executives. I glanced at Claire and at the lawyers and then slowly wrote my name. "I can't help but feel that I am signing a deal with the devil," I joked. The date was June 10, 1992, ten years to the day after my father died, and three months to the day after Watson was forced to resign from NIH.

It was difficult for me to leave NIH. Irv Kopin, intramural director and a longtime supporter, was upset when he realized that I was not negotiating but really resigning. I called a lab meeting and told my team that I would be talking to each person individually about his or her future. Some were very excited about the new possibilities; others were panicked by the thought of change. I wanted most of my team to go with me, but there were clearly some that I thought wanted the security of government employment or needed it. Only one person turned me down: Mark Dubnick, who was diabetic and blind, felt he was better off with government health benefits. I gave Mark Adams and Tony Kerlavage the job of finding a home for the new institute.

My last task, and the one that I dreaded the most, was informing Bernadine Healy of my decision. I had felt a bond with her and always found her to be an effective leader. I suspect that many of the attacks against her said more about the stodgy, male-dominated NIH establishment attitudes toward a relatively young and attractive female than about her actual abilities. Even though we were close in age, I felt as though I was telling my NIH mother that I was leaving home. Bernadine could not have been more supportive or gracious, but she did have one favor to ask: Could I disappear a few days before my departure was made public? French Anderson had just announced he was going to the University of Southern California, and Bernadine's critics had used this as evidence of a brain drain to find fault with her administration, claiming her programs were driving away the best scientists. She did not want news of my leaving to appear until after her congressional meetings regarding the NIH budget. I knew just the way to avoid even the most persistent investigative reporters: I would enter the Annapolis-to-Bermuda sailing race. At the end of the hour-long meeting we were photographed together.

At that time my boat was a Taiwanese-built forty-foot Passport 40, named *Bermuda High* to sum up that extraordinary feeling of my storm-tossed arrival years earlier in my Cape Dory 33. I was determined to do it better the second time around and had even convinced Claire to come along—no mean feat because she had found the one trip she tried, to Cape Cod from Annapolis, a trial. The rest of my crew was my uncle, Bud Hurlow; his son, my cousin Rob; and a young man named Alan from the Annapolis Yacht Club. We headed down the Chesapeake Bay in a strong breeze. The wind began to die, then shifted to blowing straight up the bay so that we had to tack back and forth, making little progress toward the ocean. As night fell we made a depressing discovery: Our freezer was not working. Claire had spent considerable time freezing some decent rations, and although we knew that her food would keep cold for at least a couple of days, after that it was going to be grim. But we were not quitters.

The wind began to build, and we started moving through the water again, this time in the right direction. *Bermuda High* had a deck-stepped mast, in contrast to one attached to the keel of the boat, and as she tacked in the strong winds, the inside of the vessel began to shift as well. The cabin that Claire and I shared was forward of the mast, and as we came down some waves, we would become airborne and hit the cabin overhead before slamming back into the bunk. Sitting on the head was also a challenge, as Claire discovered when we fell off a large wave. The force as the bow crashed into the trough was huge, and Claire was sent flying when the toilet seat that she was clinging to sheared off. My eyes still water at the thought of what would have happened if I had been sitting there.

Two days out of Bermuda we were struggling. My gloom deepened as I tracked two sails that started out on the horizon behind us and then slowly overtook us over the next few hours, both efficient sailing machines between fifty and sixty feet in length. My cousin, a primary care physician, had to get to Bermuda as soon as possible to return to his practice; we were down to emergency canned food; and we were now paying for empty rooms in Bermuda. Reluctantly, I started the engine—effectively ending my race—and found we were the third boat to make it into Saint Georges harbor. But my remorse at not finishing soon evaporated when, the next day, a severe storm hit. Two of the incoming race boats washed up on the reef, and another was demasted. In all, more than half of the fleet did not finish.

Back on dry land I started fielding the phone calls almost immediately. Healy's office was preparing press releases. Some promising potential homes for the new lab had been found. And the HealthCare Ventures people wanted to work on agreements. Claire, who had against my wishes also resigned from NIH to follow me, agreed to fly back to Maryland to take a look at the potential lab sites, relieved at not having to sail back to Cape Cod. I took to the ocean once again while my life changed around me, arriving back in Maryland for the Fourth of July weekend after an uneventful 700-mile trip. Healy's press statement was going to be issued on July 6. I had only a few changes, notably a capital T, so that my new home would be "The" Institute for Genomic Research—my institute was a Tiger (TIGR), not an Igor. The press release declared: "The innovative technology pioneered by Dr. Craig Venter in NIH's intramural laboratories has truly come of age. Now it is time for Dr. Venter to take his bold discoveries out of NIH, a great marketplace of ideas, and into the marketplace of America, private industry."

I was surprised by the amount of press coverage my move received over the ensuing days and weeks, led by *The Wall Street Journal* headline: "Gene Scientist Venter Will Leave NIH and Set Up Private Institute." As far as *Nature*'s headline was concerned, I was a "controversial NIH genome researcher," and cheeringly, it reported how my new institute "was welcomed by several researchers this week as both an acknowledgment that the genome project is ready for scale-up and that industry recognizes the long-term commercial potential." The final word in this article came from Victor A. McKusick of Johns Hopkins University, who was to become a friend and an advisor. Although there would be an "initial sense of unease," he said, "the trajectory [of the initiative] is in the right direction. This is what it's going to take to get the job done."[28] The *Biotechnology Newswatch*, a Washington policy sheet, quoted Michael Gottesman, the acting director of the NIH Human Genome program following Watson's departure. Gottesman, a former medical school classmate of Bernadine Healy, seemed to smile on the venture: "It has been the hope of the Human Genome Project for a long time that at some stage there would be enough obvious applications that the biotechnology companies would become interested."[29]

Wally Steinberg, working with a contract media firm, arranged for us to visit Gina Kolata at *The New York Times* in Manhattan. Wally, always larger than life, made me feel as if I were an exotic pet being shown off by its puffed-up owner. He deflated a little, however, when he found that he could not answer Gina's penetrating questions about my science and the broader implications of what I was doing.

On July 28 the Tuesday science section of *The New York Times* carried her feature: "Biologist's Speedy Gene Method Scares Peers but Gains Backer." Once again we had the usual huge leap in logic from patenting a few genes to owning the entire genome as she described how researchers were "aghast at the possibility that the human genome could be locked up and owned by private investors. . . . [T]hey fear a land grab for the human genome that will greatly impede scientific progress and the free exchange of information." Gina painted the strange but familiar picture of scientists who resented a breakthrough that could make their lives much easier, as if they were fifteenth-century scribes complaining about Gutenberg's newfangled printing methods: "He plucks fragments of new genes from the human gene set by a shortcut that usually does not identify its function. Critics say that the hard work is working out the full structure of a gene and determining what it does in the body."[30] But she described the science well and, to my relief, made it clear that the much-disputed patents were not my idea but that of the NIH.

Wally's dream also came true: He got his desired column space in a highly prestigious newspaper. When asked why he had agreed to fund TIGR, he replied: "Because Craig held a gun to my head. I wanted his technology and those were the only terms he would agree to."[31] That was, of course, the real answer. But I learned from this article that Wally planned to be much more than just my benefactor: He was going to be the savior of the entire American biotechnology community. "Wallace Steinberg, chairman of the board of HealthCare Investment Corp. which is financing Dr. Venter, said he had suddenly realized that there was an international race to lock up the human genome. If Americans do not participate, he said, they will forfeit the race and lose the rights to valuable genes to Britain, Japan and other countries that are in the race to win . . ." Wally's quote showed him at his selfless and patriotic best: "I suddenly said to myself, 'My God—if this thing doesn't get done in a substantive way in the United States, that is the end of biotechnology in the U.S.' "[32] I accordingly prepared to roll up my sleeves, stiffen my resolve, and help Wally save America's biotech industry.

7. TIGR CUB

Let me tell you the secret that has led me to my goal. My strength lies solely in my tenacity.

—Louis Pasteur (1822–1895)

Louis Pasteur was one of the giants of science. He made huge strides in microbiology and immunology. He discovered the principle of sterilization, which came to be known as "pasteurization" in his honor. He described the scientific basis for winemaking and the brewing of beer. He delved into the mysteries of rabies, anthrax, chicken cholera, and silkworm diseases, and he contributed to the development of the first vaccines.

Where Pasteur differed from other scientists, however, was that after having founded a new science, after having driven forward medicine, after having surrounded himself with gifted researchers, he then wanted to create a working environment that was designed for the sole purpose of helping his team explore his ideas.

The Pasteur Institute, a private nonprofit foundation, was inaugurated on November 14, 1888, and was initially funded by public contributions in the wake of Pasteur's success with a rabies vaccine. Now Pasteur found himself in the ideal environment in which to expand the use of vaccination against rabies, to develop the study of infectious diseases, and to spread his knowledge. After his death in September 1895, once his cortege had passed through crowds of mourners in Paris, his casket was eventually laid to rest in a crypt at the institute.

Unlike Pasteur, few scientists in history have had the freedom, opportunity, and privilege to start their own independent research institute. Thanks to my single-minded drive to read genetic code and my great good fortune, I was given that chance, with The Institute for Genomic Research. My motivation was clear and simple: I wanted to scale up genomics to realize the potential of the EST method by revealing the secrets of our inheritance without having to wait for lumbering government bureaucracy or to put up with the pettiness of the politics of science.

Most people still find the idea of an independent, not-for-profit research institute much less familiar than that of a university or government or industry research laboratory. When TIGR was born, the confusion was understandable. After all, why *would* a venture capital group fund a concern like this? Even my former boss, Bernadine Healy, had talked about taking my "bold discoveries . . . into the marketplace of America, private industry." Today, after more than a decade, TIGR is still often referred to as a biotech company, and we receive requests about how to purchase stock.

To add to the confusion, TIGR was created alongside Human Genome Sciences (HGS), the for-profit company that was to fund our efforts and market our discoveries. One of my critics, John Sulston, objected that I wanted to have it both ways: "to achieve recognition and acclaim from his peers for his scientific work, but also to accommodate the needs of his business partners for secrecy and enjoy the resulting profits."[1] I plead guilty, along with the rest of humanity, to committing the most heinous crime of both wanting my cake and eating it. I even thought that I had gotten away with it when, initially at least, HGS existed only on paper, with Wally Steinberg, Alan Walton, and me as its cofounders. This is the arrangement that Alan and I hoped for, the one that we wanted to remain in effect. But it would soon become clear that Wally and HealthCare Ventures Inc. had other ideas.

Now that I was beyond the grasp of the NIH, my newfound freedom was intoxicating. To counter that, however, the risks I was facing were huge, as were the responsibilities. In theory, I could have established TIGR anywhere I wanted, although I was told that the East Coast was preferred. I set my heart on Annapolis, the "sailing capital," where I thought I could combine the two great passions in my life. I found an ideal site with a large pier on the Chesapeake Bay near the Bay Bridge. There I would work near the water, live on the water, and hold sailing races during my lunch breaks.

My sailing-genomics dream ended abruptly, however, when I shared my vision with my staff in Rockville. For them it would be a nightmare. If we ended up in Annapolis, most would need to relocate, or endure hours of daily commuting. Parents faced the prospect of moving their children away from the fine public schools of Montgomery County. I was upset and tempted to ignore the howls and moans of my staff. Eventually, however, I relinquished the plan, recognizing that, if we had enacted my Annapolis fantasy, TIGR was always going to struggle to recruit the best people.

My Risk Genes

Perhaps my love of risk is due to something deeply encoded in my DNA that makes me more likely to leap before I look. One suggestion is that the thrill I get when faced with hazards has something to do with dopamine, a neurotransmitter found in the brain. Perhaps thrill seekers like me have an unusual number of receptors that process dopamine, which induces the feeling of pleasure. A team in Israel was the first to link novelty-seeking to the dopamine receptor 4 (DRD4) gene. Found on chromosome 11, different flavors of DRD4 do indeed seem to influence risk taking. Even by the age of two weeks, babies are more alert and exploratory if they have a "novelty-seeking" long variant of DRD4. There is a stretch of the gene, consisting of forty-eight base pairs, that is repeated between two and ten times—the more repeats, the more likely we are to be that toddler who likes to crash his tricycle into a wall just to see what it feels like.

The long-gened also have more sexual partners than the short-gened. In these people it seems that the receptor is less effective at capturing dopamine and, to make up for this low responsiveness, they are driven to be more adventurous to achieve the same dopamine buzz. It turns out that I have four copies of the repeated section of DRD4, which is about average.[2] By that count I am not a novelty seeker. From what we know and understand, my genes tell me I don't like taking risks. The fact is, however, that I love a little danger, so there must be more to thrill seeking than just this gene. What better example could there be to underline the absurdity of claims that there is a "personality gene"? So many genes shape our behavior that the dystopian vision of people genetically altered to have designer personalities looks more than a little simplistic.

While for the first time in my life I had the power and money to do exactly what I wanted, I realized I had to suppress my desires and think of the welfare of others if I was to be an effective leader. We soon settled on a former ceramics factory on Clopper Road in Gaithersburg, Maryland, as the home for TIGR. As a bonus, the building had substantial air-handling equipment, which we need for coping with the heat generated by the lasers in the DNA sequencers.

My next jolt of reality came at the first meeting Claire and I had with "our" lawyers. True to form, Wally Steinberg and HealthCare Ventures had hired one of Washington's largest law firms, Hogan & Hartson, and following the signing of the letter of intent, we found ourselves surrounded by a dozen of their staff, all dressed in similar blue suits and matching suspenders. We were told that some of them would represent TIGR and me, and other HGS. We were all united by the effort to work out the terms of the agreements that would govern the operation of TIGR and HGS, as well as the flow of money and intellectual property. But as the morning passed I began to feel queasy, and when it became clear that Claire shared my uneasiness, I felt even worse—sick to the pit of my stomach. Whatever they claimed, it seemed clear that the dozen suits represented Wally and the venture fund. There was not a single one present to fight for the interests of me or TIGR.

Hal Warner of HealthCare Ventures confirmed that I could have independent legal advice, and the Hogan & Hartson team referred us to some independent lawyers and smaller firms. But when we got home, I telephoned Ted Danforth, who had helped with a search for funds to start TIGR and who was destined to join my board of trustees. Ted called the senior partners of a number of Washington, D.C., law firms and set up appointments for me to interview them. All had heard of me and the $70 million grant to start TIGR and all tried to woo us, but I quickly became bored and discouraged, because they all sounded and looked the same. We then visited Arnold & Porter, where we met with a junior partner, Steve Parker, who was unlike the legal clones we had encountered so far. I felt we had something in common since Steve hailed from Maine, where his family was in the boat-building business, lived on the Chesapeake Bay, and liked to sail.

Our next encounter with Wally's legal team was pure theater—awkward, confrontational, and fascinating. Steve Parker was certain that I had been exaggerating about my earlier encounter with Hogan & Hartson and was shocked when we walked into the conference room to find the same twelve lawyers facing us. When he was formally introduced, the lead Hogan & Hartson lawyer had to confer with his colleagues and then call Wally for advice. He was not happy when he returned; Steinberg had blamed him for letting this happen but said that they would have to live with it. Future skirmishes only reassured me that by bringing in a major firm I had restored some balance to the negotiating table.

Wally decided not to play hardball on the very first issue to arise after the signing of the term sheet. Although a street fighter, he also viewed himself as a man of honor, and we had sealed the deal with a handshake. But this was just about all I had going for me: I should not have quit the NIH on the basis of the term sheet alone but only after I had a firm agreement. I had already given away my strongest bargaining chip and I could not now walk away from the deal without ruining my career.

This much was obvious when Hogan & Hartson lawyers brazenly argued that it was a conflict for Arnold & Porter to represent me as well as TIGR even though they themselves had been more than happy to represent all of us—along with HealthCare and HGS—apparently without any potential conflict. My legal team replied that TIGR and Venter were inseparable, and we thrashed out the terms and conditions. Even after these negotiations, Wally still wanted the final say, which meant man-to-man combat in an elaborate boardroom in his headquarters, the HealthCare Ventures Inc. offices in Edison, New Jersey.

I sat before Wally's partners, his senior associates, Hal Warner, and, of course, Deeda Blair, though no one spoke much other than the lead combatants. Wally was playing to his audience but did seem to want to win me over. I, on the other hand, felt the strange sensation that seemed to come over me whenever I found myself in a situation in which the stakes were high, whether at a major speech, in an intense meeting, or in the patient receiving area in Da Nang: I felt as if I could observe the proceedings with detachment, as if I were an onlooker rather than a participant. I could listen to myself and evaluate what I was saying as I spoke.

A key issue that remained was the matter of how long HGS would be given exclusive access to the gene discoveries that TIGR had made before TIGR was free to publish the data in the scientific literature. While my primary concern was the science, I did want HGS to be successful. Although some scientists are disdainful of commercialization, I was only too happy that my science could produce something of value because that also meant it had the potential for doing some good in the world.

Before our confrontation on this issue, the signals I had received from Wally on gene patents had been mixed. In the Gina Kolata story in *The New*

York Times, Wally was described as saying "he intended to be socially responsible with his investment in Dr. Venter's work. He said he would have all of the genetic information uncovered by Dr. Venter published promptly and would collaborate freely with other companies and with the NIH." The article continued: "Like others in the biotechnology industry, Mr. Steinberg said that he would be happiest if the Patent Office denied patents for gene fragments."[3]

But another article, this time in *The Wall Street Journal,* offered a more realistic assessment. Wally had told the *Journal* that HealthCare Investment had not yet established a patent policy for Human Genome Sciences: " 'We'll make the assumption that the old rules will apply, that the function of the gene must be known for it to be patentable,' he said. 'But I'm not a patent attorney and it's complicated.' " It was indeed.

Wally wanted two years of exclusive rights to the data before it could be published; I countered that six months was the standard period for federally funded scientists to hold back their data, as had recently been established by the NIH. That was the theory, at least, and I did not explain that in reality it was still a free-for-all. The NIH rule had been introduced because some teams held on to their data for years before publishing. In human genetics scientists were particularly cutthroat, blocking the access of any rivals to key data. What mattered to them was not unraveling the causes of human disease—and thus helping patients—as quickly as possible but being the first to discover them and receiving credit for those discoveries. To end such flagrant abuses of public funding, the NIH established the six-month rule. The problem was that it was never clear when the six-month period began. Was it from when data were collected or months or even years later, at the end of the determining experiment?

Six months was nowhere near enough time to develop a product or even the full set of information required for capturing the required intellectual property, argued Wally. Appealing to his ego, I said he was, of course, correct, but inserted that six months was going to be enough for someone smart, like Wally, to pick likely winners, and in those cases HGS could have more time to work on them. In any event my team was going to be discovering tens of thousands of genes, and at the very most HGS could afford to develop only a handful into therapeutics.

I eventually came up with a three-tiered approach: HGS would have six months to select genes that looked as if they could be turned into viable treatments. There would be another six-month delay before these selected genes were published. However, if any showed promise as biotech blockbusters to rival the likes of insulin and EPO, HGS would have an additional eighteen months to fully develop them. The remainder, and thus the vast majority of genes, TIGR would be free to publish when it wanted. In our meeting there was no discussion of HGS's desire to patent everything we found.

When we appeared to reach an agreement that Parker would codify as a document, there was one last item on the agenda that I knew would prove critical. Because Wally was only acting CEO of HGS, a great deal would depend on the quality of the person eventually recruited to assume the job full time. Although Wally assured me that I would have veto power over the selection of the CEO, Parker neglected to record this detail, and it never made it into the agreement. I would come to pay a heavy price for this omission.

My laboratory, with its six automated DNA sequencing machines, had been the largest DNA sequencing operation in the world. At TIGR, I put in an order for twenty machines from Applied Biosystems, by far the biggest they had ever received. With these machines in place, TIGR would be able to sequence on the order of 100,000 clones per year, around 100 million base pairs of DNA code in all, a scale unimaginable at the time, though tame by today's standards. (The J. Craig Venter Institute Joint Technology Center can sequence twice this amount in a single day.)

One of the most gratifying aspects of TIGR was that it would allow me to ignore many of the precedents, procedures, and rules followed by conventional labs. When our instruments were installed in the former ceramics factory, we decided to leave all their associated air ducts and wiring exposed. The resulting surreal look reminded me of the Pompidou Center in Paris, which caused a scandal in the late seventies when its innards were cleared of services and it was given an exoskeleton of red elevators, escalators in clear plastic tunnels, and a rainbow of conduits—blue for air, green for water, and yellow for electricity. The look of my DNA decoding facility would prove as inspirational to my team as the Pompidou in Paris was for artists and architects.

I could also ditch everything I hated about academic institutions. First and foremost, there would be no tenured positions in which one held a position on a permanent basis, without periodic review or contract renewals. Tenure actually delivers a double whammy to the organizations that endure this outmoded arrangement. The second-rate people who thrive in a tenured environment like nothing more than to surround themselves with more mediocrity and drive out those who might excel and reveal the shortcomings of the entrenched. In the nine years I had spent at the NIH, I felt that the intramural program had changed in character from a protected environment where genius could take risks and thrive, to one in which people could muddle on with science that would never flourish if it had to compete for external funds.

I wanted to employ the best, most talented scientists for TIGR, not to provide security to those who had a different agenda from the institute's or, worst of all, to the second-rate, the humdrum, the pedestrian, and those who lacked motivation. People with drive and imagination don't need tenure. More than a decade after I made this decision, I feel even more strongly that it was the correct one, and as a result I have never had a problem keeping talented people.

Another quirk of academia that I was keen to abolish was the understanding that each scientist must have his own space and equipment even if some of it is rarely used. In most institutions the status of a faculty member can be gauged by the number of square feet assigned him or the number of people working in his lab. I was very good at that game and had one of the largest and most populated labs at the State University of New York, at Roswell Park Cancer Institute, and at the NIH. But I wanted to do things differently at TIGR, where the beating heart of the organization was, after all, a vast sequencing and computational facility. Every scientist, from students to Nobel laureates, would be given a modest office or cubicle, a bench space allocation, and a share in one of the best-equipped laboratories in the world. If he or she needed specific equipment, I would provide it, but at TIGR this addition would ultimately benefit the institution, not the individual.

The culture shock some subsequently experienced on arriving at TIGR was described by an early outside recruit who likened it to being the rat immersed in fluid in the dramatic "liquid breathing" scene from the 1989 James Cameron movie *The Abyss.* Navy SEALs use the terrified rodent to demonstrate that safely taking in a lungful of a liquid rich in oxygen is possible despite the ancient reflexes that make us all panic at the very thought. (Leland Clark had shown in the 1960s that a mouse could indeed survive immersed in a beaker of oxygenated perfluorocarbon liquid.)

Among the key people I needed to recruit, one candidate was put forward by Wally, who wanted me to consider Lew Shuster, an old hand at the biotech industry, for an executive vice president position, in which he would track expenses and help establish HGS. I interviewed him and agreed that he could indeed do the job. Then came the demand for office space for Shuster, an assistant, and a third office that could be used by visiting HealthCare personnel. I objected; this represented an intrusion into the not-for-profit sanctity of TIGR, and confusion would arise about TIGR's objectives if HGS cohabited with us. I was assured that this arrangement would be only temporary, though Wally, of course, wanted a mole in TIGR to inform him about what we were doing. My relationship with Shuster was therefore awkward from the day he walked into the office.

As for the chief executive, I thought I had found the perfect candidate in George Poste of SmithKline Beecham, whom I had known since our days in New York while at the Roswell Park Cancer Institute. I had discussed ESTs and genomics with George, who was one of the few who really understood their potential. George was an opinionated character but charismatic, funny, and genial, and I felt that I could work well and constructively with him.

When George visited TIGR, he was clearly excited by the opportunities, and though in principle Wally liked the idea of bringing George on board, he seemed uneasy about actually doing so. The problem was that I was clearly not his boy, and despite his best attempts, I had held my line that I wanted TIGR to do great science and publish our data. Wally realized that George was another strong-willed and independent person, and having one to deal with already was enough. Like so many other big egos that I have encountered, Wally struck me as being more interested in control than success. Of course he was sure that success would come if he was in complete control.

Wally did meet with George on more than one occasion but clearly had someone else in mind for HGS: William Haseltine, an AIDS researcher at Dana-Farber Cancer Institute in Boston, who had already worked with Wally on founding other biotech companies. Recently married to Gale Hayman, the cocreator of Giorgio perfume, Haseltine had a taste for expensive suits, and his black thinning hair was slicked back to achieve a more corporate look. At first Haseltine was to be only a consultant, to which I did not object, because it seemed that there was not much of a plan in place for the level of his involvement. I continued to pursue George while, behind my back, Wally was wooing Bill.

By now TIGR was at last ready to do some science. I had conducted my initial EST experiments using libraries of human brain cDNA. Within the first year of its operation I wanted TIGR to sequence EST libraries made from the DNA derived from every major tissue and organ in the human body. To do this I would draw on the various NIH-sponsored banks that store tissues and organs from surgery and from donors who agree to provide their brains, for example, on their deaths. We established an Institutional Review Board to approve our research protocols and ethical standards.

We were prepared to begin what I named the "Human Gene Anatomy Project," which would define our molecular anatomy by revealing which genes form the heart, which the brain, and so on. My EST method, recall, relies on messenger RNA (mRNA) to "edit" the genome and sift the coding from the noncoding regions. The parts that code for proteins vary from tissue to tissue. Every cell contains the same genetic code, but there are some two hundred or so distinct cell types in the body (such as brain, liver, muscle), depending on which particular genes are activated. With ESTs I could use the machinery of our cells as my supercomputer to extract the spectrum of genes in use on a tissue-by-tissue basis.

But first I needed to create libraries of DNA from each human tissue, which for the most part did not yet exist. After collecting tissues and organs, we would have to isolate the mRNA from each and accurately use that as the basis for cDNA libraries. Of course, we wanted high-quality libraries, but that presented a problem: By "accurate" we meant we wanted the cDNA to be an accurate representation of the mRNA, and thus of the gene. But was it the right RNA in the first place?

What I am getting at is perhaps best illustrated with a cDNA library created from the heads of fruit flies, which led to the discovery of many genes in the fly nervous system, including the octopamine receptor that I had found. When we attempted the EST approach on this library, which was regarded as one of the very best, we found that one-half of the cDNA sequences came from the genome of the mitochondria, which power cells, and not from the nuclear chromosomes of the cell. The eyes of fruit flies contain many mitochondria—so many, in fact, that their RNA had contaminated half of this library. Using my random method, half of the selected clones would have wound up coming from mitochondria, so we had to develop new methods to ensure that the cDNA libraries were derived from cellular mRNA. We also had to formulate new computation methods to store, sort, and interpret the EST sequences.

As the trickle of data from the sequencers quickly began turning into a torrent, Steinberg wanted to launch his business plan and insisted that I accompany him on visits to pharmaceutical companies to deliver some science presentations while he sold the commercial promise of HGS. This experience gave me a new perspective on the phrase "pure academics." At one of the pharmaceutical meetings I was surprised to find Eric Lander from the Whitehead Institute, which is associated with MIT. Previously an economics professor at the Harvard Business School, and before that a mathematician, Eric was a very dramatic presenter, which, as I would come to learn, was his trademark style. He was there to promote his latest company, which had claimed to develop a new method to isolate fetal cells from maternal circulation, enabling gene studies to be done earlier in pregnancy. While this sort of activity would seem to be far removed from the world of the pure academic, in the furor about the commercialization of genomics, Lander was among those who had complained to the press about the formation of TIGR, ESTs, and gene patents. I would later learn that he was peddling his own biotech company at the same time.

I also discovered that, contrary to the plan I had drawn up with Alan Walton, Steinberg was not looking to license TIGR's discoveries to all interested biotech and pharmaceutical companies but, rather, wanted a single exclusive monster deal for the entire pool of genes that we found. Wally had by then started secret negotiations with SmithKline Beecham for exclusive rights to HGS's intellectual property and, therefore, TIGR's data. He was increasingly worried about covering his promise of $70 million to me and TIGR, and in fact, given the look of shock among his colleagues when he first blurted out this offer, I still wonder whether HealthCare ever had $70 million in the first place. While it had a $300 million fund, it was unclear exactly how much of that money had already been committed. While he portrayed himself and his fund as long-term investors, it became evident that he was anxious to conclude a big pharmaceutical company deal, then take HGS public, make a killing, and get his money back as quickly as possible.

The year 1992 ended with the announcement that Francis Collins was moving to the NIH as Watson's replacement. Described by one commentator as a "Moses for the genome project,"[4] Collins is a devout born-again Christian who believes that scientific truth provides a glimpse of an "even greater Truth."[5] When the job was offered to him, he had wondered if it was God's calling, and spent an afternoon praying in a chapel before deciding to set forth on what he called an awesome adventure. "There is only one human genome program," said Collins. "The chance to stand at the helm of that project and put my own personal stamp on it is more than I could imagine."[6] Collins and his twenty-five-person team would move into the very same precious space that I had designed for my intramural genome program. I was having so much fun at TIGR by then that I was only too pleased to leave my lab to him.

One benefit of leaving the NIH quickly became apparent: Claire and I had doubled our combined NIH income of about $140,000, which had allowed us to scrape by with our town house, two small cars, child support, student loans, and a forty-foot sailboat. We could now pay off some debts and increase our monthly housing payment. We found an unusual two-bedroom glass house in Potomac on a secluded three-acre lot for less than $500,000. The driveway was close to a quarter-mile long, and the house was suspended on four internal pillars, creating the illusion that its glass walls were floating. Our bedroom looked out onto a private meadow where deer, raccoons, and foxes would wander about.

This newfound success was enjoyable, but it was clear that our home and security depended on our future scientific success, so I rarely strayed beyond the laboratory. Heart muscle, intestine, brain, blood vessels, and more: A range of tissues continued to arrive at TIGR, where they would be processed into mRNA then cDNA, and finally DNA sequences. The resulting discoveries came so quickly that to cope with them we had to expand our efforts in bioinformatics and computing, buying one of the fastest commercial computers, a new "massively parallel" model made by MasPar. Over the next year or two TIGR would publish eight significant papers in top journals, including one on the last batch of ESTs discovered in my NIH laboratory, many of which we had located on the human genome. "Much of the work on the human genome may be done sooner than expected," conceded an editorial in the journal *Nature,*[7] which nevertheless went on to extol the old-fashioned virtues of sequencing the entire genome. Behind all our successes, however, I was engaged in a battle for survival, a struggle that would become typical of the commercial environment in which I now worked indirectly.

8. GENE WARS

The man who is swimming against the stream knows the strength of it.

—Woodrow Wilson

By the end of 1993 the controversy over the great gene gold rush rumbled on without showing any sign of abating, even though two years had passed since the first patents had been filed by the National Institutes of Health. There was now a new person who had to wrestle with the issue because Jim Watson's nemesis, Bernadine Healy, had resigned as the director of the most powerful health agency on the planet.

The post is a political appointment, and after the election of Bill Clinton, he chose Harold Varmus, a molecular biologist from the University of California, San Francisco. Varmus started out his academic life by studying Elizabethan poetry but went on to medicine, focusing on retroviruses, from their unusual life cycle to their potential to cause genetic change. He eventually shared a Nobel Prize for the insights his work gave into cancer. In November 1993 he had become the first Nobel laureate to lead the NIH. While Healy had a corporate approach that some resented, "a managerial style reminiscent of Gen. George Patton," as one article had carped,[1,2] Varmus adopted a more relaxed style. He wanted to reenergize NIH intramural research and held a strong belief in promoting basic science and the random nature of discovery. The patent issue remained a high priority, though, but unlike his predecessor, Varmus was reputedly dead-set against them, a position that soon became obvious.

Reid Adler, the official who had been behind the patents in the first place, had gone ahead and filed them in an attempt to clarify what to do with all those pesky uncharacterized gene fragments. At the end of 1993, Alder paid the price for doing the "right thing" when Varmus removed him from his job. The journal *Science* pointed out how many patent experts, even Francis Collins, had come around to Adler's view that filing the applications had cleared the air: "Varmus opposed those patents, but top NIH officials say that his decision to transfer Adler to an unspecified policy position is based less on that than a desire to improve general OTT [Office of Technology Transfer] operations."[3]

I, too, was about to feel the effects of a management shake-up. Steinberg had by then begun negotiating an unprecedented $125 million deal with SmithKline Beecham for exclusive rights to TIGR's gene sequence data. Only when I found out about the deal did I grasp the real significance of an earlier conversation I had had with him.

At that time Wally had grilled me about my immediate publication plans. By now some of TIGR's data were already six months old and could, according to our agreement, be published. But I felt that rather than release a little at a time into the public domain, I would finish the initial phase of the Human Gene Anatomy Project and publish a major paper that would contain at least half of all human genes. Wally asked for a timetable for this plan, and when pressed, I estimated that it would take eighteen months. If I would agree to this in writing, Wally said, I could have a $15 million bonus, bringing the total TIGR budget to $85 million over a decade. In retrospect it was clear that he needed that cast-iron commitment from me to close the SmithKline agreement.

The HGS–SmithKline Beecham deal for exclusive rights to my gene data and 8.6 percent of HGS stock was announced just prior to HGS's going public on December 2, 1993. The IPO price was $12 for each share and rose rapidly to the $20 range. But any warm glow I felt about this development quickly disappeared when Wally announced that Haseltine was now going to be the chief executive officer of HGS. I knew they had been talking, but I did not realize how serious Wally had been about Haseltine. I reminded him that he had promised I would have a major say in the selection of CEO and that he had granted me the power of veto. My preference was for someone who would mind the business while I got on with the science, which is why I thought a heavyweight from the pharmaceutical industry, such as George Poste, was a better fit for the job than an antagonistic AIDS researcher who had already started up a few biotech companies. I accordingly exercised my right to veto Haseltine. Wally agreed that he had indeed offered that option but added, "Too bad that you did not get it in writing." Haseltine was obviously smart, slick, and aggressive to the point that, by reputation at least, he was given to stalking his competitors, even with 3:00 a.m. phone calls. But my gut instinct was that he was also not the partner I was looking for.

From the time of his arrival Haseltine made it clear that he was running the company and had little interest in my plans, leaving me to feel that I was merely the "booster rocket" that would launch him into orbit as a biotech tycoon. The problem for him was, of course, that the agreements I had thrashed out with Wally did not allow Haseltine to direct TIGR's research. To make matters worse, SmithKline was troubled by our rights to publish our data. And what bothered SmithKline bothered Haseltine. The solution was obvious: From the outset Haseltine wanted to eliminate TIGR, thus doing away with any obligation to fund it.

Haseltine ordered a number of DNA sequencers of his own. By late 1993 he had set up a lab to compete with TIGR. ("There was no reason for me to compete with him," protested Haseltine. "I already owned everything he did.")[4] I was consequently forced into shifting a significant portion of my science budget—provided, of course, by HGS—to Steve Parker and his legal colleagues at Arnold & Porter. Was it really only a year since I had told *The Washington Post:* "It's every scientist's dream to have a benefactor invest in their ideas, dreams, and capabilities"?[5]

Haseltine was not the only one bothered by my intention to publish my data and discoveries in the scientific literature. By that time TIGR had completed the sequence of the smallpox genome with our colleagues in the Centers for Disease Control in Atlanta. Given the significance of this achievement, we were, of course, writing up the work for publication in *Nature.* Not so fast, warned senior government officials. Even before I had left the NIH, talk of classifying our data had led to a debate that at times became heated. Eventually the row was settled in a surreal fashion when our colleagues in what had been, until the end of 1991, the Soviet Union announced that they were going to publish their version of the smallpox genome. Since we could not be outdone by our former Cold War enemy, we published the genome analysis in the journal *Nature;*[6] it went on to win a medal for its scientific significance.

Among the widespread press coverage was an article in *The Economist,* which considered a *Jurassic Park*–style effort to use the genome to resurrect the smallpox virus (which "gives a whole new meaning to the term 'computer virus' ").[7] The article argued that assembling the DNA a letter at a time would be a tedious process in today's laboratories, a position with which I disagreed. From what I knew of DNA technologies, I believed that resurrection of the virus from the genome would be possible within a few years. I made this point to urge federal officials not to create a false sense of security by staging a public execution of the smallpox virus.

I thought that the act of destruction was naïve at best and argued that there were multiple remaining sources of smallpox. The United States and the Soviet Union had stopped immunizing their populations against the disease, but no one could be certain that there were not hidden or forgotten freezers containing vials of the deadly virus, let alone virus present in the bodies of smallpox victims buried in the permafrost. This is hardly a far-fetched concern: The resurrection of the 1918 influenza virus in 2005 would depend on finding the body of a woman who had been lying in the Alaska permafrost since November 1918, one of the estimated 50 million who had been killed by that particular strain of pandemic flu. In addition to being able to synthesize the DNA of smallpox from its DNA sequence, there was also a multitude of pox viruses that infect other species, including our close relatives, that might evolve to again infect humans.

Eliminating smallpox was therefore highly unlikely, and the public announcements giving the impression we could do so were misguided. I can date my change of mind to the beginning of 1994. That January *The Washington Post* ran a full-page story with a large pullout quote from me: "If you believe in capital punishment, then smallpox should fry."[8] In fact, I don't believe in capital punishment and sent a letter to the editor arguing that the virus should not be destroyed. Smallpox would continue to play a role in my life for some time to come, from discussions with the CIA and even to being the subject of a briefing to the president and his Cabinet in the White House.

By that time my success was beginning to create its own problems; my personal finances had suddenly become interesting—so much so that their details were published on the front page of the Sunday, January 3, 1994, *New York Times.* If scientists tend to get a little bitter and resentful when one of their peers attracts significant press attention, the one sin they can't forgive is when a rival makes money, too. Like most human endeavors, science is driven in no small part by envy.

Around this time we held a party at TIGR to celebrate a milestone in DNA sequencing that I considered historic: We had successfully produced our 100,000th sequence in less than one year. More important, we were producing very high quality data, which was crucial if we were to keep ahead of any competitors, who I felt put too much emphasis on simply counting sequences and genes. In addition to HGS there was the company Incyte, which was the first to begin using ESTs after my NIH publication in *Science* in 1991. Randal Scott, its chief scientist, recognized immediately their power and changed the entire focus of his company from drugs to ESTs. Both Incyte and HGS would essentially copy the original NIH patent applications put together by Reid Adler, in an attempt to patent every EST they could find. I found it ironic that such plagiarism is tolerated by the U.S. Patent Office when companies pursue intellectual property protection on their "unique inventions."

To his credit, Harold Varmus had decided to talk to anyone and everyone with a stake or strong views of any persuasion about the EST patent quagmire. I had a "spy," a member of the TIGR staff who had worked in the NIH and was kept informed of developments by friends in Varmus's office. Eventually it was my turn to be summoned for a meeting with Varmus, which was cordial and to the point. Once again I said I was against the patents being filed in the first place but that I understood Adler's view that the NIH could do a great public service by rapidly pursuing the patents, establishing once and for all the lack of patentability of ESTs, and removing uncertainty from the community. By now Varmus had decided to drop the patents but, as ever, the matter was not that simple. Government rules said that if the NIH did not attempt to apply for the patents, the inventors themselves could still claim the rights. In an irony to end all ironies, the NIH patents could become the "Venter patents," just as Watson had warned. Would I take over the patent claims if the NIH abandoned them? Varmus and others were keen to know my answer.

I had already discussed what to do with the patents with Mark Adams, who was listed as coinventor. We did not want them, and we did not want to profit from them, either. We decided that we would take ourselves out of this messy situation by assigning our rights to the patents and any royalties resulting from them to a good cause, one that I had already supported: the NIH Children's Inn, which housed the families of children being treated for cancer. When I told Varmus of our decision, he seemed surprised and flustered. His expectations of me were clearly low, presumably because he had bought into the characterizations of me put about by Watson et al. He would have been even more shaken if he knew the whole story: By then I had been offered a million dollars in cash by Haseltine and Steinberg to assign the patents to HGS—an offer I had refused. On my return to TIGR less than an hour after seeing Varmus, my spy delivered a full report. Varmus was "stunned that Venter was not in it for the money." Unfortunately, it was not apparent that Varmus made any effort to share that appraisal with others.

While I was understandably pleased that Varmus appeared to understand I was not driven by financial gain, I was frustrated, too, with his apparent lack of understanding of NIH rules. No one can get rich from an NIH patent because the royalties are capped by law to $100,000 each year. I knew of only one NIH scientist who had received the maximum amount (Robert C. Gallo, for his work on the AIDS test), and the extra $100,000 merely brought his salary in line with his better-paid peers at academic institutions. A few weeks after our meeting, Varmus announced he would withdraw the NIH patent applications. None of the resulting press stories mentioned Mark and my decision not to pursue our rights or our offer to transfer them to the NIH Children's Inn. *Science* pointed out that "NIH's decision leaves unresolved the question of whether uncharacterized gene fragments can, in fact, be patented."[9] The news depressed HGS and Incyte stock prices, and the reason that Adler had been so concerned about the effects of uncertainty was by now crystal clear.

To aid our analysis of the new human gene sequences, we had developed a computer system that exploited one of the great truths of biology: When the process of evolution has yielded a protein that successfully performs a critical biological function, Mother Nature tends to use the same protein structure over and over, whatever the species. Chris Fields, a computational wiz who had an obsession with eating the hottest chiles, used computers to take advantage of how, as scientists say, genes or protein sequences are "highly conserved" to determine the function of the ESTs we had discovered. Our computer would take each new human sequence (around three hundred base pairs of genetic code) we sequenced and compare that string of base pairs against a database of all currently known genes. If we found, for example, that it strongly matched a DNA repair gene in fruit flies, there was a good chance that the human counterpart had a similar function. Now we could realize the full power of the EST method by using the automated DNA sequencing to pump out massive amounts of raw data and the computer searches to mine these data for new gene discoveries.

One of our early triumphs dated back to December 1993, when I received a call from Bert Vogelstein, the leading colon cancer researcher at Johns Hopkins University in Baltimore. Cancer can occur when mutations accumulate in the genes that regulate cell division, leading to uncontrolled growth. Bert wanted to search for faulty DNA repair genes after his team had found that a mutation in one gene of this type, a DNA mismatch repair enzyme, was associated with about 10 percent of nonpolyposis colon cancer cases. This in its own right was a major discovery, but Bert felt certain there had to be at least one more DNA repair enzyme that would account for more cancer cases, and wanted to know if we had seen any new DNA repair enzyme genes in our Human Gene Anatomy data set. I thought that we had, in fact, and told him we would review our data. To help the search he e-mailed me some unpublished DNA sequences of mismatch DNA repair genes he had isolated from yeast.

As soon as I received the yeast gene sequences, I told the bioinformatics team to start looking through the TIGR database of human DNA sequences for something similar. In a short time three new human DNA repair genes appeared on my computer screen. It was an exciting moment, and I called Bert right away. He was excited as well and wanted to map the new gene sequences to the human chromosomes to see if they lay in the three regions that his studies of patients had linked to colon cancer. We attached fluorescent dyes to the genes and used them to probe the chromosomes. By looking through a microscope to see where the sequences attached, we could see that they did indeed map to the identified colon cancer regions.

It was now clear to Bert and me that we had a major discovery, one that not only would improve our understanding of colon cancer but would also show the immense value of my EST method. While the first DNA repair enzyme had taken Bert several years to find, we had uncovered another three after a quick search through the EST database. But, of course, there was one outstanding issue: HGS had the commercial rights to all TIGR discoveries. I told Bert that I would not let Bill Haseltine stand in the way of a discovery that could be of immediate benefit to cancer patients.

We decided that we had to bring Haseltine and HGS on board now to prevent this dramatic advance from being bogged down in future battles. Bert was backed by a major pharmaceutical company, so he knew the score as well as having constraints of his own. I called Haseltine, explained what we had discovered, and told him that HGS should immediately establish a collaboration agreement with Bert's team at Hopkins. Bill, to his credit, could sense this was a significant breakthrough. The deal was signed in March 1994 and announced in the press. Within one week we had each of the new genes completely sequenced.

Vogelstein's team, led by Kenneth Kinzler, took DNA from colon cancer patients and from controls. They used a DNA amplification method called PCR (invented by fellow Californian and surfer Kary Mullis) to make copies of the DNA responsible for the repair enzyme from each patient. We then sequenced the repair genes from the patients to see if mutations were associated with the cancer. All three genes were linked to the presence of the cancer. Hereditary nonpolyposis colon cancer is the most common inherited disease in humans and accounts for almost 20 percent of all colon cancers. The two resulting papers rapidly became some of the most highly cited in the literature and marked a turning point in the credibility of the EST method. What was most gratifying was that the establishment of the link between mutations in the DNA repair genes and this disease would rapidly lead to new diagnostics for colon cancer: Basic science and some of my very own discoveries were now helping doctors and patients.

As we pursued this remarkable work, my own colon would take center stage in my life when I was struck by wrenching pain, nausea, and fever, and rushed to the hospital. It was initially thought that I had a ruptured appendix and peritonitis, a potentially life-threatening infection of the peritoneum, the membrane that lines the wall of the abdomen. The doctors put me on high doses of antibiotics, and within a few days I had almost recovered. They wanted to remove my appendix, but I refused surgery because my symptoms did not match what I knew about appendicitis. Within a few weeks the condition returned, and once again I was given massive doses of antibiotics, plus a CAT scan, barium enema, and X-rays to try to reveal what was going on in my guts. The diagnosis became clear: I had diverticulitis, a disease in which weak spots develop in the colon. Diverticulitis results when they become perforated and gut bacteria leak into the abdominal cavity, causing peritonitis.

I was told that this was the new executive disease, caused by high stress. "Are you under any pressures?" the doctors asked. Well, there were those daily attacks by my peers in the press. I was also involved in a struggle to establish the credibility of a new method in science. And, of course, there was my difficult relationship with Bill Haseltine and the endless attempts to destroy my institute by the company that I had helped start. Around that time I also remember feeling a moment of paranoia as I was about to board a private helicopter bound for lunch with an HGS stockholder in New York, hesitating when I remembered that the $85 million contract between HGS and TIGR was in effect only as long as I remained alive. Apart from all that, I told the doctors, I was feeling relaxed. But although my mind could handle the stress, my body clearly couldn't.

Stress, Impulsivity, and Thrill Seeking

The ability to deal with pressure and a propensity to seek thrills have been linked to a gene on the X chromosome responsible for the enzyme monoamine oxidase (MAO). One form of this gene is particularly related to sensation seeking and to the regulation of messenger chemicals (such as dopamine and serotonin). The resulting low MAO levels are linked with impulsive tendencies to seek immediate rewards without any regard for the consequences. One unusual variant of this gene was even tracked through three generations of criminal males in a Dutch family, providing a remarkable link between molecules and malice.

The gene encodes an enzyme that is found in the synapses between brain cells and sweeps up excess messenger chemicals. A common variation of the gene produces a less active enzyme, which isn't as efficient at clearing away the excess, and this form is present in high sensation seekers.[10] Meanwhile, the high activity version of the gene seems to have a protective effect against stress. A scan of my genome reveals that I have that high activity form and thus a lower risk of antisocial behavior. Some may find this hard to believe: I have been called the bad boy of biology, mischievous, and even satanic. However, I doubt that even my sternest critic would deny that I can cope with a great deal of stress.

I was warned that I could develop peritonitis anytime, and because this can be lethal, I would need surgery as soon as possible. First, however, I had to be free of infection. I ended up being treated with Augmentin, an antibiotic made by SmithKline Beecham, the company that had helped create the stress in the first place. My life continued at its hectic pace, until toward the end of the year when I went to Monaco to speak at a major international symposium.

I began to feel myself burning up as I developed a fever. I barely made it through my talk, almost collapsing because of the pain. I called my surgeon, who told me to take some Augmentin and get home. On my arrival in Maryland I went to the hospital to have sections of my large intestine removed.

I made what I thought was a remarkable recovery, having agreed to try a new pain management method in which, at the slightest twinge of pain, I would trigger a pump to inject small doses of narcotic into my spinal cord. Given what I had seen in Vietnam, I was amazed that even with a six-inch incision in my belly, I was soon walking up stairs and around the hospital. Everyone, including myself, thought I was superman. After two days I wanted to go home, though the doctors thought that a week or more would be a better idea. I insisted: The hospital reminded me too much of my time in Da Nang. The spinal pump was removed, and by the next morning I lay tight-lipped in my hospital bed, pale and weak and no longer bounding around; the slightest movement caused extreme pain.

When Claire came to take me home, she was shocked at my appearance and argued that I should stay in the hospital. But by now I had set my heart on going home, and she ended up driving at jogging pace: Every little bump and shake gave me a jolt of pain. In the days that followed she threatened several times to send me back to the hospital in an ambulance. But I was soon back at work.

If it was at all possible, SmithKline Beecham and HGS were by now becoming even more concerned about TIGR's right to publish its data. The pressure seemed to mount proportionate to the discoveries from our Human Gene Anatomy project. While I thought that I had pacified them by agreeing to wait eighteen months before publishing a paper on half of all human genes, this proved to be another strategic error on my part: A slow bleed of sequences into the public domain probably would have caused them less grief than my grand unveiling of thousands in a single paper, which was explosive. As my team continued to analyze EST data and to attempt to put it into some biological and medical context, SmithKline Beecham launched a new round of discussions on how they could let us publish while limiting the access of commercial competitors to the data. I did not see how it was possible to achieve the greater goals of science without making my data broadly available in the literature. Of course I wanted new drugs and tests to result from my work, but I felt that HGS and SmithKline had the necessary safeguards in place with their patents and armies of researchers.

Paradoxically, the torrent of data generated by TIGR, which should have been an achievement to be celebrated, was the source of the problem: HGS was simply overwhelmed. By now it was also generating its own sequence data, which, by agreement, they had to hand over to us. Had I delivered them a single gene that was linked to a disease, they would have known how to mount a major discovery effort to turn the find into a test and new drugs. But I had given them thousands of genes over the course of a few months. As HGS complained that to exploit the data "was like trying to drink from a fire hose," I did not understand why HGS and SmithKline wanted to restrict access to it when they themselves could work on only a few dozen genes at most. Ultimately, it was a matter of pride: They did not want to be embarrassed when someone else made a discovery using the DNA sequences that they had passed over.

We eventually replaced the earlier three-tier approach, which gave HGS up to two years from when it was found to exploit a gene. In the new deal, more than 11,000 ESTs, representing 7,500 genes, would be deposited in GenBank and released on publication of our paper, while the rest of the data, thought to represent more than 100,000 sequences, would be available on both TIGR and HGS Web sites with additional restrictions. Eighty-five percent would be "level 1" data that would be freely available to all researchers at universities, U.S. government laboratories, and not-for-profit research institutions upon the signing of a liability waiver agreement. The waiver may sound unnecessary, but SmithKline was unrelenting; pharmaceutical companies have a more cautious view of litigation than most organizations. Due to their deep pockets, payouts imposed in lawsuits can run into the billions of dollars. The company was reasonably worried that someone would use the TIGR data to develop diagnostics and therapeutics that could inadvertently cause harm. A complaint against TIGR or HGS had little value, while SmithKline's link to the data created a legal bull's-eye with the possibility of a huge payout. Like many scientists I thought this approach overcautious but figured it would not matter because researchers would be happy to sign anyway. (I was wrong.) For the final 15 percent of the data, the so-called level 2 data that was being actively studied at HGS or SmithKline, scientists were required to sign an agreement granting HGS the option to negotiate a license to commercialize potential products derived from the use of the data. All the level 2 data had a time stamp on it providing that within six months it would automatically convert to level 1.

After much discussion I submitted our Human Gene Anatomy data to *Nature.* Because our paper was about twenty times longer than its typical features, the journal planned to publish it in a special issue, "The Genome Directory," which would also include the first papers to roughly map the human genome, on which, in effect, our ESTs were key landmarks. While all this news was very exciting, it led to only more grief from my funders. The SmithKline executives proposed to hold a one-week "summit" somewhere inconspicuous—a second-rate local hotel—to clear up any pending issues associated with the *Nature* publication.

I walked into the dark, drab conference room with Steve Parker and a few TIGR scientists and was taken aback to see massed before us the assembled ranks of SmithKline and HGS, about twenty-five people in all, and more than a dozen lawyers. After four days of tense, detailed arguments I finally understood how a multimillion-dollar divorce agreement eventually falls apart over who owns a silver picture frame or why the purchase agreement on a house founders over who gets to keep the towel rack. The final sticking point in this case was Table 2 of the 174-page manuscript, which showed what organs, tissues, or cell lines were used for the EST analysis. Under the category "bone" were five subcategories: bone marrow, condrosarcoma, fetal bone, osteosarcoma, and osteoclasts. The osteoclasts, which erode bones, became the subject of intense debate.

Osteoclasts are thought to play a role in osteoporosis, a condition characterized by weakened, brittle bones that afflicts roughly 50 percent of Caucasian and Asian women over age sixty-five. We had isolated several new protease genes from the osteoclast cDNA library, and SmithKline, with the backing of Haseltine and HGS, insisted on removing what became known as the "O" word from the table and replacing it with the nondescript "bone." Unknown to us SmithKline had already been intently perusing the proteases in order to develop inhibitors to treat osteoporosis. For four hours we chewed over this particular bone; by the fourth emotional day we began to scream at one another. I wanted my paper to go out, and, more important, I just wanted to escape this insanity and oppression, so I gave in and left. I doubt a single reader has noticed the word "bone" on Table 2 of page 18 of the Genome Directory. But I did, and so did my colleagues. This compromise did not make us feel proud, but at least we now had the go-ahead to publish our *Nature* paper.

That bone of contention was the least of our PR problems. The terms by which HGS and SmithKline allowed researchers to use our data also came under scrutiny. *Science*'s headline ran, "HGS Opens Its Databanks—for a Price," and *Nature*'s "HGS Seeks Exclusive Option on All Patents Using Its cDNA Sequences." George Poste, research director of SmithKline, said that the terms of the agreement were little different from those that any commercial sponsor would attach to its support for a university-based group of researchers. He argued that free access to the cDNA database should be considered a form of support for university scientists comparable to a grant that they might receive directly from a pharmaceutical company. Along with Haseltine he defended the aggressive stance that HGS took in driving hard bargains with commercial competitors who sought access to the TIGR data: "We have committed more than $100 million to Craig Venter's work at TIGR, as well as an extra $8.5 million in making the database available to the academic community and feel that we have a reasonable right to profit from any results that emerge from something that we have made an investment in."[11]

The scrutiny of the HGS/SmithKline terms by journalists was diverted temporarily by a controversy over another high-profile gene, this time of a mutation in a gene linked with breast cancer. It had been known for some time that in rare cases breast cancer was hereditary, caused by mutant genes passed from parents to their children. Estimates of the incidence range from 1 to 3 percent of all breast cancers. In 1990, Mary-Claire King of the University of California, Berkeley, and her team, after detailed work mapping the breast cancer families, identified the approximate region of one breast cancer gene on chromosome 17, launching a major quest to pinpoint its location.

I was fond of Mary-Claire and found her to be not only a good scientist but an intellectual leader, so much so that when Bernadine Healy was looking for a replacement for Watson, I had suggested Mary-Claire. When Francis Collins was selected instead, I invited her to join the TIGR board of trustees. But I would have to rescind the invitation—the price I had to pay for offering to help her find the breast cancer gene with high-throughput DNA sequencing.

Dozens of groups worldwide became involved in what was then the most high profile gene hunt of all. If the earlier quest to find the Huntington's disease gene was anything to go by, it would take a decade or so of work using traditional methods, so Mary-Claire was very enthusiastic about our participation. Then, understandably, she became worried about HGS's patent rights.

After extensive discussions with Steinberg and Haseltine, I got HGS to state in writing that it would not pursue patents on the breast cancer gene if TIGR discovered it by sequencing the region. With that agreement in place we were prepared to join the hunt, which could begin as soon as Mary-Claire sent me the DNA clones from the region of chromosome 17, where the gene was thought to lie. First, however, she had to consult one of the scientists who had helped isolate the clones, Francis Collins.

When I did not hear from Mary-Claire for a while I telephoned her to find out what had happened to the clones. Mary-Claire explained that the problem was Francis. He had supposedly told her that, unless she guaranteed that he would be an author on any resulting breast cancer paper (which she felt was inappropriate), he would stop funding her grant if she sent us the DNA for sequencing. All things considered, she did not want to put her funding at risk and preferred not to send the clones. Claire (who was also on the call) and I were outraged. I knew there were tensions between Mary-Claire and Francis, but if she reported what he had said accurately, it would mean that he was abusing his position for scientific credit. Even so, I was more outraged at Mary-Claire for yielding to this blackmail, if that was indeed what it was. Our frustration was all the greater because this particular gene hunt could affect the lives of many women—it promised a test for those at risk within a few years. And since HGS had waived its intellectual property rights, there would be nothing to prevent that from happening soon. (I learned later from a venture capitalist that Collins had been approached by King for funds for a new biotech company based on the breast cancer gene.)

A few months later the National Cancer Institute, part of the NIH, decided that sequencing the breast cancer region should be made a priority, to speed the search for the gene, and to put out a request for proposals. We once again attempted to obtain the DNA and once again were rebuffed. I called Sam Broder, the NCI director, and explained that we did not even need any funding, just the DNA. While sympathetic, he said that he could not order researchers to make the clones publicly available.

Once again politics came before science. While Collins was attacking HGS and TIGR for not making their data more widely available to the scientific community, he was reportedly blocking free access to DNA clones that had been created with taxpayers' money so that he would gain a competitive advantage in the scientific world's game of credit and fame. In September 1994 the race was eventually won by Roger Wiseman, of the National Institute for Environmental Health Sciences in North Carolina, and Mark Skolnick, of the University of Utah. In total, forty-five scientists had participated in the isolation of what came to be named BRCA1, the Breast Cancer 1 gene.

As members of Skolnick's team posed for photographers from *Time* magazine, the intellectual property issue was about to flare once again. Skolnick had set up a company called Myriad Genetics to patent and market a test for the gene, which led to inevitable attempts by the media to draw a parallel with the "Venter case." But I thought the reality was best expressed in a quote from a researcher at Massachusetts General Hospital who told *Nature*[12] that it felt as if a new era in genetics had begun: Industrial support was providing the power and the opportunity for high-quality experiments to proceed quickly. After all, no one was attacking Francis Collins for his patents on the cystic fibrosis gene, the pyrin gene for Mediterranean Fever, the neurofibromatosis gene, and the Ataxia-telangiectasia gene.

Science and *Nature* continued to publish reports on HGS and TIGR and gene patents, now throwing Myriad into the mix. Labs around the world wanted to use the breast cancer test but Myriad prevented them from doing it commercially. The ultimate irony is that if we were not blocked from receiving the clones, the breast cancer gene would not have been patented, and the test could be freely carried out in any lab.

By now, there were other major forces in sequencing on the world stage, including the Sanger Center, established in 1993 as a partnership between the Medical Research Council, the Wellcome Trust, and the European Bioinformatics Institute, which was created to provide fast and easy access to the latest sequence data for biologists across the continent.

Britain had a special reason to take genomics more seriously than most nations. The country had, of course, helped launch the DNA revolution with the elucidation of the double helix structure of DNA by Watson and Crick in 1953, and Sanger had shown us all how to read the code. Yet the nation seemed to fail when it came to the application and commercialization of its science.

Frustrated by its own attempts to gain value from genomics, the Wellcome sponsored a private meeting in Washington in the fall of 1994 to discuss the growing commercial control of genetic data. The gathering of about thirty people was spearheaded by Michael Morgan, program director of the Wellcome; Francis Collins; and Tom Caskey, the Baylor geneticist who had left basic science to join Merck when he could no longer obtain government funding.

The major players had much in common. Wellcome and NIH were being outcompeted by a small institute with less than 1 percent of their resources, while Merck was being left in the dust by SmithKline. To deal with the competition they would cast themselves in the role of spoilers rather than potential winners.

Nature, which was then handling my Human Gene Anatomy paper, ran a contemporaneous editorial[13] referring to how I had "put the cat among the pigeons" but concluded: "Rather than seeking to cramp Venter's style, the genome community might usefully seek to emulate his methods." Although Watson and others were still threatening to boycott the journal if it published my paper, Morgan seemed to give the meeting a positive spin, suggesting that it could showcase a conciliation between SmithKline, HGS, and their academic critics[14] and get the community talking about a possible new genetic map based on ESTs, marking official recognition of their importance.

That four-hour encounter took on quite a different and predictable purpose, however. On October 14, 1994, my birthday, I was greeted with a three-page story in *Science:* "A Showdown over Gene Fragments."[15] The participants at the meeting, it reported, had vented feelings of "intense anger" and "frustration" that had built up in the academic community on the gene patent issue. As *Science* put it, "If any outfit was cast in the role of Frankenstein that afternoon, it was The Institute for Genomic Research." (One of my TIGR colleagues complained in response, "They're coming after us with torches and pitchforks.")

Merck, a major competitor of SmithKline, was eager to break its rival's lock on ESTs for reasons that were not hard to discern: Merck had no internal genomics program, lacked even primitive bioinformatics capabilities, and had a director of research who was effectively antigenomics. The company rejected licensing data and clones from either HGS or Incyte. But it had become rattled after HGS announced its exclusive deal with SmithKline Beecham. If Merck was not going to be a major player, then no one would: it paid Bob Waterston at Washington University in St. Louis to sequence human ESTs and dump them as fast as he could into dbEST, the public database established while I was at NIH as a special repository for my EST sequences.

Merck could market its spoiler exercise as a gesture made for the good of humanity, and Waterston was only too happy to go along even though only a few years earlier he had tried to stop the EST method from being published in the first place. The media love stereotypes, and the picture that was emerging was simple and seductive: There were those who were working for the public (the good guys), and there were those who were working for the companies (the bad guys).

The Merck management must have laughed long and hard in their executive boardroom: A decision made for cold-eyed strategic business reasons had been successfully rebranded as an attempt to save the world from rapacious capitalists. Sulston summed up the prevailing view: "Merck's action was a great thing for science and a triumph for the principle of free access to genomic information."[16] Their fellow "good guys" were Wellcome, the world's second-largest charity, which was funded by the profit and largesse of a drug company and blessed with tax-exempt status, and a U.S. government organization that had failed to seize the initiative when ESTs had first showed promise. Why the latter should be involved in a British-led attack on a U.S. biotechnology company still makes my head reel. The charges were all the more baffling because we were the only team generating ESTs that was prepared to make many of them available.

I had now stepped to center stage of a soap opera as the villainous public face of, as a *Science* article put it that December, "the company that genome researchers love to hate."[17] The usual themes were played out: I was "cream skimming" and "land grabbing." But there were the occasional compliments, too, and from some unexpected quarters: "TIGR and Venter have done a wonderful job, and their data are of high quality," stated Eric Lander.

My favorite quote of all came from Varmus: "I wish we had pushed for getting someone to do [a similar] database in the public domain some years ago." I wish they had, too. Perhaps Varmus did not know that I had developed ESTs at NIH, had lobbied for them to be done there, and had finally had to leave the NIH to get support for them. Collins himself denied "that we've been asleep at the switch and hadn't thought about this." Although they had turned their backs on this opportunity, they now demanded my results in Washington, even though no taxpayer money was involved in my research.

The gut- and brain-twisting media attention was, ironically, helping my cause in many ways. *The Wall Street Journal* headline of September 28, 1994, "Plan May Blow Lid Off Secret Gene Research," brought even more attention to my method and ideas. Along with the millions now being poured into EST sequencing, this was seen as a ringing endorsement of my method. Second, my goal remained to publish our work in journals, so it was a great gift to my team when the various "public" efforts dumped their data without any intellectual content or analysis. HGS could not argue that they had to keep my data secret if similar sequences were already in the public domain.

But I was still compelled to keep one step ahead of the game. HGS and SmithKline wanted us to continue to sequence ESTs, an effort I felt was pointless, given that HGS had duplicated our DNA sequencing center and TIGR had already sent a manuscript to *Nature* outlining ESTs corresponding to more than half of all human genes. Because I was driven by the science, I had other ideas. After all, one of the reasons I wanted TIGR to be an independent not-for-profit institute was that I could exploit new scientific opportunities as they arose and not be bound by the strategic objectives of a commercial organization. In this I had only partly succeeded, as evidenced by my ongoing turbulent relationship with SmithKline and HGS. While HGS was doing all it could to close down TIGR to save about $10 million annually, and while Haseltine was trying to take credit for my work, I felt the time had come for me to do something new. I was trapped in a doomed relationship and now lacked a stretch of my intestine because of the intense stress. I was ready to move on.

9. SHOTGUN SEQUENCING

If you cannot—in the long run—tell everyone what you have been doing, your doing has been worthless.

—Erwin Schrödinger, 1933 Nobel Prize winner in physics

Although we were by now discovering human genes at an unimaginable rate, this achievement only whetted my appetite for even more ambitious projects. Now I wanted to get back to a comprehensive view of the entire human genome—to read every one of the 6 billion base pairs of genetic code that comprise all the chromosomes in each of our cells. Despite my earlier arguments supporting ESTs as a viable alternative, it had always been my intention to eventually sequence the entire human genome. To do so, I had to develop and test new approaches. I remained certain that there had to be a better way than the one that government-funded scientists around the world were embracing, an approach that felt like something out of the Dark Ages.

My critics had often grumbled that my expressed sequence tag method to find genes was a cheap and inadequate substitute for the expense and grind of sequencing entire chromosomes. I can understand where they got this idea: Out of frustration, and in response to the way Watson and others had tried to belittle my approach, I had indeed said that ESTs were a bargain compared to the estimated $3 billion price tag of the human genome project. But I also believed that ESTs were no replacement for reading the entire genetic code; a point I made in my first paper describing them, where I concluded that ESTs would provide the ultimate way to annotate the human genome, serving as crucial landmarks to reveal where genes lay on the vast tracts of incomprehensible DNA.

From the time of my involvement in genomics in 1986 and my use of the first automated DNA sequencer the following year, I had dreamed of having a factory where row upon row of machines automatically read the DNA code. I now had the first such scientific facility in history and was determined to use it. The alternative to what I wanted to do was the government-backed genome project, which was crawling slowly down a long path that made no sense to me. From the outset they had viewed it as a massive labor effort; the prototype of this approach was the yeast genome project, which took a decade or so and the sweat and grind of more than one thousand scientists and technical staff spread over dozens of countries.

The challenge that we all faced was to figure out how to read the sequence of the entire code when the available technology could provide only a few hundred base pairs of code per sequence read. If you were a monk and faced the massive task of sequencing millions of base pairs of genetic code, you would logically break the DNA into smaller manageable fragments. To handle them you could use various methods to grow these fragments of DNA. Small pieces of only a few thousand base pairs could simply be bound and multiplied in standard plasmids. For pieces of DNA up to eighteen thousand base pairs, a bacterial virus or phage, called lambda, was utilized; and for what were then considered extremely large sections of DNA, measuring around thirty-five thousand base pairs, a special plasmid, called a cosmid, was used. In early genomics almost everyone used cosmids. The process was very logical, but the logical way is not always the fastest way forward; randomness can sometimes help as well.

In his time-consuming, labor-intensive, and expensive project, a monk would first want to carefully place all the cosmids in the proper order, the same order found in the book of life. The result was a cosmid-based map of the genome. Only after this mapping phase had been completed did their superiors provide monks with the money and the blessing to begin sequencing the cosmids, one at a time. This crucial step of establishing a map before sequencing could be done but it would take time, lots and lots of time. Frederick Blattner's work on the gut bacterium E. coli, whose genome is close to one thousand times smaller than the human genome, required three years to line up the lambda clones into a map of the genome before he was allowed to begin sequencing. For the human genome, more than a decade and $1.5 billion had been consumed in an attempt to create chromosome maps, but they were never completed. As one biologist remarked, "Several good careers would be built sequencing the human genome bit by bit, clone by clone."[1]

Watching these projects in progress, I was certain that there had to be a better way to carry them out. I put my faith in randomness, not order, by using the EST method on a grand scale. I had also learned from establishing my pioneering DNA sequencing center the value of the DNA sequence itself. In a strange way scientists of the day seemed afraid to actually launch into DNA sequencing. Its message of Cs, Gs, As, and Ts was complex, and it was tedious to generate with commonly used methods. Much of the genome mapping phase seemed aimed at actually avoiding sequencing the DNA. But the EST data clearly demonstrated how much information was contained in only a few hundred base pairs of DNA code: It not only provided a unique signature of the fragment that could be used to map it on the genome but often contained enough information to find the gene structure and function as well. Why not use the power of this sequence information? Why not eliminate the tedious clone mapping and the distributed monk work?

I had already thought of an alternative approach when I proposed shotgun sequencing the smallpox genome several years earlier, fragmenting its genome into thousands of shards of DNA that could easily be sequenced and then using the sequences of the individual fragments to reconstruct the genome by finding specific overlapping sequences. To me this was like the first step in solving a jigsaw puzzle when you spread out all the pieces and then select one that you compare with the others until a match is found, a process that is completed over and over again until the puzzle is completely assembled. However, with thousands to millions of genome puzzle pieces involved, this matching had to be done with a computer. At the time of the smallpox genome effort I had to abandon this approach because I lacked the necessary computational tools to stitch the sequence together again. All this would soon change due to advances in the EST method, such as new mathematical algorithms, and with a chance meeting in Bilbao, Spain, in March 1993.

I had been invited to speak there at a conference organized by Santiago Grisolía, a leading Spanish geneticist who was also a distinguished lecturer in the biochemistry department of the University of Kansas Medical Center. I was the last to address the conference, and many in the audience looked shocked by the advanced results of our EST effort and the nature of TIGR's discoveries, including the colon cancer genes. The questions inevitably turned to gene patents, and a Catholic priest and theologian told the meeting it was immoral to patent human genes. I asked him if it was also wrong to patent genes from other species. No, he replied—just the answer I had been waiting for. I told him that TIGR had just sequenced a human gene that was identical to one from a rat, so the proteins each described were one and the same. Wouldn't patenting the rat gene be equivalent to patenting the human gene? He was taken aback, having been certain that the human genome would be unlike that of any other species.

Once the crowd that had come up for a one-to-one talk with me had melted away, I was confronted by a tall, kindly-looking man with silver hair and glasses. "I thought you were supposed to have horns," he said, referring to my demonic image in the press. It was Hamilton Smith from Johns Hopkins University. I already knew Ham through his huge reputation in the field and his Nobel Prize. I took an instant liking to him; he was clearly going to make up his own mind about me and my science and not have his opinion dictated by others.

Ham had discovered restriction enzymes, the molecular scissors that are used to cut DNA at a precise location. Today, hundreds are known, each of which slices the DNA at a precise sequence. Some recognize four base pairs, such as GTAC, and will dice DNA wherever they encounter a GTAC in the sequence. Other enzymes recognize eight base pairs uniquely, and these sites will occur only every hundred thousand base pairs or so—the more an enzyme is keyed to, the rarer the site. The applications of Ham's discovery are endless, and molecular biology could not have progressed to its present level without them. In 1972, Paul Berg used restriction enzymes to induce a bacterium to make a foreign protein, launching the modern biotechnology industry. The first genome maps were even called "restriction maps," based on the size of the fragments that one obtained with a given enzyme. Today these maps are used for the genetic fingerprinting of individuals in forensics, among other things.

Ham and I went to a bar for a drink, and it soon became apparent that this self-effacing man wanted to bask in new science, not the glory of earlier achievements. As Ham sipped a Manhattan and I drank beer, he grilled me about our sequencing, its accuracy, the robotics, and the genes that we were discovering. I invited him to dine with me and some friends. He explained that he was scheduled to attend a formal dinner where he was to be paraded as a Nobel trophy, but then said, "Oh, what the hell," and we joined a small and festive party at a local restaurant, which, in true Spanish fashion, lasted into the wee hours.

After dinner we went back to the hotel to talk further. Although Ham is more than a decade older than I, we would come to discover that we had much in common in terms of our upbringing. Both of us loved building things as kids, were inspired by an older brother (sadly, in Ham's case, his brother had been hospitalized for mental illness), and had medical training. Ham had been drafted and stationed in San Diego. He had even had a run-in with Bill Haseltine in which Ham had suspected him of trying to hold up a paper written by a competitor. The following day I invited him to join the scientific advisory board of TIGR.

Later that year Ham attended his first board meeting, during which he raised his hand and said: "You call yourself The Institute of Genomic Research. How about doing a genome?" Then he told us about *Haemophilus influenzae,* which he had been working with for twenty years, and explained how it had a smaller genome than E. coli as well as other attributes that made it an ideal candidate for genome sequencing. I had been looking for a good genome to test my idea of whole-genome shotgun sequencing and had toyed with the idea of rapidly sequencing E. coli and competing with the public program (one that would take the monks thirteen years to complete), as a test. But I liked the idea of sequencing *H. influenzae* better. *H. influenzae* had many virtues as a subject for a test shotgun sequencing project, including that it had a similar composition (G/C letter content) to human DNA. Here was a chance to sequence the first genome of an organism, one that Ham knew inside out.

Our first collaboration got off to a slow start. Ham explained that there were problems with producing the libraries of clones containing *H. influenzae* genome fragments. Only years later did he reveal that his colleagues at Johns Hopkins were less than impressed with our project, viewing me with suspicion because of the attacks by Watson and others, and fearing that his association with me would ruin his reputation. Even though many of them would spend their careers studying *H. influenzae,* they did not immediately see the value of having its entire genome sequence. One of Ham's postdocs even asked him, "What's in it for me?" Their lack of vision and interest forced Ham to sidestep his team, as I had done years earlier with ESTs.

Nevertheless, Ham thought that he could make a library from *Haemophilus.* Although the computer algorithms of the day would choke on as few as one thousand sequences, we now had a better program to reassemble the pieces. Ham had done some modeling to simulate the assembly, and he thought it would be possible to achieve with about twenty-five thousand pieces. While the TIGR team was enthusiastic, Granger Sutton, who had designed the TIGR "assembler" algorithm, was uncertain if his code was up to the task of putting all the sequenced DNA back together into an entire genome consisting of 1.8 million base pairs. Granger was as quiet as he was modest: His assembler had in fact just joined more than 100,000 EST sequences into clusters of related DNA, and I was certain his algorithm could handle the *H. influenzae* genome.

We set about applying for an NIH grant, submitting the request to try our new method in the summer of 1994. Naturally, I was worried that because of the politics involved, the NIH would not fund our new approach, and Ham and I were impatient to get started. The yeast and the E. coli genome projects had already had years of funding, and if we trumped them by using this new method, it would mark a very significant milestone: By reading the code of this human pathogen with nearly 2 million base pairs, we would be the first to decode the genome of a free-living organism. Rather than wait nine months for a likely rejection from NIH, I made the decision to redirect a portion of TIGR's budget, about a million dollars, to the *H. influenzae* genome project, a gamble that I was confident would pay off.

Four months later we had the sequences of the twenty-five thousand pieces of *H. influenzae* DNA, and Granger's team got going. Within a few weeks the data looked promising, with several very large sections assembled from the pieces. But various small fragments were unaccounted for, and it was not at all clear how they fit into the circular chromosome.

These results fell short of the great genome dream, in which all the DNA clones from the genome were grown in E. coli and sequenced, and then these sequences were compared and assembled in a computer until, finally, the entire chromosome popped out. There are good biological reasons why this sort of outcome is rare. One flaw inherent in molecular biology is the dependence on growing foreign DNA fragments in the bacterium E. coli. Some DNA is clearly toxic to E. coli, and those particular fragments get deleted by its cellular machinery. Restriction enzymes are used by bacteria to defend themselves from foreign DNA attacks, as DNA is moved around constantly in our environment, including by viruses.

Still, the missing pieces of the genome puzzle helped me appreciate that a genome map would help order the sequence and assembled fragments, in the same way that having a picture of a finished jigsaw puzzle helps in its assembly even if pieces are missing. Just as sailors used crude and primitive navigation tools to find their way, so geneticists have used various kinds of maps over the years: for example, they could make what is called a functional, or linkage, map. During reproduction, genes in one parent organism are often—but not always—transmitted together to the offspring. The farther apart the genes are on the chromosome, the less likely it is that such a transmission will occur. By studying how often two genes are transmitted together down the generations, scientists can estimate how close they are on the chromosome and create a linkage map. The first chromosome mapped this way dates to pioneering work on the fruit fly in the early 1900s by the American zoologist Thomas Hunt Morgan. (A genetic unit, the centimorgan, was named after him and consists of roughly one million base pairs of genetic code.) A centimorgan-resolution map had long been a dream of geneticists.

Another form of genetic cartography rests on finding the physical location of a given gene: determining on which chromosome it resides, who its neighbors are, and where approximately on the chromosome it is found. This is known as a physical map.

But I did not want to make either a linkage or a physical map a prerequisite of sequencing, as had my government-funded rivals. Fred Blattner's team had spent three years developing a lambda clone map of E. coli, and the end result was an elegant feat of traditional genetic craftsmanship, 18-kilobase clones tiling the genome like overlapping pieces of Lego. But I did not have to chart such a map. As anyone who has assembled a jigsaw puzzle knows, you can proceed without knowing what the bigger picture is if you take advantage of edges and other recognizable features to reconstruct the jigsaw from the bottom up. After all, the DNA sequence itself is the ultimate physical map in which all the base pairs of the genetic code are known in their exact order.

In the absence of any map of the *H. influenzae* genome, we developed several novel methods to organize large assemblies of fragments to re-create the genome. In one we used a technique called PCR to copy DNA from the genome. Two chemicals, called primers, determine the beginning and end of a given region to be copied. We would use primers derived from the sequences toward the ends of the assembled fragments. Then we would try to use PCR between every combination of primers, using a PCR probe from the end of each sequence in turn with all other PCR probes from the ends of all other assemblies. If any DNA fragment amplified from the genome, we rapidly sequenced it. The sequence would then link up and order two of the fragments. By doing multiple combinations simultaneously, we were able to order much of the genome relatively quickly.

The PCR method did not work with every gap so I came up with a novel idea that would change how we did sequencing, particularly when it came to the human genome. Once we had used the computer to assemble the full complement of twenty-five thousand fragments of the *Haemophilus* genome as best we could, we had ended up with larger pieces, called contigs (from contiguous), that consisted of a set of overlapping DNA segments. To assemble the contigs into the genome I thought we could compare the sequence from both ends of a few hundred random lambda clones. If the end of one lambda clone matched one contig and the other end matched another contig, then we automatically knew the correct order and orientation of those two contigs. We had to work out some new methods for sequencing just the ends of the lambda clones but that work proceeded rapidly. From even the first few paired end sequences we were able to link up the sequence assemblies in the correct order. This "paired end" strategy is like knowing the exact number of pieces separating two features on the genetic jigsaw and became the key to the whole genome shotgun method. We were soon down to only a few sequence gaps in the entire bacterial genome and were confident that we had found the winning strategy.

The genome sequencing conference was rapidly approaching, and I wanted to present our results there. Although we were proud of the success we had achieved and I looked forward to the meeting, I would have preferred to finish the job in Rockville completely before someone could beat us to this important milestone. We had gone from testing a wild idea to nearly reaching a breakthrough, the first genome of a free-living organism to be sequenced in history. Now that we were so close to actually achieving that, I did not want to miss the opportunity of doing so.

That September, Robert Fleischmann outlined our results before the Genome Conference in Hilton Head, South Carolina. We thought the presentation was well received but were stunned when Bob Waterston got up to attack the approach as useless. He argued that it would never work and that all we had ended up with was eleven fragments that could not be put into any kind of order. Ham was particularly upset and even to this day makes reference to Waterston's attack in 1994.

Shortly after we returned to Rockville, we received the inevitable and expected reply from NIH on the *Haemophilus* grant application that Ham and I had submitted earlier that year. The score was low, and it had not even come close to being funded. The verdict of the reviewers reflected that of the genome community: Like Waterston, they thought what we were proposing (and unbeknownst to them, already putting into practice) would not work and was not even worth attempting. I did take some comfort from the NIH response in the (highly unusual) form of a minority report from a small group of peer reviewers who dissented from the majority view and believed that our program should be funded.

I tacked the rejection letter onto my office door. By then I had no doubt that we would succeed. Ham and I decided to file a rebuttal to the critics and to appeal to Francis Collins directly to fund the project. We included the latest data that showed we were likely to have the first genome in history within a short time. But of course more was at stake here than science. I called Francis to inform him of our probable success and to assure him that our goal was not to embarrass his NIH program but simply to receive funding from it. We were shocked when we received a letter from the NIH Genome Center a few weeks later that supported the review. The letter was signed by Robert Strausberg, then the head of the sequencing grant section. When he later joined TIGR, Bob indicated to me that his position required him to write the rejection letter even though he thought that we would succeed.

Rather than discouraging us, the letter motivated us to prove our critics wrong, and a short while later the final gaps in the *Haemophilus influenzae* sequence were closed. We had become the first to sequence the genetic code of a living organism, and of equal significance was the fact that we had done so by developing a new method, "whole genome shotgun sequencing," whereby we could rapidly (twenty times faster than any other project)—and without a genome map—sequence and reconstruct in the computer an entire genome. We owed a great debt to Sanger, certainly, but there were important differences in what we had accomplished. The viruses sequenced by Sanger in his pioneering work were not living but complex chemical structures that needed to pirate another creature's cells in order to multiply. To sequence them, Sanger had broken up the viral genomes into manageable pieces with restriction enzymes, so his shotgun method was not truly random. And although Sanger had used computers to put the pieces back together again, his software would have choked to a halt on the amount of data we handled.

While Sanger's work was pioneering and represented a milestone in DNA sequencing, his methods had to be extended and adapted to tackle the genomes of living species. His own attempts to do so had been thwarted by the territorial instincts of his coworkers and a lack of automation. After Sanger retired, his protégés started using sonication, a good random method, but they still applied it to clones of restriction fragments as they moved to larger viral genomes. Others, such as Clyde Hutchison of the University of North Carolina (now at the Venter Institute), had looked into the shotgun method but were daunted by the problem of manually sequencing and reassembling the random pieces of DNA, which got exponentially harder as the genome grew larger.

In short, Sanger's methods had been as critical for genetics as the invention of the wheel or the first steam-powered cars in the seventeenth and eighteenth centuries had been for the automobile industry. But the Sanger approach predated the age of genomics, just as the wheel and steam cars predated the age of the automobile. To help launch large-scale genomics, my team used a combination of random coverage of the genome, the paired end sequencing strategy, a blend of mathematics and new computational tools, and pragmatism in the form of novel methods to fill any gaps. Above all else our successful combination of a number of methods was carried out in the context of a factory environment, where the scientists involved in sequencing expressed their natural territorial instincts by producing the best libraries and the smartest algorithms, rather than by staking out claims to particular pieces of the genome. And that is why the champagne flowed freely at a party to celebrate the sequencing of *Haemophilus influenzae.* This marked our work as the first full-fledged demonstration that the shotgun method could be used to read a whole genome. This also marked the start of a new era when the DNA of living things would be read and compared and understood.

The first unveiling of our success to our peers and rivals would come in England, where Richard Moxon of Oxford, a key collaborator on the *Haemophilus* genome project, organized a four-day meeting. Moxon had worked at Johns Hopkins for many years, regarded Ham Smith as his mentor, and had himself been "absolutely flabbergasted" by TIGR's progress. He felt confident that the eventual success of the project was assured even if the genome assembly still had some rough edges or was incomplete. The gathering was backed by John Stephenson, a senior official at the Wellcome Trust.

The Wellcome bureaucrat Michael Morgan, who had a reputation for being a bruiser among his colleagues, had clearly bought the Watson line that I was jeopardizing the scientific world order and posed a major threat to the Sanger Center. He seemed very unhappy with the prospect of my being the star attraction of a Wellcome Trust meeting with the unveiling of the first genome in history. Although the gathering was scheduled to take place before publication of the paper, I was advised to bring the sequence on a CD-ROM so that Morgan and others could verify its existence. Assuming that I would be tied up by commercial secrecy requirements, a Wellcome Trust official had cockily asserted that I would not show up, or if I did I wouldn't bring the sequence, or if I *did* bring it, I wouldn't allow anyone to see it.

Ham and I accordingly decided to up the ante. At that time Clyde Hutchison had realized that a species of *Mycoplasma genitalium,* a parasitic bacterium that lives in the genital tract, would make an attractive candidate for genome sequencing because it was the smallest genome of a living organism. Ham knew that with our new methods and tool set we could sequence the genome very rapidly, and he took great delight in calling Clyde from my office and inviting him to the British meeting in a few months . . . and oh, by the way, did he want his genome sequenced prior to the meeting? That would be interesting, Clyde replied with his wonderfully dry sense of humor, and accepted the offer. (He later remarked, "We might have finished it [*Mycoplasma*] before 2000 if you had not come along.") We submitted a proposal to the Department of Energy review team containing the same data that we had shown to the NIH and Francis Collins, and it provided an immediate grant to sequence *Mycoplasma genitalium* and several additional genomes that it would help us select.

Even though we had by then completed the first genome, I chose to delay making our triumph public. I wanted more than just the DNA sequence; I wanted to analyze the genome to determine, for the first time in history, what a sequence could tell us about a species, and then write up a key scientific paper that would set the standard for the field. Interpreting the genetic code and specific genes is not an easy process and had not been done before at this scale and *in toto* for a free-living organism. We had 1.8 million As, Cs, Ts, and Gs that needed analysis and translation into English, and for this we needed new software, new algorithms, and new approaches.

We were most interested in finding the organism's genes, the blocks of genetic material (typically around nine hundred base pairs of code, equivalent to three hundred amino acids) that are the actual blueprints for proteins. Called open reading frames, they contain the stretch of the genetic code that describes all the amino acids comprising a single protein. Bacteria have no introns (meaningless DNA) to break up genes and complicate things, so we were able to look for all the open reading frames in the genome and then search for what protein those sequences coded for by trawling through public databases for similar genetic sequences. Once again, because Mother Nature is conservative, we knew that if a protein did a certain job in, say, E. coli, it probably did the same job in *H. influenzae*. But the latter contained about two thousand genes, and this strategy took time. Because of the limited information in public databases, this strategy worked only for six out of every ten genes. The rest did not match any known proteins or genes and therefore were listed as new genes of unknown function. We then constructed a giant metabolic chart of all the identified genes and their likely pathways, showing how one gene "talked" to another to enable this bacterium go about its daily life. This was thrilling work, because we were able to fill in more details about how this species functioned in its metabolic map every day. But I still wanted more.

Here we were, the first people ever to see the complete set of genes necessary for basic life, and the story it told of this organism remained frustratingly incomplete. If we could fill in all the gaps, we would lay bare secrets of the evolution of that species and so much more. But Ham and I were forced to agree that those goals lay beyond the current level of analysis or understanding and that we would have to fight that battle another day. We decided to write up our results and send the paper to *Science*. I called Barbara Jasny, an editor there, and told her what we had, and she and the other editors were obviously extremely excited. I negotiated the cover as well, assuming that the paper would pass peer review.

The final product took forty drafts; we knew this paper was going to be historic, and I was insistent that it be as near perfect as possible. The number of people mentioned as authors of a paper is usually a tricky issue, and even more so when it comes to Big Biology, which depends on a small army, from molecular biologists to mathematicians and programmers to sequencing technicians. Two positions on the paper really count, first author and last, and attached to either position is the status associated with being the author for correspondence. The best combination when you are young is to be first author and author for correspondence. Being last author and author for correspondence signals that the paper is a product of your lab, that you are the main person responsible for its contents, and that a younger colleague has made major contributions. After juggling the many contributors several ways, we decided to stick with Rob Fleischmann as the first author, because no one other than Ham and I had contributed more. In the end, everyone was just happy to have taken part in the tremendous accomplishment and to be an author on so key a paper. We sent it to *Science* to be reviewed by our peers, the last hurdle before publication.

We could not have been more pleased with their feedback. Reviews usually get very critical very quickly, but this time we had compliments, some of the most gratifying I had ever seen. We made some changes requested by the reviewers that we felt strengthened the paper and sent it back to *Science,* where it was scheduled to appear in June 1995. But, of course, rumors of our success were circulating weeks beforehand. As a result, I was invited to deliver the president's lecture at the annual meeting of the American Society of Microbiology, which was being held in Washington, D.C., on May 24, and accepted with the understanding that Ham would join me on stage.

Because scientific journals are a business and make money on subscriptions and advertising, leading publications such as *Science* and *Nature* attempt to prevent news of their papers leaking out before they are published, in order to maintain their substantial influence. Papers are "embargoed," and journalists who write or broadcast the results before they formally appear are punished by being cut off from future prepublication news releases. Scientists who break the embargoes by discussing their work openly before publication can likewise see their paper rejected or lose the coveted spot on the cover. This system benefits the journals, but such rules do, of course, run counter to the fundamental principle of open and free communication that is supposed to be the foundation of science. Ham and I did not want to miss the chance to present the first free-living genome in history to several thousand microbiologists (more than nineteen thousand attended the meeting), who were probably the audience most likely to appreciate what we had accomplished. *Science* initially objected, but the rules do permit scientific presentations as long as no press interviews are involved.

Ham and I arrived that evening in suits and ties, and I gazed around the giant ballroom and its thousands of seats. As I hooked up my computer and tested my slides on the vast screens, I began to feel nervous. Not only was the size of the meeting daunting, but I was delivering my first paper in microbiology to the crème de la crème of microbiologists. I also feared the usual questions about patents and hostility from my peers in the public genomic community. But I reminded myself that as long as I had allowed for sufficient preparation, I would bear up under pressure, however unnerving. As always, I had that peculiar out-of-body feeling in which I can evaluate what I just said as though I myself were listening in the audience.

The pressure really came to bear on me when the society's president, David Schlesinger of Washington University in St. Louis, announced what he described as an "historic event." Ham introduced me in his typically warm manner. As my computer fired up, I began to speak with confidence and clarity. I described how we had constructed DNA libraries from the *Haemophilus* genome and how critical it was to shatter the DNA into specific-size fragments so that when twenty to thirty thousand fragments were randomly selected from a set of millions, they would statistically represent the entire DNA in the genome. I illustrated how we developed the paired end method of sequencing both ends of every fragment to aid in the assembly. I discussed how we used the new algorithms developed from the EST method and a massively parallel computer to assemble twenty-five thousand random sequences to form large contigs covering much of the genome, then paired the sequences from the ends of these contigs and closed a few remaining gaps. The result was that the 1.8 million base pairs of the genome were re-created in the computer in the correct order. We had transformed the analog version of biology into the digital world of the computer.

But although we had the first genome of a living species, the fun was only now beginning. I described how we had used the genome to explore the biology of this bacterium and how it causes meningitis and other infections. There was more. We had in fact sequenced a second genome to validate the method, the smallest one known, that of *Mycoplasma genitalium.* When I ended my speech, the audience rose in unison and gave me a long and sincere ovation. I was almost overwhelmed since it was so unexpected, and I had never seen so big and spontaneous a reaction at a scientific meeting before.

Science was right to be concerned about my talk: The meeting triggered an avalanche of news coverage in advance of the paper. The journal itself carried the headline "Venter Wins Sequencing Race—Twice" and carried a quote from Collins, who called it "a remarkable milestone."[2] As *Time* magazine put it, "Having been turned down for government support because his method was thought to be unreliable, Venter used private money to beat federally funded scientists in achieving what his rivals acknowledged as a significant milestone."[3] In *The New York Times,* Nicholas Wade wrote: "As if to prove that the *Haemophilus* sequence was no fluke, Dr. Venter at the end of his lecture produced another rabbit from his hat, the sequence of a second free-living organism."[4] Fred Blattner, the head of the first genome project funded by the NIH (E. coli), called it "an incredible moment in history." This was most gratifying, since I admired Fred's work, and I now admired him even more for his graciousness. Wade went on to say, "Dr. Venter's achievement threatens to make him part of the scientific establishment with which he has long been at odds because of his liking for short-cut approaches to genome sequencing that other experts say are unlikely to work."

It was a wonderful moment for me, for Ham, and for the rest of the team. We all knew how much grief we had endured to come this far: the politics of the NIH, the cold-shouldering, and being ignored by the public genome community. While our achievement did help some academics forgive what they viewed as my previous crimes, their opposition and opprobrium had been child's play compared with what rained down on us from our "partners" in the EST sequencing effort. Behind the triumph, my already bad relationship with HGS and Haseltine deteriorated further.

The refusal of Haseltine and SmithKline Beecham to let me publish our EST data had made me so resented and loathed by some in the genome community that when it came to *Haemophilus* and our whole-genome sequencing efforts, I was determined to do things differently. I managed to find a loophole in the agreements between HGS and TIGR, and realized that I could exploit the fact that they were focused on single EST sequences and had never anticipated the assembly of entire genomes.

My aim was to prevent them from once again interfering with publication. HGS, recall, had six months (from the moment TIGR transferred the data) to select genes for commercial development, after which the rest could be published. To set this clock running for *Haemophilus,* I began to transfer raw sequence data to HGS before it had been reassembled. Over the course of four months twenty-five thousand bacterial sequences poured into the computers at HGS, creating puzzlement rather than curiosity. As we began to stitch the sequences together into the genome, and the significance of what we were doing became clear, the attitude changed from bafflement to outright hostility.

Part of the reason was that as HGS's competitors began churning out human ESTs at an ever increasing rate, Haseltine was mortified to discover that we were sequencing a mere bacterium. "I'm going to get you," he growled at one TIGR board gathering but then changed tack when SmithKline realized the sequence could have commercial value and help to develop new vaccines and antibiotics. The now familiar wrangle over the release of data started in earnest.

Haseltine began to claim that the commercialization clock would not start ticking until HGS received the complete genome sequence. And then, of course, it would invoke the clause that allowed it to keep the genome secret for another eighteen months because it was a single sequence. The last thing I wanted was for HGS to file a patent on the genome or slow down our publication and risk our coming in second in the race to sequence the first genome. This particular battle was not about the money, however, but control. Haseltine recognized that if my team sequenced the first genome in history, there would be a power shift, and I and TIGR would clearly survive without HGS. Haseltine threatened to file an injunction in court to block the publication of the genome and hired attorneys to do just that.

I knew that compromise could prove fatal to TIGR and to my career, even as I received new memos and new demands daily. I devoted more time and money to Steve Parker's team of lawyers at Arnold & Porter, and Parker himself was by now spending half of his time in my office or on the phone with me or the HGS attorneys. Haseltine upped the ante by bringing in an even bigger Washington attorney who had just stepped down as the counsel to the U.S. president. Now he planned to file an injunction as well as a patent application on the genome. But it soon became clear that to win an injunction HGS would have to show in court how my publication would actually harm its commerce—a tough proposition for a company with no interest in microbes.

Through the president's former attorney HGS sought a last-minute compromise: If I would provide the complete genome sequence before I sent the paper to *Science,* then HGS would give in. Feeling that I had won the right to publish, I agreed, and the data was transferred to HGS as we submitted the paper. I had not counted however, on the actions of Robert Millman, a patent attorney who had once worked in a biotech company that Eric Lander had helped set up. A peculiar figure, with his red hair tied into a ponytail, a beard, and a bizarre dress sense, Millman was to patent law what geeks are to computing and had a background in molecular biology. With his help HGS managed to file a patent before my paper was published, albeit at great expense. The application included twelve hundred pages containing the 1.8 million base pairs of the bacterium's genome. As would also become true for thousands of patent applications that HGS filed and that Millman would continue to file as the Celera patent attorney after he left HGS, their only real value was for the patent attorneys themselves. The only tangible consequence of this aggressive patent policy was the incredible resentment it stirred up among the scientific community.

The *Haemophilus* genome paper was published on July 28, 1995, in *Sci-*

ence[5] and included forty authors, with Ham and me as senior authors. It was featured on the cover and in a centerfold, which carried a detailed gene map, marked with colored bars on the organism's circular DNA: Green corresponded to genes involved in energy metabolism, yellow for copying and repairing DNA, and so on. Almost half had no color because their role was unknown. The article itself not only described what was in the genome but, importantly, what was missing. We had decoded the Rd laboratory strain, which is not infective in humans, and found that it lacked a complete cassette (set) of genes associated with infectivity. We discovered that some of its metabolic pathways were incomplete, and notably that the "TCA cycle" associated with cellular energy production lacked one-half of its enzymes. As a consequence, the species required high concentrations of the amino acid glutamate in order to grow. On seeing this detail, a prominent biochemist at Stanford said that we had clearly screwed up because everyone knows that each cell will have a complete TCA cycle. In fact, since pioneering this microbial sequence, we now know that every combination is possible, from cells that have no TCA cycle to those that are completely dependent on a TCA cycle for energy production.

We published a second paper in the same issue[6] of *Science* that described how *Haemophilus* could speed up its evolution by exchanging DNA with its peers, as if it were installing a software update to its genome. Ham found the key to this mechanism in a unique sequence of nine base pairs, 1,465 copies of which were scattered throughout the genetic code, in the middle of genes. Molecules on the surface of the bacterium bind to the sequence and transport the DNA into the cell. There was little variation, suggesting that it was too important to the bacterium to allow the sequence to change without causing harm. Strikingly, there was much less variation in this software update mechanism than in the software itself; it is as if the quantity of the new software, not quality, was most important for the survival of the bacterium.

One of the most exciting findings came from Richard Moxon's team at Oxford University. After studying a gene that codes for an enzyme that helps make molecules on the surface of the bacterium, called lipooligosaccharides, they had discovered why our bodies find it so hard to fight microbes. Later, Moxon recalled how "Derek Hood and I had identified more than twenty novel (hitherto unrecognized) genes in the pathway of lipopolysaccharide synthesis within a space of a few weeks, more progress in this short time than we and others had made in several years."

His team found a DNA sequence repeat in front of the gene that created errors when the gene was copied to daughter cells by an enzyme called DNA polymerase. The enzyme slipped on these repeats, and when we looked across the genome, we found that they were associated with a number of the genes responsible for producing cell surface molecules. This was a clever way for the bacterium to continually change its cell surface antigens so that new strains could keep one step ahead of the immune defenses of the body, a process one can see at work in the respiratory tract: As the body picks off the familiar strains, a different version of *Haemophilus* takes up permanent residence. We now know that similar mechanisms are built into the genetic code of many different human pathogens, which is one reason that we will never win the war against infectious disease, and the best we can do is stay one step ahead of bacterial evolution.

By then, of course, Richard had discovered to his cost that this work had a political dimension. The Dormy House meeting that he had organized, between April 23 and 26, had been a triumph. One attendee remarked: "Craig bounced up to the podium and described how the *Haemophilus influenzae* genome had been assembled—the impact was amazing and immediate—it was clear to all that microbiology was going to change, and of course it did." Not only did I show up and speak (it was unusual for me to attend four-day meetings in full) but I brought with me a CD with the *Haemophilus* sequence and that of the *Mycoplasma* genome as well. At the meeting, the scientists pored over our data for hours. "That's it," remarked one. "This is what this organism really is."[7] But Morgan himself did not turn up and seemed unaware of the meeting's significance, even though it had been sponsored by the Wellcome Trust. Richard was disappointed, since he regarded it as one of the most successful meetings of its kind (Wellcome called them Frontiers of Science meetings) and yet there had been little senior representation from the Trust, save John Stephenson, who had helped organize it and was blown away by how it eventually turned out. A report on the proceedings concluded that our approach was the way forward in genomics and was a significant factor in the later decision of the Trust to develop a program of bacterial genome sequencing at the Sanger.

While TIGR was now running short of money, endless possibilities beckoned. Richard wanted to apply for funding from the Wellcome so that his Oxford lab could work with TIGR to tackle the genome of the major cause of meningitis in children, *Neisseria meningitis.* He and I had one awkward meeting with Morgan in which it emerged that the Wellcome genome czar had not even read the *Science* papers. Still, the Wellcome's Infection and Immunity panel recommended it as a project of the highest priority, given the suffering, death, and disability caused by meningitis. Approval by the Trust usually follows as a formality. But this time there was a problem, a technicality to do with TIGR's not-for-profit status not having been ratified by the American authorities and concern that charitable money could end up benefiting HGS. Morgan vetoed the meningitis proposal on the grounds that there could be legal problems with the U.K. Charities Commission. I had even started sequencing the bacterium but had to halt progress at that point.

The *Haemophilus* paper soon became the most cited in biology. Lucy Shapiro, a Stanford University professor, described how her team stayed up all night poring over its details, thrilled by the first glimpse at the complete gene content of a living species. Hundreds of congratulatory e-mails arrived with such sentiments as "Now we understand what genomics is all about" and "This is the true start of the genomic era." Fred Sanger even sent me a nice handwritten note on the publication of the *Haemophilus* genome, saying he always believed my approach would work, but he never got the chance to test it because his colleagues all wanted their own piece of DNA.

Articles about the paper continued to appear. Our work was billed as a "colossal feat with enormous potential for twenty-first-century medicine." In *The New York Times,*[8] Nicholas Wade waxed poetic: "Life is a mystery, ineffable, unfathomable, the last thing on earth that might seem susceptible to exact description. Yet now, for the first time, a free-living organism has been precisely defined by the chemical identification of its complete genetic blueprint." He quoted George Church of Harvard and one of the leading intellectual voices in genomics: "It's a really beautiful story, because they made everyone wait until everything was done." Even Jim Watson declared it "a great moment in science." Had Watson read the *Science* paper to the very end? I wondered. There I speculated that "the methods described here have implications for sequencing the human genome" and the journal ran an accompanying news story[9] that highlighted a similar quote from me: "The success with the *H. influenzae* sequence has raised the ante worldwide for sequencing the human genome."

Not long after the *Haemophilus* paper appeared, we published in *Science,*

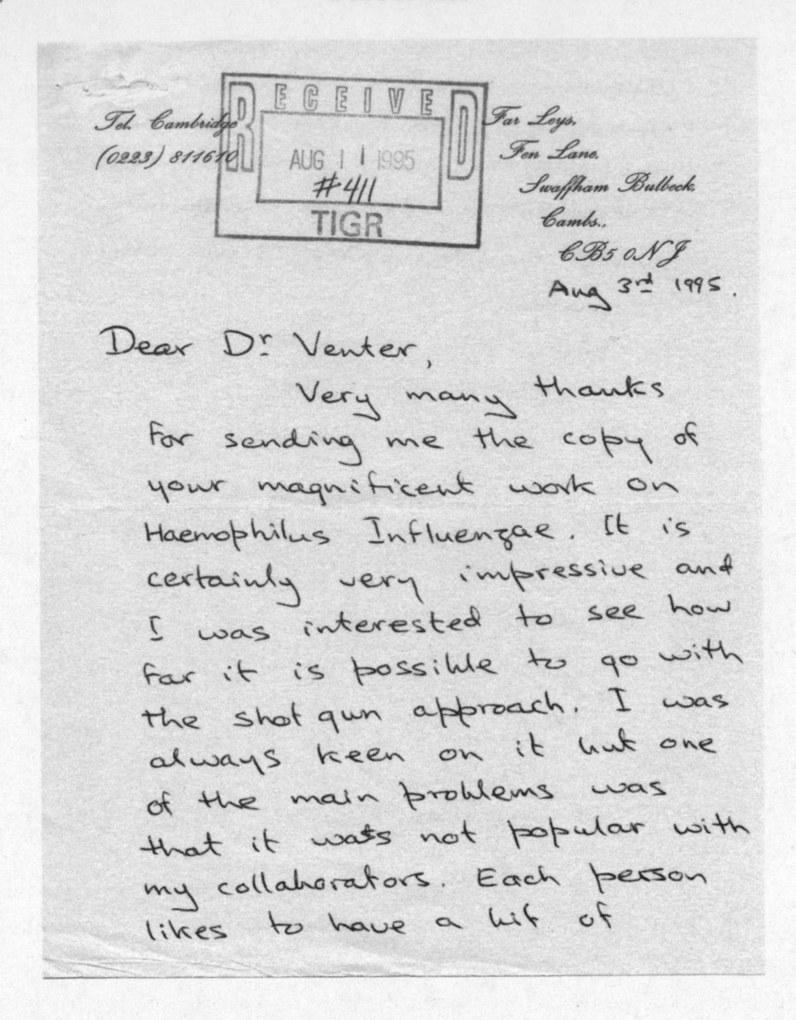
RECEIVED
AUG 11 1995
#411
TIGR

Tel. Cambridge
(0223) 811610

Far Leys,
Fen Lane,
Swaffham Bulbeck,
Cambs.,
CB5 0NJ
Aug 3rd 1995.

Dear Dr Venter,
Very many thanks for sending me the copy of your magnificent work on Haemophilus Influenzae. It is certainly very impressive and I was interested to see how far it is possible to go with the shot gun approach. I was always keen on it but one of the main problems was that it was not popular with my collaborators. Each person likes to have a bit of

as promised, the minimal genome of *Mycoplasma genitalium.*[10] In an editorial Andre Goffeau, the leader of the international effort to sequence the genome of yeast,[11] reminded its readers how for years it had been thought that the first genome to be sequenced in its entirety would be that of the bacterium *Escherichia coli,* "but to everyone's surprise" the race was won by an outsider, who had now gone on to sequence a second genome. He went on: "One of the most impressive features of the sequencing effort for the *M. genitalium* genome is its efficiency, a testament to the power of the TIGR sequencing and informatics facilities." Clyde Hutchison had sent us the *M. genitalium* DNA in January 1995, and we had submitted the manuscript August 11 of that year.

sequence they can call their
own. I suppose this is not
a problem now with all the
automation. It has certainly
come a long way since the
λ work.
with best wishes &
congratulations
Yours sincerely
Fred Sanger.

With the availability of the second genome of a free-living creature we were able to launch a new discipline, that of comparative genomics. Some of the reactions to that development were carried by *The Scientist*:[12] "I started reading the portion of the *Mycoplasma* paper on the comparison of the genomes and all of a sudden it dawned on me—Wow, this is going to be a whole new field of biology with enormous implications," said David Smith of the DOE. Though they had backed our efforts, even they were only just coming to terms with the ramifications of what we were doing. As was Elke Jordan, the deputy director of the NIH genome institute, first under Watson and then Collins: "I think we're going to take our baby steps on the microbial genomes and then—when the bigger, more complex genomes like yeast, *C. elegans* and *Drosophila* become available—we'll transfer some of the experiences to them." In the *Scientist* article, Ham summed it all up perfectly: "Craig built this whole thing over the doubts of virtually everybody else in the country. It seemed like everyone was expecting him to fail and fall on his face. But he's delivered more than anyone ever thought and quicker." I was just getting started.

My yearlong struggle to have the TIGR EST work published finally succeeded in September 1995 with a special 377-page Genome Directory issue of *Nature.*[13] The previous month saw a turning point in my battle for recognition when the editor of *Nature,* John Maddox, wrote an unusual editorial[14] that discussed the "irksome" conditions on the use of my EST data. It opened with the memorable line: " 'If you publish this Venter stuff,' said the distinctive voice on the telephone a few months ago, 'I can promise you that nobody in the US genome community will ever send you anything ever again.' " Maddox continued: "The source of the voice, who will recognize himself, is one of the most distinguished geneticists in the United States." A *Nature* editor later told me that the distinctive voice was, of course, that of Jim Watson.

Maddox had always had an eye for headlines as well as for science and decided to publish the paper despite the threats: "There are several good reasons (other than bravado) for publishing this material. The chief, as will be seen when the Genome Directory is distributed, is that the work described is excellent science with great intrinsic interest. The scale of the enterprise is also remarkable. Venter's group will be reporting that the total length of the ESTs now sequenced amounts to 5 million base-pairs, or some 0.15% of the human genome. . . . [T]here are more than 55,000 ESTs that correspond to authentic genes, of which only some 10,000 are at present logged in public databases."

The scientific community and the press applauded our efforts with front-page stories and headlines that marked them as the starting shot of the human genome race: "Gene Pioneer Opens His Databank";[15] "New Directory Is First Atlas of Ourselves";[16] "Rapid Gains Are Reported on Genome";[17] "Scientists Glimpse Gene's Division of Labor";[18] and "Study Details Progress in Massive Human-Gene Project."[19] As one observed, the directory marked a significant step toward learning what makes us human.[20] The *Nature* biological sciences editor, Nicholas Short, told *The New York Times* that the terms of access to the data "had been widely misrepresented and were now 'really rather liberal.' "

I made the cover of *Business Week*[21] and was profiled in *People* magazine.[22] *U.S. News & World Report* pointed out that despite the scoffing and carping of my critics, "Craig Venter had the last laugh."[23]

My success with the first two genomes brought in fans, collaborators, and money. The DOE Office of Science now gave us funding to sequence a number of other microorganisms. After much discussion we chose as our third genome project an unusual representative, *Methanococcus jannaschii,* which lives where a hot, mineral-rich liquid originating in the bowels of the Earth spews out of the seabed like smoke in what is called a hydrothermal vent. This organism was isolated in 1982 by the deep submersible *Alvin* out of the Woods Hole Oceanographic Institution in the boiling waters where one such "white smoker" lies 8,000 feet deep in the Pacific Ocean, one hundred miles off Cabo San Lucas, Mexico. It was like finding a little alien on Earth: This creature was tough and strange enough to thrive on other planets.

At that depth there are more than 245 atmospheres—equivalent to the crushing pressure of 3,700 pounds per square inch. The temperature at the heart of the white smoker was more than 329 degrees Celsius (752 degrees Fahrenheit), while the surrounding water was only 2 degrees Celsius (35.6 degrees Fahrenheit). *Methanococcus* likes to bask somewhere in between, in waters of about 85 degrees Celsius (185 degrees Fahrenheit). Living off mineral substances, not organic matter, *Methanococcus jannaschii* uses carbon dioxide as its source of carbon, and hydrogen for energy, belching out methane as a by-product of its metabolism.

Methanococcus jannaschii was thought to be from the third branch of life, an idea put forward by Carl Woese of the University of Illinois, Urbana. I liked Carl very much and found him to be a tremendous thinker. He had suggested that all living species are divided into three general classes: eukarya, creatures such as humans and yeast whose cells contain a control center housed in a compartment called a nucleus; bacteria; and archaea, microorganisms that share some features of the other groups but are nonetheless distinct and have no nucleus to contain their genomes. Traditionally, bacteria and archaea were lumped together into a single kingdom called the prokaryotes, and Carl had been denounced and ridiculed for his attempt to separate them.

Carl, who took such attacks even more personally than I did, had become somewhat of a recluse but had agreed to collaborate with me. As the *Methanococcus jannaschii* genome sequencing proceeded, Carl became more and more excited, which I could certainly understand given how much he had at stake with the outcome. He was anxious to get his hands on the fragment data, while I urged him to wait until we had the whole chromosome assembled. Fortunately, he did not have to wait long. Only a few genes had been characterized from high-temperature organisms, so we were all very curious to see what distinguished them. At the temperatures endured by this organism, the structure of most proteins denatures (falls apart), a process that often occurs at around 50 to 60 degrees C, so I expected to see proteins that had been substantially transformed by the evolutionary process to cope with higher temperatures. One change I particularly expected was a greater amount of the amino acid cystine. Cystine can lock the complex three-dimensional structure of a protein into place by forming strong chemical bonds with other cystine amino acids. But we were stunned to see no obvious differences in overall amino acid composition. We often found that a *Methanococcus* protein that was very similar to ones from other species had only a few specific differences, and these could not readily account for the temperature stability. The random mutations that drive evolution apparently only have to fine-tune the structure of a protein to prevent it from being denatured by high temperatures. But those similarities did not mean that this bug was otherwise familiar. Only 44 percent of the proteins of the creature, the first of the archaea to be studied, resembled anything that had been characterized before. Some of its genes, including those associated with basic energy metabolism, resembled those from the bacterial branch of life. However, in stark contrast, many of the genes associated with the copying of the chromosome, information processing, and gene replication had their best matches in the eukaryotes, including human and yeast genes. This was a remarkable vindication of Woese's theory.

By the time the paper on *Methanococcus* was about to appear in *Science*, NASA had published some tentative evidence of microbial life on Mars. This primed the media interest, and we had a packed press conference at the National Press Club in Washington, D.C. Carl Woese was ill and could not travel, so I arranged for him to be there by video conference, since he had to be center stage. I also wanted to honor the original expedition team that had discovered the organism and cultured it in the laboratory, so I brought down from the Woods Hole Oceanographic Institution Holger Jannasch, the expedition leader after whom the species had been named, and Dudley Foster, who was the pilot of the *Alvin.* The Department of Energy sent the deputy secretary. The TIGR genome team, including Ham and me, as well as the *Science* editor completed the group that sat before the reporters and cameras to discuss the publication of our paper.[24]

Our genome study appeared on the front page of every major paper in America and much of the rest of the world: "Microorganisms Confirmed as 3rd Branch of Life," said *USA Today;*[25] "Evolution of the Species, Microscopic Life Unlike Any Other" was the headline in the *Christian Science Monitor.*[26] *The Economist* settled on "Hot Stuff,"[27] while *Popular Mechanics* announced "Alien Life on Earth,"[28] a theme also pursued by the *San Jose Mercury News* with "Something Out of Science Fiction."[29] An earlier conversation with my mother gave me a glimpse of one theme that would emerge again and again: When I tried to explain that our findings proved that the third branch of life was real, she asked whether it was animal, vegetable, or mineral. I gave up trying to explain it to her out of frustration, but on the night of the announcement, NBC news anchor Tom Brokaw posed the same question. *The Washington Post* returned to the theme: "Not animal, vegetable, or bacterial. Forget Martian organisms—the real ruckus is over a genetic code of another life form on Earth."[30]

We now had the first genome from two out of the three branches of life, as well as the first three genomes published in history. (The first eukaryote genome, that of the brewer's yeast, was announced prior to our publication of the *Methanococcus* genome but was actually published in *Nature* after our paper appeared.) In the background the EST work continued apace: We had worked with Brazilian scientists to advance the research on schistosomiasis, also known as bilharzia, a disease caused by parasitic flatworms that leads to chronic ill health in developing countries. We had studied how genes use changes in nerve cells, uncovered genes associated with Alzheimer's disease, and, as I had predicted in the original EST paper in 1991, used ESTs to map genes to the human genome.

My interests never strayed very far from figuring out ways to speed up sequencing the human genome. Hundreds of millions to billions of dollars were being spent by NIH to "map" the genome so that serious sequencing could eventually begin. As with the E. coli genome, mapping meant breaking sections of the human genome into more manageable (now 100,000-base-pair)-size clones called bacterial artificial chromosomes, or BACs. (Daniel

Cohen's feat of growing bigger chunks of sequence in yeast, so-called Mega-YACs, had run into problems because the pieces would break up and re-arrange.) The worldwide genome community was then going to line up all the BACs in the correct order and begin to sequence them. I believed that if they were going to insist on using BACs, years and a great deal of money could be saved by sequencing 500 to 600 base pairs of genetic code from each end of hundreds of thousands of BAC clones to create a large database, just as we had done with the lambda clones and the *Haemophilus* genome. When any group randomly selected a BAC clone and sequenced all 100,000 base pairs of its code, the next step would be simply to compare the sequence against the BAC-end data set. Any overlaps would be immediately obvious, and efforts could be focused on sequencing clones with the least overlap, so a map and sequence could be generated at the same time. Ham really liked the idea, and we polished it up. Lee Hood heard me discuss our proposal and also became an enthusiastic supporter, and the three of us wound up publishing the method in *Nature.* As with ESTs, the BAC-end sequencing method has become standard.

At last, our science was becoming respectable, and even our critics were coming around. At a conference in Hilton Head, South Carolina, the senior Wellcome Trust official, John Stephenson, remarked on how, only two years earlier, "everyone was very skeptical that Craig Venter could do what he did." Now the world of microbial genomics had been "changed overnight." The sentiment was echoed by Anne Ginsberg of the National Institute of Allergy and Infectious Diseases, who spoke of how the *Haemophilus* genome had changed the landscape of research.

Money from NIH and DOE flowed into TIGR, and the genomes flowed out apace. *Methanococcus* was followed by *Helicobacter pylori,* a bacterium of the stomach that infects more than half of the world's population and is linked to gastritis, ulcers, and cancer. Then came a second archaea, *Archaeoglobus fulgidus,* followed by *Borrelia burgdorferi,* the organism—called a spirochete—responsible for Lyme disease, the most common tick-borne disease in the United States. (Spirochetes are so named because of their spiral shapes and their ability to corkscrew their way through tissues.) Soon we had sequenced a second spirochete, one that causes syphilis, as well as the first malaria chromosome.[31]

We were hot, the science was hot, and we were beginning to attract even more money for our work. Despite this surge of interest, however, the funding was far from sufficient to fulfill my ambitions. One aspect of TIGR still rankled most scientists and funding agencies: our link to HGS and Bill Haseltine. I had nothing but sympathy for the view that funding TIGR would also benefit Haseltine and HGS. I still felt trapped by the constant legal hassles as HGS blindly went on attempting to patent the data flooding from my institute, whether they had a use for it or not. This was crazy; something had to give.

10. INSTITUTIONAL DIVORCE

It is a cursed evil to any man to become as absorbed in any subject as I am in mine.

—Charles Darwin

The beginning of the end had actually come two years earlier, on the morning of Wednesday, July 25, 1995. I was still at home, and the voice on my telephone was both tense and resigned as it informed me that Wally Steinberg had died in his sleep at age sixty-one from an apparent heart attack. I greeted the news with disbelief, and I also felt a knot of pain as I recalled a similar phone conversation thirteen years earlier when my mother told me how my father's life had ended in the same way. Even though Wally and I had often clashed, as I had with my father, I would miss him. Wally had backed my work when many others seemed pitted against me.

Wally had always been larger than life. He lived a Great Gatsby existence in a sumptuous New Jersey mansion with lavish entertainment and endless tennis and dancing. More than most, Wally wanted to live forever, and I am sure he believed that the various health-care companies he owned would help give him a better shot at longevity than most. Even though his motive seemed to be pure profit, he was more aware of the vital potential of genomics than most molecular biologists.

Wally's early death reinforced the main lesson I had taken from my Vietnam experience: Life must be lived to the full, something Wally had done for as long as he could. His death reminded me of how I wanted more from my own life than endless battles with my supposed backers. From day one my relationship with HGS had failed to fulfill my vision of a cooperative venture out of which basic science discoveries could rapidly make their way into the clinic. In reality, it was all about greed and power, not health. I knew that Wally's death was going to mark the end of the relationship between TIGR and HGS, since that strained relationship had been held together in no small part by his desire to be linked to me and my research. With his passing, all that was left was Haseltine, who wanted to do away with both.

My Fickle Heart

I carry higher-risk versions of various genes linked with cardiac disease, notably GNB3, which is linked to high blood pressure, obesity, insulin resistance, and a kind of enlargement of the heart, and MMP3, which is linked to heart attacks.

GNB3 has a natural variation that we have known for some time to be associated with high blood pressure, particularly in people with left ventricular hypertrophy, a thickening of the heart muscle that can be life threatening. Variations in the gene also affect how patients respond to a frequently used medication for high blood pressure, e.g., the diuretic hydrochlorothiazide (HCTZ). Animal and laboratory cell studies have given some insights into why the gene is harmful: It raises the activation of G protein, a messenger system within cells, which leads to a greater production of fat cells. That discovery led German researchers to find that people who inherit two copies of the gene, one from each parent, have a high risk of obesity, which itself is a risk factor for many common diseases.

MMP3 is an enzyme known as a matrix metalloproteinase (MMP), which is important during an organism's development and during wound healing. This family of enzymes plays a role in epithelial tissues, which are specialized for managing the flow of substances into and out of the body and for protecting underlying organs. The matrix metalloproteinases are digestive enzymes that normally act as bulldozers to clear the way for building new organ structures or repairing old ones. The identity of the letter at one position in MMP3, also called stromelysin-1, influences the rate of production of an enzyme that degrades the extracellular matrix that makes up the walls of arteries. In this way the MMP3 protein plays a role in regulating the elasticity and thickness of blood vessels. People like me, who have the low-expression version of the gene, are slightly more prone to atherosclerosis, the narrowing of arteries associated with the accumulation of plaque in the arterial walls.

To be fair, Haseltine and company had different goals, which arose from regarding patients as future customers and sources of revenue. That is acceptable as long as a new treatment helps, but it becomes objectionable when industry provides false hope or deceives patients with smoke and mirrors.

Although I knew that the genomics revolution was going to lay the foundations for the future of medicine, I also felt that its short-term potential was being enormously oversold by the biotechnology industry as a whole.

My unease had come to a head a few weeks before Wally's death when I held a meeting at TIGR with a few of my trustees, Wally, and some of his advisors. At the top of the agenda was my decision to sell any of my HGS stock that I had not already given to the TIGR foundation. I knew this would be interpreted as a vote of no confidence in Wally's commercial venture, and Wally, understandably, took it personally and was extremely upset. This was more than merely a thumbs down for HGS, it was a vote of no confidence in him. He liked being rich, and I think he wanted to make me rich, too. Why was I selling my HGS stock when it would be worth so much in the future?

But I could not see a future for my relationship with HGS when it was so clouded by the problems of the present. This meeting took place in the middle of the battle to publish the first microbial genome paper, which marked a new low in the troubled relationship between HGS and TIGR. I explained to Wally how I had had to carry the burden for Haseltine's antics, which had alienated so many in the academic community. Because of Haseltine I now had to sell and move on. I upset Wally but knew that he would have done the same had he been in my position. Unfortunately, that was the last time I saw him alive.

Wally would have liked his three-column obituary in *The New York Times,*[1] which appeared the day after the *Haemophilus* genome paper was published in *Science.* The article memorialized his accomplishments, from the Reach toothbrush to the biotech companies he had started, and when it came to me, the obit declared that "the gamble appears to have paid off. Dr. Venter has now deciphered the first full set of genes from a living organism." Wally would also have adored his funeral. Opulent excess was the order of the day, and I felt as if I had wandered into the final scene of a movie that combined elements not only of *Gatsby* but also of *The Godfather.*

Wally was hardly settled in his grave before Haseltine, now free of his reasonable oversight, began to exert his new power and independence. I had told Haseltine that the time had come for HGS and TIGR to part company even though HGS still had $50 million left to pay us—a message I sent directly and through the lawyers. But Haseltine did not want to dissolve our relationship, for however strained it had become, it was part and parcel of the HGS agreement with SmithKline Beecham. HGS would never agree to a split because it would have to pay back tens of millions to SKB if it lost TIGR.

Skipping a Heartbeat

Many factors, genetic and environmental, contribute to an irregular heartbeat and other maladies that may lead to sudden cardiac death, a condition that affects about 300,000 Americans each year. It is well worth taking a look at my genome because if I find something badly amiss, current medicine does offer me a chance to fix it with beta-blockers and other drugs that regulate heart rhythm, or even by implanting an automatic defibrillator.

Using a new strategy to survey the entire human genome and identify gene variants that contribute to complex diseases, Dan Arking and his colleagues at Johns Hopkins, together with scientists from Munich, Germany, and the Framingham Heart Study in the United States, found a gene that may predispose some people to abnormal heart rhythms that lead to sudden cardiac death.

Called NOS1AP (nitric oxide synthase 1 [neuronal] adaptor protein), the gene was missed by more traditional gene-hunting approaches and appears to influence the so-called QT interval length. QT interval measures the period of time it takes the heart to recover from the ventricular beat when the two bottom chambers pump. Corresponding to the *lub* part of the *lub-dub* of the heartbeat, an individual's QT interval should remain constant.

But there is evidence that having a QT interval that is unusually long or short is a risk factor for sudden cardiac death. The interval, which can be measured with an electrocardiogram, is the length of time from electrical stimulation of the heart's pumping chambers to recharging to the next heartbeat. But in more than one-third of all cases, sudden cardiac death is the first and last sign that something is wrong.

The study revealed one particular single-letter spelling mistake in the code (SNP) correlated with QT interval. That SNP was found in the NOS1AP gene, which makes a protein that is a regulator of an enzyme (neuronal nitric oxide synthase) that has been studied for its function in nerve cells and was not previously suspected to have a cardiac role. The research team found, however, that the NOS1AP gene is turned on in the left ventricle of the human heart, in the proper place and time, underlining its role in the QT interval.

Further studies revealed that approximately 60 percent of people of European descent may carry at least one copy of this SNP in the NOS1AP gene. With two copies of one variant, one ends up with a shorter QT, while two copies of the other variant mean a longer QT interval. I have a blend consistent with the least variation in the QT interval. This is reassuring, but, as ever, this particular SNP is not the whole story, being responsible for up to 1.5 percent of the difference in QT interval.

I decided to appeal to George Poste, who by then was the director of research for SKB and was directly responsible for the HGS–SKB relationship. We had often met on the SKB scientific advisory board, of which I was a member, and it was clear that George and the SKB management had grown annoyed with the problems between TIGR and HGS. SKB's primary interest was access to TIGR's data, both human and microbial, which would be useful for developing antibiotics. George and I accordingly hatched a plan. George would waive the tens of millions of dollars that HGS owed him if he obtained a major concession from HGS on using the data he already had to test for disease genes and for other diagnostics, a right that SKB did not currently have. George promised that if he and SKB were granted those rights, he would not veto a divorce between HGS and TIGR. Haseltine seemed to accept the plan, and in discussions with the HGS president I was told that if I was willing to forfeit all the money owed to TIGR over the next six years, then HGS would end our agreement and give us our freedom.

What I had craved for so long was now within my grasp. My critics had dwelled on how I was only in research for the money. They got it backward: I was interested in money only to have the freedom to do my research. The funding from HGS came with too many unexpected strings attached; it was restrictive and damaging my reputation. But would the TIGR board support my turning my back on the millions of dollars that HGS owed us so I could pursue an independent existence for TIGR? I telephoned key members, and every one of them was nervous—very nervous. TIGR had barely enough money to last one year. In truth, I was very anxious, too, because my decision involved the one hundred scientists and colleagues who worked for me. I had many sleepless nights.

But there was an optimistic logic to my proposal. Our link with HGS had discouraged public funding from government agencies because the reviewing scientists did not want any money to go to TIGR that would benefit HGS. We had also been denied tax-exempt status, a critical element for a research institute. My hope was that once the link to HGS was cut, the research funds would follow; after discussions with the Internal Revenue Service, it was also likely that we would be granted tax-exempt status. I had put into the TIGR endowment approximately $30 million in HGS stock. TIGR could receive the full amount, not half, but only if it had tax-exempt, not-for-profit status. I called a special board meeting to vote on the issue, and despite the serious misgivings, its members made it clear that they were there to back me, betting on my intuition and me. It was a big bet, one that could either make or break my career and my organization. If the grant money and the tax-exempt status did not come through, TIGR would be forced to close down, and we would all be out on the street. I realized that I needed more time to think about one of the biggest decisions of my life. I needed to get away, and had the perfect opportunity with a major sailing challenge, a race that would test my abilities across the Atlantic Ocean.

Ever since I had sailed to Bermuda and back, I knew that I wanted a bigger, faster boat that could handle the extremes of the ocean. When I began to sell my HGS stock, I began to plan for the perfect ocean-sailing craft. For years I pored over the sailing magazines, and based on my experience, I began to picture a 55- to 65-foot sloop that I could sail single-handed, that was relatively light and fast, and that used water ballast for stability. I began to work with an Annapolis yacht designer named Rob Ladd who specialized in light cruising boats, and we came up with a 65-footer—at the limit of what could be handled solo—with water ballast and a small pilothouse that seemed like the perfect compromise between living space and performance, styling and cost. All that I needed now was a builder to give me an estimate of how much my dream would cost. After much research I made an appointment to meet Howdy Bailey. With the captain of the 65-foot steel North Sea trawler that I had purchased through TIGR to begin marine research, and an Annapolis yacht broker who was interested in my design as the basis of a production boat, I flew to Bailey's yard in Norfolk, Virginia. We examined the drawings and then toured his yard. We discussed aluminum. We discussed welds. We discussed the construction time (up to two years). Finally came the cost: around $2.5 million. I almost fell over and told Bailey that I needed to think about it for a while.

By then my dream had a rival. In *Yachting* magazine I had been following one of the most beautiful yachts I had ever seen, pictured in one advertisement after another. On the flight to see Bailey in Norfolk I noticed in the latest issue that this stunning 82-foot yacht—now in Fort Lauderdale, Florida—had just had a major price reduction: The sleek sloop was for sale for $1.5 million. She had more than 55 tons displacement because of the 55,000 pounds of lead in the keel alone, and she had been designed by German Frers, the Argentinean naval architect who was famous for his fast hulls, including some America's Cup boats. The very next day I flew to Florida to take a look.

When I saw the low freeboard sleek blue hull with the teak deck and low-cut deckhouse, it was love at first sight. As I went down the companionway into the warm cherrywood cabin, I was truly smitten. The boat was laden with features including bow thrusters, hydraulic winches, generators, water makers, and so on, a far cry from the simple, fabulously expensive yacht that I was designing. And there was a raw visceral beauty to her that would make any sailor drool. The broker described how she had been built three years earlier by the Palmer Johnson yard in Wisconsin for Gary Comer, former Olympic star-class sailor and the founder of Land's End, the catalog and Internet specialty clothier known for its conservatively styled apparel. Comer used her to circumnavigate the globe and, once the feat had been completed, put the boat up for sale. I decided to make an offer that same day for under $1.25 million. A few days later it was accepted, and I knew that a major new chapter in my life was about to begin.

The boat was called *Turmoil,* which I thought was a strange choice for such a beauty, and I was relieved when Comer wanted to retain the name for his new power yacht. I spent a great deal of time searching for a new name that would sum up the magical effect exerted on me by being on the water with just the power of the wind to move me. I settled on *Sorcerer,* as a nod toward how science seems as mysterious as magic to the average person, toward my son's middle name (Emrys, which is Welsh for Merlin), and toward the deep links between science and magic—how astronomy had its origins in astrology, chemistry in alchemy, and so on.

During our first year together *Sorcerer* did not exert quite the enchanting hold on me that I had expected. Most of her systems failed; they had apparently been held together with any fix that would get the boat safely back to Florida. While getting things shipshape was expensive in terms of time and inconvenience, as I repaired each fault I began to know every nook and cranny of the boat. I had by then sailed to New England and down to the Caribbean, but I felt that such a megayacht should not sit in a harbor and be used for day sailing. It had global potential, and had already successfully made one circumnavigation. I wanted to step up to transocean sailing, but I needed a good excuse to leave research for a while. Then I saw a posting for the Great Ocean

Race, sponsored by the New York Yacht Club, a repeat of one that was last run in 1905 when the transatlantic sailing record was set by the 185-foot schooner *Atlantic.*

The race, which was scheduled to start on May 17, 1997, was three thousand miles from New York to Falmouth, England, across the stormy North Atlantic. The challenge both frightened and excited me, but the minimum size for the race was 85 feet, and *Sorcerer* was only 82. I looked into various ways to extend the hull, until I noticed that the rules said the length requirement included all extrusions, including bowsprits, and so on. Richard Von Doenhoff, the race coordinator, historian, and official measurer, measured *Sorcerer* in Annapolis from the tip of her flagpole, which extended from the transom, to the tip of her bow rail. *Sorcerer* was 85 feet 1 inch overall, which would be a disadvantage when it came to the rating handicap. That was not a concern, though, because I was now in the race and, in any case, I had no illusions about winning against the 130- to 212-foot-long vessels with experienced crews that had also entered.

The Washington Post ran a story about this plucky entry from the only "local" crew,[2] and I began to receive letters from people who were inspired by what I was doing and wanted to join my crew. Among them was David Kiernan, a Washington attorney at Williams & Connolly who was also an M.D. A former surgeon could be a valuable addition to the crew, and after I arranged a dinner to meet David, I knew I could survive weeks at sea with him. I now had a crew of eleven, including my seventy-plus-year-old uncle, Bud Hurlow.

Thanks to my discussions with potential industrial partners for TIGR, I was able to find some money to cover the $20,000 I needed for a new spinnaker. By then the company Amersham, which had bought a DNA sequencing concern, knew that my relationship with HGS was troubled. I enjoyed meeting its new CEO, Ron Long, who was keen that we join up to harness microbial genomics to develop new antibiotics, which were desperately needed, given the rise of resistance to conventional drugs. Ron agreed to cover the expense of the sail—which bore the image of a sorcerer, complete with beard and pointed hat—and we carried a large banner announcing Amersham's support.

The race festivities began with drinking and partying in New York at the yacht club and on the waterfront. Some of the owners had hired elite crews to captain their yachts across the ocean, and I was stunned to learn that a few of them were not even going to race themselves but were going to fly to England to attend the party after the finish. Out of twenty entries there was only one other "owner-captain" in the race, a 130-footer. And the smallest boat after *Sorcerer* was a 100-foot British entry captained by Sir Robin Knox-Johnston, who was the youngest person to have sailed around the world solo. I had read his books and admired what he had done but found a somewhat arrogant man in the wake of these inspiring feats. He told us all how he was going to win because he knew the ocean better than the rest of us. This would be his twentieth transatlantic crossing, after all.

When our handicap ratings were posted, I was downcast to find *Sorcerer* in the middle of the fleet, tied with boats 30 to 50 feet longer. I even had to give time away to Sir Robin. Though discouraged, I reminded myself that I was there for the adventure (not that I had given up on winning). Sir Robin had announced he was going to go low across the Atlantic to avoid the worst of the weather and pick up what he was certain would be the best winds. I was puzzled by this; I had done extensive research, and my sources told me to go as far north as possible and into the gales. Perhaps Sir Robin knew something I did not.

But I also had some technology at my disposal. This race was testing a new safety feature that required all boats to report their positions automatically every six hours via satellite. We wrote a computer algorithm on *Sorcerer* so that we could take the updated information about the speed and position of each boat and calculate, based on their ratings, how they were really doing.

The race started just outside of New York Harbor in dreary weather but with good winds. Then came the first storm, a full North Atlantic gale along with seas that topped 20 feet. I found that by steering *Sorcerer* like a giant surfboard I could surf down the ever increasing waves at speeds up to 21 knots.

We had to change the person at the helm every twenty minutes to keep the concentration level high enough to avoid making an error that could prove tragic in such conditions. We had too much sail up for a gale, and for a few terrifying minutes found ourselves in a breach lying on our side, spinnaker in the water, at the mercy of the big waves. There were many other hair-raising incidents. Drew Donald, an Annapolis sailmaker, went up the mast in the boatswain chair while we were surfing the big seas at 18 knots to make a repair so that we did not have to drop the sail and slow down. My brand-new spinnaker exploded into hundreds of pieces. And in gale-force winds using the backup spinnaker (which resembled a giant American flag), the spinnaker sheet wrapped around the main two-foot-diameter winch drum, snapped, and was pulled through at such high speed that it melted the line cover onto the drum, making a terrible noise. Three different gales carried us over the first two thousand miles, during which we moved up on some boats while others dropped out due to the storms. We had to save every last drop of water, and one of my crew, a TIGR sequencing technician named Cheryl (our only female crewmember), offered to shower with her husband despite teasing from the crew.

With one thousand miles to go, a major low developed near England. We changed from sailing downwind to sailing right into the wind, which was blowing right out of the British Isles. *Sorcerer* began to earn her penalty rating, because she could sail upwind better than most of the larger yachts. By the end, constant headwinds had forced a third of the original fleet to pull out.

When we finally crossed the line off The Lizard just before 1:00 P.M., fifteen days and a few hours after starting, *Sorcerer*'s rigging was damaged, and we were spent. I knew that we had won, but by then we were so exhausted that we had only enough energy to drop anchor and sleep. Three days later Sir Robin Knox-Johnson motored into the harbor in *Sapphire,* one of the last to announce her retirement. Pressing into strong to gale-force headwinds for much of the crossing was not what she was designed for. A few hours later Sir Robin walked up to me, stiffly offered congratulations, and then turned his back on the festivities. According to the local media, our win was considered remarkable.

The days after our triumph were memorable for other reasons. A special luncheon was held in the stateroom of *HMS Victory,* the flagship of Horatio Nelson, the greatest British naval hero. He had died on board in 1805 during his crushing defeat of the French and Spanish fleets at Trafalgar, after being hit by a musket bullet from a sniper on the French ship *Redoutable.* After the battle, the *Victory* was towed to Gibraltar with Nelson's body preserved in a barrel of brandy. As part of the return trip and as a homage to this seafaring legend, we set sail for Gibraltar, only to encounter awesome seas off the Bay of Biscayne, including a wave that stood three times as tall as the *Sorcerer.* For what seemed like an age, the wave submerged us as it broke over our heads. As we entered the Straits, Africa was close on our starboard and Spain to port. I tried to imagine how Nelson sailed through the Straits to do battle in the days before global positioning satellites and engines. As we entered

Gibraltar, the historical and military importance of the "Rock" was clear. I felt more alive than I had in a very long time, as though I could take on the world—including HGS.

On my return to TIGR I announced to the board that I was going to terminate our relationship with HGS. The members spontaneously passed a unanimous resolution to support me. In *The New York Times,* Nicholas Wade reported that "a long-expected divorce between the oddest couple in genome sequencing . . . took place Friday."[3] *The Washington Post* warmed to the odd couple theme, depicting me as a "surf bum-turned-biochemist who prefers khakis to a suit and tie and drives himself to work from his modest home in Potomac," and Haseltine as a fifty-two-year-old biophysicist "so impeccably dressed that he looks like he stepped off the pages of GQ magazine. Each morning, he is chauffeured to work from his swanky Georgetown home."[4]

Despite having a chauffeur Haseltine did not make it to the formal signing of the termination agreement in my office at TIGR. We were now free, and although we had no financial support, our data were our own, and I could decide independently what to do with them. My first decision was to put every last letter of genetic code I had collected into GenBank, the public database. It was the largest single deposit ever made, comprising 40 million base pairs' worth of data, including sequences for twenty thousand new bacterial genes from eleven species. Several editorials appeared in response. One, entitled "$38 Million of Altruism," concluded with a heart-warming payoff: "[G]iving up millions of dollars in order to serve science and humanity is not something witnessed every day, and deserves high praise."[5] In light of this new data, scientists around the world began changing their experiments, according to David Lipman, director of the National Center for Biotechnology Information at the NIH.[6] "This is the best news we . . . have had in years," said the Stanford University biologist Lucy Shapiro. "It allows us to get at the genes used by bacteria to invoke virulence."[7] But I needed more than praise. Now that I had my freedom, I had to secure enough funds for my staff of 140 people. That meant tapping new sources of funding if TIGR was to survive.

11. SEQUENCING THE HUMAN

What do you say if you are trudging towards the top of an unclaimed mountain and you see another group of climbers scrambling up a parallel path? When doing science, you can suggest working together—cooperation seems so much more creative than competition—but in the DNA case that did not seem possible.

—Maurice Wilkins, *The Third Man of the Double Helix*[1]

My prospects improved no end after my research institute struck out on its own. Now that I had walked away from Haseltine and HGS, various companies lined up to engage in meaningful discussions with TIGR. Chiron wanted my team to tackle the genome of the major cause of meningitis in children, *Neisseria meningitis,* for vaccine development. In lieu of a fee I agreed that TIGR would share the risk and take royalties on the vaccine if it worked. (Gratifyingly, not one but two vaccines would go on to enter trials.) Corning, Becton Dickinson, Amersham, and Applied Biosystems also joined the line.

On my return to TIGR after the race, I discovered that Steve Lombardi of Applied Biosystems had called several times. When I eventually got back to him, Steve was his usual highly exuberant self, only more so. The fall genome conference in Hilton Head was coming up, and he and Mike Hunkapiller wanted to bring along Tony White to meet me. White was the new head of PerkinElmer, which was now the parent company of Applied Biosystems. Fine, I said. But I was not enthusiastic; ever since ABI had been bought by PerkinElmer I had heard nothing good about White. Steve himself had said that both he and Mike were close to quitting. And by then I found myself more and more drawn to Ron Long of Amersham, which wanted to put up $30 million to start a joint program to develop new antibiotics.

I did not give Steve's calls another thought until September, before the conference session in the Hyatt on human genome sequencing. The Hyatt's grand ballroom was dotted with large round tables where the audience would be served dinner before we discussed the status of the human genome. (It was, frankly, pitiful, and completion of the human genome looked to be decades away at best.) Steve had rushed up and excitedly told me that White was on his boat in the harbor but was about to join us. My first thought was that this was going to be a nuisance.

The Applied Biosystems group sat at a single table, and as I approached, there was only one person I did not recognize: a short, rotund man in a polo shirt, slacks, and loafers with no socks and a highball glass in his hand, sitting in a classic fat cat pose. I assumed that this was none other than Tony White, and knew it to be so when everyone but Mike was duly deferential to the great man in their midst, a Cuban American who spoke with a southern drawl. I immediately wanted to turn and walk away, but by then I had been seen. The toadies jumped up to be the first to introduce me to Tony. I said a polite hello before I set about the important business at the meeting of ensuring that TIGR would carve out a significant role in the human genome project.

My Waistline and Diabetes

Adult diabetes is now reaching epidemic proportions in the West as a result of the rise of obesity. Although the expanding waistline and diminishing activity of the population are key factors, another is genetics. Does my genome have anything to say about my own risk? Analysis reveals that I do have something to worry about. Two genes—ENPP1 and CAPN10—have been linked with susceptibility to diabetes, and it turns out that I have a variant of the former, called K121Q, that has been associated with earlier onset of type 2 diabetes and heart attacks. In the case of CAPN10 I lack a variant distinguished by single-letter spelling mistakes—SNPs—linked with type 2 diabetes in studies of Mexican American and Finnish populations. But our understanding of genetic influences is far from clear, and yet to be determined is how the apparently contradictory results from these two genes influence my overall type 2 diabetes risk. Nor are genes the whole story, for activity and obesity play a huge role in the development of adult diabetes. So far I have not succumbed to a disease that can have many serious complications, such as blindness, impotence, and amputations.

By that time the NIH had announced it was going to fund test centers for human genome sequencing and I had made certain that TIGR applied. Because of the visceral antipathy toward me from some in the genome community, I appointed Mark Adams the principal investigator on the grant. I was pleased when TIGR and Mark got an award as one of the test centers, but as I had now come to expect, the funding was only token despite strong scientific reviews and despite our having completed the first genomes of living species. Francis Collins wanted to fund only the usual suspects—Washington University, MIT, and Baylor—and the rest was down to politics, spin, and window dressing.

In December 1997 the challenges that Collins faced had become clear at a meeting of the leading scientists on the project held in Bethesda. Maynard Olson of the University of Washington, a genome purist, had estimated that to do the job properly might cost as much as $20 for each base pair, and claims and counterclaims about the true expense of reading sequences generated much acrimony. Then followed a wave of fear that estimates claiming it could be done cheaply might undercut funding from Congress. Although I was not present at this gathering, I had seen the spin and sugar coating the year before in Bermuda at the first "international strategy meeting" of what I liked to call the "liars' club." They were at one another's throats so fiercely that it was "the low point of the Human Genome Project," according to one of the participants.[2]

I had tracked the progress on Mark's program and it was clear that the path to the sequence taken by NIH Genome Institute remained slow, painful, and expensive. Each mapping lab was running into the same problems and surprises as it created libraries for each BAC clone, and then sequenced thousands of clones from it. Attempting to close every tiny sequence gap, most of which consisted of meaningless sequence repeats, drove the cost up tenfold, since repeats constitute about one-third of the genome. After the pilot studies were completed, the plan was to give a few of the centers more money to scale up the human genome sequencing. Mark and I made it very clear that our intentions were to scale up greatly and that we would be competing to become one of the major centers. However, in every discussion with the Genome Institute officials at NIH, particularly Jane Peterson and Francis Collins, they were cool to the idea.

A month or so after my brief encounter with Tony White, Steve Lombardi was on the phone again. He adopted a conspiratorial tone when he told me that PerkinElmer was contemplating putting up $300 million to sequence the human genome. What did I think? Given what I knew about the state of current technology, I just dismissed the notion with, "You guys are crazy." If they were serious about the subject, I thought, why send the man who sells machines and reagents to raise it with me? Surely that was a job for Mike Hunkapiller, not the head of sales.

At the end of the year I was back at the same hotel in Hilton Head for the New Year Renaissance Weekend with President Clinton and two thousand others. Before we could get into the New Year's Eve dinner, Claire and I were pulled out of the security line and asked if anyone had called us about the seating arrangements. To our surprise and great pleasure, we found ourselves sitting next to the Clintons. I enjoyed their company and found them keenly interested in our work. The First Lady was like a sponge eagerly absorbing what I had to say about the genome.

Most of the sessions that weekend were short talks followed by a discussion. After one such session the $300 million PerkinElmer project surfaced once again, this time through Mark Rogers, its newly hired executive vice president of business development. Mark told me how Tony White was looking for a way to change PerkinElmer into a more dynamic, forward-looking company. They wanted to sequence the human genome using a new instrument they were building. Would this be possible using my method? I answered him by posing more questions since their plan still seemed hazy and vague. But then he told me they were forming a high-level science advisory board and agreed to pay me $50,000 for consulting time—an unusual offer, because ABI was known for being frugal. If they were serious, I told him he should send me a letter outlining the terms of the consulting.

Back in Maryland I was once again focused on sending a proposal to the NIH and called Lombardi to see how soon I could test one of ABI's new sequencers so that I could include them in my grant application. Steve again gave me the $300 million project pitch. Was I interested? I asked him to tell me about the new technology, and then I would let him know. A few days later Mike Hunkapiller invited me to see his new machine, still only a breadboard prototype. By the way, he added, they were serious about talking to me about doing a whole-genome shotgun sequencing of the human genome, but I could not discuss their proposal with anyone except a few of my closest associates such as Mark Adams or Ham Smith.

I decided to fly out immediately with Mark to see what they had and proceed from there. At that time Mark and I could not settle on a strategy to deal with the NIH. Mark wanted to give them what they wanted: sequencing of a map based on BACs. And because I wanted to give the NIH something better, this new ABI machine might be the trick we needed. I don't think I told Mark about the $300 million plan, because I did not believe it was real. But I did do a few back-of-the-envelope calculations to figure out what the new device would have to deliver for a single center to be able to tackle the entire genome.

Mark and I arrived in Foster City in February 1998, and I felt a twinge of nostalgia for the days when I sailed down the bay from Redwood City in my 19-foot folk boat and tested my hydroplane at nearby Coyote Point. The Applied Biosystems complex is a series of one- and two-story warehouse-type buildings situated just before the San Mateo Bay Bridge, at the end of a sandy road adjacent to the San Francisco Bay. We were ushered into a conference room with cheap tables and plastic chairs decked out in the ABI colors of gray and purple. The whiteboards suggested the room was used for brainstorming and not sales presentations. The tall figure of Mike Hunkapiller loomed among the group of engineers and software people standing before us, and he began to outline the challenges.

At that time ABI pretty much dominated the market for DNA sequencers with its Prism 377 model. While called "automated" DNA sequencers, they were far from being shining examples of robotics and automation. Each pair of sequencers had to be tended by three people, so that any major scale-up would require thousands of technicians just to run the machines. The time people had to tend them had to be cut from twelve hours every day to twelve minutes. The chemical reagents involved were also costly and would have to be scaled down ten to a hundred times to even begin to be economical. The data had to be of better quality, too, if tens of millions of short sequences of DNA were to be assembled back into the genome. In short, the hurdles we had to overcome looked daunting if not impossible.

And they looked all the more so once we started our tour. We were not about to see a prototype machine but prototype *components,* each under test in a separate work area. First came the capillary array, a set of hairlike fibers about 18 inches long that separated out the DNA molecules by size. These were going to replace the sequencing gels that were not only time-consuming and expensive to make (someone had to mix the chemicals, then pour a liquid gel between large glass plates separated by thin spacers) but also a weak link because the gel was never uniform, which undermined the quality of the data. We were shown some sample runs of the capillary array, which worked successfully. It was an encouraging start, for it solved many problems at a stroke. In the next room a team was developing an automatic loader so that lab technicians would not be required to load DNA samples manually onto each gel. We wandered to a lab where new reagents were being developed and tested. In another building a team was writing new software that would run the machines and process their data. We were impressed.

Put all the pieces together, and you might just end up with a truly automated DNA sequencer. Apply row upon row of them to the task of the human genome, and you eliminated the need for thousands of workers while improving the quality of the data. I made constant mental notes and calculations as we walked, looked, listened, and asked questions. I like to think that I am better than most at assembling parts and complex systems in my head, and I became convinced that if they could deliver all that they were showing—better still, all that they were implying—then this was the technical breakthrough I had been waiting for in order to make a run at the human genome.

Back in the conference room Mike asked me what I thought. I went to the whiteboard and started to work through the possibilities. I knew the answer I wanted: a small team that could sequence the entire human genome in two or three years for 10 percent of the cost of the federal program. But would the figures back my instinct? When I finished, I thought that it was doable, but only marginally so. Mark Adams, who always took a conservative view, still thought it was impossible. I always listened to Mark, because I would end up counting on him to help make things happen at TIGR. But when he challenged me on whether the machines were really up to generating the huge number of sequences I had claimed, I looked more closely at my calculations and found a tenfold error. Once fixed, this decreased the DNA sequences we would require by that factor. Mark immediately said that it would not be easy, but that it might be possible. As Aristotle observed, a likely impossibility is always more preferable to an unconvincing possibility.

I wish I could say that my miscalculation had been deliberate, so I could claim it as a master stroke in corporate psychology and boardroom manipulation. But even though it was no more or less than a simple mistake, it had helped prepare the minds in that conference room for what was then a revolutionary idea: that one team could tackle the entire human genome within the time frame of a few years. Had I presented the real number first, there would have been a kneejerk response of "No way!" The fact that it was now ten times smaller than the original one made the goal seem plausible. Because of the incremental improvements in machines, software, chemistry, and enzymes, the crazy idea was now sane (possibly). Mike reminded me that if I thought it was doable, they were prepared to fund the project to the tune of $300 million.

Mark and I flew back to Maryland to consider our options. But the truth was I had already made up my mind: I wanted to go for it. That is an understatement. This new technology offered me precisely the kind of way that I wanted to deliver the genome—with a quick and aggressive campaign. There would be no better way to accelerate human medicine and science, but there were many issues to consider. I wanted to do the project at TIGR despite there being mention of starting a new company. And not everyone was convinced that it was doable. For reassurance I consulted Ham, who did not give me any. After a brief pause my scientific muse told me he did not think it would work and then added, "But if you are going to try, then I'm going with you." When I raised the idea with Claire, she made it clear that she thought I had taken leave of my senses.

Over the next few weeks we had more discussions with Mike and others at ABI, and the idea began to look a little crazy again. The idea of sequencing the human genome with their money and new machines again began to look half-baked at best. After what I had been through with Haseltine and HGS, I was determined to pin down all the details, and some remained frustratingly elusive. But there was one detail on which I was clear. Even before I left Foster City, I had told Mike that there was a single nonnegotiable term for my involvement: If I were to sequence the human genome, then I had to be able to release the data and to publish a major paper on our analysis. Mike said that he did not see that as being a problem.

I was asked to give a presentation on the project to the PerkinElmer management, which was holding a retreat in Arizona to discuss the future of the company. I had not been impressed with Tony White on our first meeting, and this one did nothing to change my perception. Once I had finished my overview of genomics and an estimate of what it would take a single team using the new ABI sequencer to tackle the human genome, Tony White aggressively took command of the question-and-answer session. "How would I make money by sequencing the genome?" he blurted. I had not given it much thought, I told him; my role was to get the sequencing done and publish the data. The response was inevitable, an echo of the bad old days with HGS: "If you are going to sequence the human genome with my money and then give it away for free, then you'd better have a plan for how you are going to make money." I did not think it could get any worse—until, that is, Tony White declared that he viewed TIGR as a threat to the commercial success of his new enterprise. I must have looked stunned because Mike Hunkapiller intervened, and White finally backed down.

Although the basic idea of the plan was not crazy, I left Arizona questioning my own sanity for even thinking of getting involved with White et al. Desperate men swinging wildly with the hopes of hitting a home run were not likely to be ideal partners. Even if I hit a home run, which I fully intended to do, would they even realize it? My inner voice was screaming, "Run, and run fast," but the lure of the genome was so seductive and strong that I could not yet turn my back on this opportunity.

In the days that followed came the phone calls: The meeting had gone well; PerkinElmer was set to go forward; Tony White was prepared to build a new company to do it. Would they fund me to do it at TIGR? On that point White was firm: "No. I'm in the business to make money, not give it away." If I wanted to sequence the genome and offer free access to the data, I had better come up with a business model to convert my scientific largesse into financial common sense—and before their annual board meeting in Florida. To help me do so, PerkinElmer assigned Peter Barrett, a senior vice president who had been with the company for more than two decades, to visit me at TIGR and thrash out the terms of my involvement with a new company and the all-important business plan to ensure a return on a $300 million investment.

I began talking to close friends and colleagues about whether we could come up with a strategy to release data that would keep both PerkinElmer and the scientific community happy. My working assumption was that my data would be essentially the only real data from the human genome for several years to come: The government-led competition was advancing at a crawl.

In December 1995, Eric Lander predicted that the human genome would be sequenced by 2002 to 2003, "plus or minus two years." By the spring of 1998 only 3 percent had been completed,[3] and not one of the six pilot centers NHGRI funded in 1996 to encourage faster, cheaper DNA sequencing methods had achieved the production rates they had promised, including TIGR, because the methods were not scalable. We were by then midway through the fifteen-year Human Genome Project, and teams were only just starting large-scale sequencing. Some advisors to the project had private concerns that Collins was not even serious about the genome, since less than half of the funding from his genome institute over the next decade was earmarked for sequencing. A long article on the delays that appeared in the journal *Science* ended with a quote from Rick Myers, codirector of the Stanford Genome Center: "A lot of us are in for the long haul."

In that context, it seemed to me that anything I did could only help propel science forward. I was given encouraging feedback from Richard D. Klausner, the director of the National Cancer Institute, the largest institute at the NIH, and Aristides Patrinos, a senior DOE official and long-term TIGR supporter and funder. Both endorsed my idea of releasing data every three months into GenBank. To help win over the PerkinElmer people and validate my arguments, I enlisted the help of David Cox, the head of the genome center at Stanford University, which was also funded by Collins, to come along when I was introduced to the board of directors at PerkinElmer.

Compared with Tony White, they were an agreeable bunch. Many told me privately that if all they accomplished with the money was to get the human genome sequenced well ahead of the public schedule, then they believed the risk of a few pennies per share was well worthwhile. The board gave its approval to move toward starting the genome company, which I would head. I had feared another cold-eyed financial grilling and walked out feeling better about this plan than I had for a long time. The warm, fuzzy glow did not last long, for when I walked into a private meeting with Tony White, he in effect said that he did not buy into or understand this science crap: He wanted a winner, and "It is clear that you are a winner." At least we had one important thing in common: Neither of us liked to lose.

After days of negotiation we settled on a term sheet: I would head a new independent company funded by PerkinElmer, and the human genome sequence it produced would be published and placed in the public domain on completion. I was offered 10 percent of the stock in the new company. I would remain chief scientific officer of TIGR and made it clear that if I was going to temporarily leave TIGR, I wanted half of my stock to increase its endowment so that one day I could return to do the research I wanted to do for the rest of my life. White did not want me to take less stock personally (it served as a set of useful "golden handcuffs") but left it up to me to decide if I wanted to give it away (which I did).

Now all that was left was to turn the term sheet into a definitive employment agreement. I hired a highly recommended New York attorney to represent me in the final drafting, and we had copies of the employment agreements of White and other senior PerkinElmer employees as a template for my own. I was to be president of the new company and a senior vice president of PerkinElmer—the same status as Mike Hunkapiller. I would receive a 5 percent stake in the company that resulted, and TIGR another 5 percent. Although I was now one of the three top people at PerkinElmer, the agreement was modest by their standards; I was stunned to discover that Tony White had the private use of a $25 million aircraft that mostly saw service flying him and his wife between Connecticut and South Carolina, where he had a vacation home.

White and his attorney came back with so many onerous terms that it was hard to understand their motives. I was familiar with the Maryland employment terms and decided I would be better off as an "at will" employee without a contract: While I could be fired at any time, and for any reason, at least I would have limited legal rights to fight my termination. By then even my hardened and experienced New York attorney had become exasperated, remarking that they were "some of the biggest assholes [he had] ever encountered." He offered me a simple piece of career advice: "Get the hell away from them while you still can." It had also become obvious that although White had agreed to a separate company in the term sheet, he wanted to make it a division of PerkinElmer.

I asked Claire for advice, though we had not talked much about all this. She was furious; how could I be so stupid to even think of doing this after all the problems with HGS? To me the answer was obvious: The genome was the biggest prize in biology. I had never been very good at working out what would anger her, but I thought that she would understand what was driving me. I reminded her that I had known Mike Hunkapiller for fifteen years and felt that he had integrity and would not let Tony White renege on the promise to publish the data.

She was not mollified and said that she felt I was walking out on her, effectively beginning a trial separation at a professional level. But I had one more bomb to drop on her: Would she take over as acting president of TIGR for the three years I would be away? While she was reluctant and clearly frightened by the challenge, I felt that she was the only one I could trust in that interim position.

To guarantee the best chance of success at sequencing the genome, I wanted to bring along a select group of people from TIGR, including Ham, Mark Adams, Tony Kerlavage, and Granger Sutton (who wisely asked, "Can I think about that for a moment?"). I urged anyone who played a key role in the microbial genome efforts, which had by then become the mainstay of TIGR, to remain in place to ensure TIGR's continuity. I also made it clear that I would one day return to TIGR and that I was building the endowment to guarantee our future. After I assembled the staff and told them of my decision, I broke down, overwhelmed by the emotion of the moment.

I had often been accused of creating a cultlike following at TIGR, and if that meant I had motivated my team to look beyond their monthly paycheck, to believe in a mission, and to undertake a scientific crusade, then I plead guilty. But now, of course, I was off on a new and bigger crusade, to sequence the human genome on my own, and I was not taking everyone with me. Claire was not the only one to feel abandoned, and I believe I understood the complicated emotions involved. However, what I still can't understand is how that feeling of abandonment turned into hostility from some of my closest friends and colleagues, people I had hired, encouraged, and supported.

My immediate concern was to find a place to do the human sequencing work and to hire a team and build the infrastructure. There was no way to keep what we were doing a secret, for as a public New York Stock Exchange company PerkinElmer had a requirement to announce publicly anything substantive regarding their business. A $300 million effort to sequence the human genome was certainly in that category.

There was considerable debate as to how to unveil our ambitious plan to the world. Some wanted to do it by press release, but I preferred to touch base with key leaders in the genome community first to see if we could have real cooperation in one effort. I was cautiously optimistic this would be worthwhile after my various discussions with Ari Patrinos and Rick Klausner, as well as with the PerkinElmer board members Arnold J. Levine, then president of Rockefeller University, and Caroline Slaymen of Yale University. I also wanted a top scientific advisory board to advise the program. Arnie Levine agreed to participate, as did Richard Roberts, who had won a Nobel for the codiscovery of introns, Victor McKusick of Johns Hopkins, the father of modern medical genetics, Norton Zinder, a pioneer of molecular biology, and Arthur (Art) Caplan, a distinguished bioethicist from the University of Pennsylvania. I also thought that reaching out to Jim Watson could be constructive and could lead to synergy with the public program. One of the advisory group offered to call him, and when he did, Watson was clearly surprised. Our project rested on a new technology that Watson did not seem to know anything about. He would have considered advising us except for the fact that just two weeks earlier he had testified before a congressional committee that there was nothing at all new on the horizon and that the human genome was going to have to be sequenced with the existing technology. I was told that Watson ended by saying: "I would like to pretend that this call never happened so that when you make your announcement, I can act surprised like everyone else." (Which, of course, is precisely what he did, telling Sulston that he wondered why I had not called him myself to tell him of my plans.)[4]

His reaction was all the more baffling because we had tried to tell Watson what we had in the works weeks earlier when Ham Smith had been invited to the same congressional briefing with Watson. Ham and I had discussed whether he should even turn up: A no-show might be safer than being compromised by not being able to openly discuss all that he knew about the new ABI machine. We decided, however, that Ham should attend so that he would have a chance to give Watson a hint of what was to come. As they sat around a table, Ham mentioned that a new technology was available that could change the landscape. Jim looked up and asked in a desultory way: "You mean that capillary stuff?" Jim dismissed it with a wave of his hand: "Everyone knows that it doesn't work." As Ham later put it, "I have the greatest respect for Watson, but sometimes he is just damn wrong."

Ham had assumed that Watson knew what he was talking about, but in fact Watson was referring to a capillary machine made by a company just purchased by Amersham that did have serious problems. Watson was apparently ignorant of what ABI was hoping to produce, which was a much firmer prospect given its established track record in the field. Ham was stung by Watson's reaction and decided not to respond, even though we were weeks away from the announcement, and I was to learn later from Rich Roberts that Watson was upset with Ham for not persisting.

Although the U.S. Securities and Exchange Commission would frown on what we were about to do, we all felt that before we went public we should at least discuss our plans with Harold Varmus, the current NIH director, and with Francis Collins. It made the most sense to start with Varmus, in the hope that he would be reasonable and therefore help bring Collins around. Mike Hunkapiller accompanied me to Building 1 of the NIH, which looked like a grand southern mansion, complete with a larger circular drive, pillars, and a grand stairway that led to the NIH director's office. Unlike the days when I climbed up the stairs to see Bernadine Healy, the building seemed to have taken on a different vibe, that of cold bureaucracy.

Harold greeted us cordially. His office was stark with the exception of the bicycle that he rode to work when he could. There were no intermediaries or supporters present, so we were able to have a real discussion. The meeting went well, I thought, and it seemed as if Varmus could not only live with what we were doing but wanted to be constructive. I planned to do a test project first, and when I told him that I was thinking of the fruit fly, he replied, How about a second nematode (worm)? That December the rough draft of the *C. elegans* genome had been presented by Sulston, Waterston, and colleagues. And, he asked, if the pilot project worked, what then? I answered that I thought we could continue to do the human, and for the sake of efficiency, the public program could focus on the mouse, bringing huge benefits through comparative genomics. (In terms of genetic code, humans are equivalent to big, tailless mice.)

Varmus responded in the same cool logical, scientific manner in which the idea had been offered, and without any territorial, political, or emotional agenda. He understood that having human and mouse together would make the effort that much more valuable—while we share most of our genes, one can conduct experiments on mice to figure out what the genes do. There was only one real issue, and that was the familiar one: data availability. Mike and I explained that we were going to publish the data and establish a high-end database with tremendous added value, along the lines of LexisNexis, the legal and news base that repackages public information and presents it in a friendly format that allows for fast searches. We liked to think of our venture as comparable to a business that sells laboratory reagents, such as enzymes: While you could make and purify them yourself, it was so much quicker to buy them from a company that specialized in them. Few had complained when companies like New England Biolabs started selling restriction enzymes to the scientific community. (In fact, Rich Roberts moved to New England Biolab in 1992 to help make restriction enzymes and do basic research—a good precedent for what I wanted to do.) Varmus said he thought it was a reasonable argument and agreed to keep an open mind. We shook hands, and before we said good-bye, I told him we were also going to discuss the issues with Collins before making any public announcement.

That same day we caught up with Francis at Dulles Airport, en route to San Francisco. I had not told him I was bringing Mike Hunkapiller along, otherwise he would have been able to figure out, with the help of a few well-aimed phone calls, why we had to see him to discuss a matter of the utmost importance. I wanted him to be aware of our plans, but I did not want to give him the chance to play politics, which had already ended one potentially important collaboration with the Department of Energy.

By that time an attempt to cooperate with the DOE on sequencing—which had gone as far as a memorandum of understanding—was already dead and buried. The DOE's rationale was simple enough: Big projects dominated by physicists had recently been in the driving seat of science, commanding the lion's share of resources for big experiments with atom smashers and reactors. The multibillion-dollar scale of the genome effort suited a department with a huge track record and equally huge ambitions.

But at a meeting held on December 3, 1998, Varmus, Morgan of the British Wellcome Trust, and Collins had warned Ari's representative, Marvin Frazier, that lending its support to a Venter-led sequencing effort would undermine cooperation between the NIH and the DOE and turn the latter into pariah.[5] As Morgan put it, the DOE would be ostracized by the other labs in the public effort, and as one participant observed, "It offended everyone royally. If this was an attempt to bring about peace, it wasn't a very well thought-out one. It was dead on arrival." Ari remembers how the memorandum was "violently squashed."

That experience did not set an encouraging precedent for my encounter with Collins. Mike and I cleared security, rode on a people mover out to the terminal, and found Francis waiting in one of the business rooms of the United first class lounge. While we had had an informal meeting with Harold Varmus in formal surroundings, the converse was true here. Collins had his senior staff with him, including Mark Guyer, the deputy director. He was surprised to see I had brought along Mike Hunkapiller, which made the speed of sequencing that I proposed much more tangible. Collins referred to him as "Craig's mystery guest"[6] in a way that signaled his disapproval.

Mike and I tried to foster the cool and constructive tone of the discussion that we had had with Harold Varmus, but as I had both feared and expected,

Francis reacted with his heart, not his head. To suggest that a privately backed effort sequence the human while the government program focused on the mouse was nothing less than an insult. Even though it was the most efficient way forward for the science, Francis wanted nothing to do with a collaboration, even one that would deliver benefits to humanity much faster than if we competed. For us to share leadership of the genome effort was "vastly premature." I left with the impression that Francis cared primarily about getting the credit for the human genome. For his part, Mike did not think the meeting went so badly, but then, as he sat next to Francis on the flight to San Francisco afterward, he did agree to sell NIH the new machines so the public program could compete with me.

Rather than issue a press release, we decided to offer the story on a plate to a reporter who we thought could get the space to do it justice. Nicholas Wade of *The New York Times* and Rick Weiss of *The Washington Post* had both written numerous detailed stories on genomics, were astute about the politics involved, and would do a good job of explaining that our effort was not an end run around the NIH. Eventually we came up with a plan: We would issue a press release before the markets opened on the morning of Monday, May 11, 1998, while Nicholas Wade's story would appear on the front page of *The New York Times*[7] the day before—two days after I had met Varmus and Collins. The story read, "A pioneer in genetic sequencing and a private company are joining forces with the aim of deciphering the entire DNA, or genome, of humans within three years, far faster and cheaper than the Federal Government is planning."

Although Wade's article was a red flag to the public genome bull, it still managed—with the benefit of years of hindsight—to hit the right note. Both Collins and Varmus had told Wade that our plan, if successful, would enable the desired goal to be reached sooner. Remarkably, Varmus also seemed to have swayed Collins, for Wade reported: "Dr. Collins said he planned to integrate his program with the new company's initiative. The Government would adjust by focusing on the many projects that are needed to interpret the human DNA sequence, such as sequencing the genomes of mice and other animals. . . . Both Dr. Varmus and Dr. Collins expressed confidence that they could persuade Congress to accept the need for this change in focus, noting that the sequencing of mouse and other genomes has always been included as a necessary part of the human genome project."[8] The spirit of collaboration was reinforced when Wade described how even I ("known for his direct approach") had stated that I wanted to work closely with the NIH and not "with an in-your-face kind of attitude."

The problem with journalists, of course, is that they love to add spin and twist knives. Wade could not resist either, in this case, and pointed out how my effort could to "some extent make redundant the government's $3 billion program to sequence the genome by 2005" and how Congress might ask why it should continue to fund the public effort if the new company was going to finish first.[9] Although that final journalistic flourish was entirely complimentary, the suggestion that I could not fail in this venture was guaranteed to enrage my powerful enemies.

The day after the press release was issued, Wade returned to this theme with vigor, suggesting that I had "snatched away" the historic goal of the human genome from the government[10] and observing that my "takeover of the human genome project is a venture of unusual audacity." By now John Sulston and Michael Morgan had also begun to panic, to worry that the announcement would torpedo a proposal they had to ramp up their own genome effort.[11] ("We were no longer going to be the largest genome center in the world," said Sulston. "That was disturbing.")

On Monday, May 11, I attended a press conference at the NIH with Mike Hunkapiller, Harold Varmus, Francis Collins, and Ari Patrinos. Sulston remarked that it was "the first of what was to become a series of bizarrely staged shows of unity,"[12] an assessment that proved accurate. That day Francis adopted an aggressive new line: The government would remain on its current course for the next twelve to eighteen months, by which time it would be clearer whether the project should change its approach to accommodate me. Collins also changed tactics and called into question the shotgun sequencing method: "The government considered switching to the approach Venter will use a few years ago and roundly rejected it as too problematic."[13] He predicted that any human genome sequence I might produce "is likely to be peppered with more and larger holes than that produced by the federal program."

My old foe Bill Haseltine rubbed salt into the public project's wounds with a quote in *The New York Times* that was a thinly veiled attack on Watson and Collins: "There have been serious problems of organization and management both at DOE and at NIH, together with internal dissension among the senior scientists involved."[14] Other media accounts only added to the discomfort of any federally funded sequencer. Haseltine was quoted again saying that I had dropped a bomb on the Human Genome Project: "All of a sudden somebody is going to pull a $3 billion rug out from under you? They must be deeply shocked."[15] A front-page story in *The Washington Post* declared: "Private Firm Aims to Beat Government to Gene Map.[16]

The annual meeting on molecular biology and genomics was just about to begin on Watson's home turf of Cold Spring Harbor. All the recent publicity was going to be damaging to the public project, particularly because it had come only a few days before the journal *Science* revealed that none of the genome centers funded by Collins had come near its goals.[17] When they gathered in Cold Spring Harbor, the leaders were by one account "in various stages of shock, anger, and despair."[18] Lander and others were so goaded by these taunts that they pressured Collins to compete, not cooperate, according to James Shreeve in his book *The Genome War*.[19]

While I was annoyed by the backtracking, I was still optimistic that we would find some middle ground for cooperation. Mike and I had arranged to meet the heads of the public genome centers and officials from the DOE and NIH genome programs on the Tuesday morning of the conference on the Cold Spring Harbor campus. We found ourselves in a new building that had been funded by millions in donations from pharmaceutical and other companies. The Plimpton conference room was just outside Watson's private office. Unlike the tatty and functional room at ABI where we had put together our plan, this was a bright space with rich oak panels and large windows. Only a few of the forty or so people sitting at the U-shaped table—the few friends we had—acknowledged our presence. Reading the faces around us, we felt we had walked into a cross between a funeral and a lynch mob. Their anguish was palpable.

I gave an overview of the science, the accelerated time scale—a finish in 2001, four years ahead of the public program—and why I thought our approach would work. I explained that we would use a model organism for a dry run before tackling the human. I discussed the computational power developed at TIGR to put a shotgunned sequence back together again. I cited an equation, the Lander-Waterman model, in support of our approach. A cry went up from a familiar red-haired figure: "But, Craig, you've completely misapplied Lander-Waterman. I should know. I'm Lander." I described how we were going to release the data on a frequent basis, assuming that we were the primary source of human data and that we planned to publish a paper on the whole genome when it was completed. But when I suggested they focus on the mouse to complement this effort, the atmosphere became so hostile that I was not sure anyone was listening anymore. (One of those present recalled that "I almost punched him in the fucking mouth.")[20]

As Mike began to describe the details of the new ABI machine that would make it all possible, I had an idea. I asked Gerald M. Rubin, the head of the fruit fly (*Drosophila*) genome project at the University of California, Berkeley, to join me for a chat in the hallway and motioned for Francis to join us. No one seemed to notice us slip out of the room. I did not know the bearded and cherubic Gerry except by reputation as one of the brightest scientists in his field. I got right to the point and told them Varmus thought I should sequence another worm as the pilot project to debug my approach but that I wanted to do the fruit fly. Would Gerry like to help me sequence the *Drosophila* genome instead? Years later Rubin recalled: "I didn't know whether I was going to belt him or whatever, but I immediately said, 'Great, anyone who wants to help finish *Drosophila* is my friend, as long as you are going to put all the data in GenBank."[21] Anticipating his answer, I promised that we would publish the genome as soon as it was sequenced and analyzed. Gerry's positive response was gratifying, and he explained that if he turned me down and I sequenced the worm instead, the "fly groups" toiling over *Drosophila* genetics would have him shot. He added that he hoped Francis would not cut his grant so that he could perform complementary work. This plea was a wise move, for Francis was by then looking a little pale. He meekly said that it was probably workable but that he would have to understand what Gerry had in mind. Gerry and I shook hands warmly. With a minimum of fuss and politics, I had initiated one of the best—if not *the* best—collaborative projects of my scientific career. The three of us walked back into the room, and after Mike finished his presentation, we informed the group what the test project would be. There were a few questions but mostly statements of a hostile nature from Waterston and Lander. "Everyone is rushing around like headless chickens," as one of those close to the events recalled.[22] "Jim is calling foul, Francis is apoplectic." And there were "mutterings that Gerry is collaborating with the Devil."[23]

As soon as Mike and I left, Watson joined the group. Watson refused to be in the same room as us, so incensed was he by the mouse proposal. As he expressed it, "It is an understatement to say it was done in an insulting fashion."[24] Watson had also apparently likened me to Hitler ("Craig wanted to own the human genome the way Hitler wanted to own the world"[25]), and during breakfast that day he had called out to Francis Collins to ask if he was going to be Winston Churchill or Neville Chamberlain;[26] in conversation with Gerry Rubin later that day he had declared: "I understand the fruit fly is going to be Poland."[27] I was scheduled to deliver a major address at the Cold Spring Harbor meeting, but even without being aware of what Watson had said, I decided that I had had enough anger and hostility for one day and left for Rockville.

That evening my opponents began to fight among themselves. Members of the smaller sequencing labs were frozen out of crisis meetings held by what were then, in effect, genome businesses run by the likes of Waterston and Lander. It was Eric who was pushing the hardest for the public program to get out a draft sequence at the same time as we did, even if it was full of gaps, a proposal that was anathema to purists such as Maynard Olson. For his part, although Collins found the idea of collaboration with me repugnant, he had at least to appear as though he were taking my proposal seriously, or Congress would attack him for turning down a public-private partnership that could save taxpayers hundreds of millions of dollars.

Not everyone in the public program was infuriated. Sulston remarked on how that evening he caught grins on the faces of symposium participants who were not involved in large-scale sequencing as they saw the members of the genome labs huddled grimly in corners. . . . [T]hey were enjoying the thought that this very small group, who had taken so much in funding to produce the human sequence, were finally getting their comeuppance."[28]

Shortly after returning from the meeting I received a call from David Cox, codirector of the Stanford University Genome Center. He would not be able to join my advisory board after all, he explained; just after I had left, Francis asked him to go for a stroll down Bungtown Road, the lane that runs past the pine trees and stone walls of the Cold Spring Harbor campus. This is the path of choice for any scientist who wants to have a confidential conversation.

In his book James Shreeve gives a colorful description of what transpired between them.

> *"It will be sad not having you part of the genome project," Collins said. His implication was clear. If Cox accepted Venter's offer, the funding his lab enjoyed from NIH would cease. "Why does it have to be one or the other?" Cox said. . . . Collins shook his head and smiled wistfully, as if regret had overcome him for the ideal world that could*

not be. But there would be no appeasement. "If you join Craig's board," he said, in patiently enunciated syllables, "nobody in the program is going to want to work with you anymore. You have to choose."[29]

Within twenty-four months Collins cut the funding to his lab anyway, and David ended up leaving Stanford to form Perlgen to map variations in the human genome sequence. My dream of working on the human genome together with the public program had lasted little more than a weekend. I was on my own again and anxious to get started.

By then, after some discussion, we had come up with a logo, a little dancing figure whose limbs formed a double helix. And we had a name: Celera, chosen because it was derived from the Latin for swiftness (the same derivation as "accelerate," for example). That was by far the best of a huge list that included such clunky alternatives as Biotrek and Sxigen. (And it was certainly better than the nickname used by our rivals, who called the company the Venter-Hunkapiller Proposal—often shortened to Venterpiller or VentiPede.)

Celera needed people from a range of disciplines to make shotgun sequencing work. The process starts by extracting the DNA from human cells (either blood or sperm). Converting this DNA into easy-to-handle and easy-to-sequence pieces—creating what is called a genomic sequencing library—is a crucial step. In whole-genomic shotgun sequencing the library is made by breaking the DNA into pieces using sound waves or other methods to apply shearing forces to large DNA molecules and chromosomes. Using simple laboratory procedures the fragmented DNA can be separated according to size. We can then take a piece of a given size—say two thousand base pairs long (2 kilobases, or 2 kb)—and insert it into a "cloning vector," which is a set of bacterial genes that allows the piece of DNA to be grown in the bacterium E. coli. Repeat that process with all the pieces and the resulting genome library will have all the sections of the human genome represented in millions of 2,000-base-pair pieces. Given that the fragments were made in the first place from millions of complete genomes, many pieces contain overlapping sections of DNA due to the random way the chromosomes are broken up. Then, by random selection of DNA clones from such a whole-genome library, it should be possible to sequence them and match the overlaps by computer to stitch together a copy of the entire genome again. Like playing a piano, the description of the basics is simple enough but can be done well only with a great performer. I needed the best people in the world to make this work at a scale that had never been contemplated before.

One was Ham Smith, who has golden hands and can handle and manipulate DNA molecules better than anyone else I know. Ham was with me from day one and already wanted to construct new cloning vectors that would improve the efficiency of sequencing. Mark Adams was another member of my dream team. No one was better at getting complex technology to work and to work quickly and I could also implicitly trust his judgment in hiring top people to work with him. Mark had a knack for reaching out to the best and bringing the best out of them. I asked Mark to take on building the DNA sequencing core of the Celera operation. We had built three previous ones, but the scale of what we were now proposing was beyond the imagination of most scientists in the field.

We needed more robotics to scale up all the steps in the DNA sequencing process, and from the TIGR sequencing core I pulled out Jeannine Gocayne, who had worked alongside me in getting the very first DNA sequencer to work while at NIH in 1987. I always admired her skill and dedication, and I knew that I could count on her 100 percent.

I still had a number of concerns. I had not actually seen one of the new machines work, and at that time they did not even exist. But even when the inevitable initial troubles emerged, I figured we would have the benefit of the engineering team at ABI to fall back on. What worried me even more was how to handle the torrent of genome data we would produce. Every calculation we did on the processing and assembly of the entire human genome came to the same conclusion: We would have to build one of the biggest computers ever—perhaps even the biggest on the planet. I formed a team to cope with this challenge around Anne Deslattes Mays, my head of software engineering at TIGR. Also on the team was Tony Kerlavage, who had been with me from the early NIH days and played an important role in bringing in new computing as the trickle of data became a stream and then a river.

Our problem quickly got the interest and the attention of the computer industry. Soon we were bombarded with presentations by the major companies: Sun, Silicon Graphics, IBM, HP, and Compaq, which had just acquired Digital, the makers of the Alpha chip, then the most powerful computer chip available (and liked by Anne). They each tried to convince us that their computers were the only ones that could possibly do the job, as they all wanted to provide *the* computer that assembled the human genome. As the other computer manufacturers fell away, lacking the hardware or the performance, Compaq and IBM began to draw ahead in the field.

But the more that I learned, the harder it became to decide. We visited Compaq (the former Digital labs) in California, where I found impressive work done by a team that had been demoralized by the merger. We visited IBM research labs in New York where they were building the ASCII computer for the DOE to upgrade the biggest computer in the world, which it used to simulate nuclear explosions. I was very taken with what IBM was doing and with its senior management team, particularly Nicholas M. Donofrio. But we seldom had a meeting with IBM that included fewer than ten or twenty people, and they would not mix and match the best on offer, pressing instead a single solution to all our computing needs, whether PCs or databases: IBM. I had to negotiate a computer deal worth anything between $50 million and $100 million and needed a system that would work well, would work quickly, and would run complex computer code that had not yet been written.

To make sense of the claims and counterclaims by the competing manufacturers, I decided to do an experiment with the assembler that was being used routinely for sequencing at TIGR. With the Sun computers that we ran it on, a simple (relative to human) genome such as *Haemophilus* required several days to assemble. By extrapolation, it would require years for it to assemble the 3 billion base pairs of the human genome. How would our various suitors do in coming up with more efficient hardware?

Only Compaq and IBM agreed to the challenge. In the first run of Compaq's Alpha chip, we went from taking days to nineteen hours, and then finally nine hours. The best that IBM could manage was thirty-six hours. My programmers wanted the Alpha chip and the results of the challenge left no doubt that was the one to go with. IBM knew it had not done well and asked me what it would take to get the contract. I bluntly told them that IBM would have to provide the system for free and include a development team to get it up and running. While IBM was contemplating my reply, it dawned on me that I might not have made a particularly good proposal: I had no leverage if the system was free, and I could hardly withhold a payment for poor performance. I realized that I had to pay for what I really wanted, for what had the best chance to work in an area that no one had ever been before. The CEO of Compaq flew out to meet with me and to pledge his commitment to make every effort to guarantee the success of our effort. He wanted his computers to be the ones that carried out the largest calculation in the history of biology and medicine to date.

After a few days I signed the Compaq contract and called the CEO to tell him that I had accepted his offer. A half an hour later I was about to phone Nick at IBM to give him my decision when he called me instead. He had just met Lou Gerstner, the IBM CEO, who authorized him to provide the entire system gratis. I responded that I would have been interested thirty minutes earlier but had just signed the deal with Compaq. He wished me well and said that when the Compaq computers failed, they would be waiting. But I could not let his offer even enter my head, for I could not countenance failure: For this project there were no second chances.

I learned that the number-one guy at Compaq for building a system of the scale we envisaged was Marshall Peterson, who was then in Sweden working for Ericsson. Peterson had done three tours of duty as a helicopter pilot in Vietnam and had been shot down several times, earning him the nickname "Mad Dog." I liked him immediately and offered him the job on the spot, which he accepted. The supercomputer hardware was now under way, but it would only be as good as the software we ran on it. Granger Sutton, who had built the only complete whole genome assembler to date, knew that it was doable in theory, but the TIGR assembler could not simply be extrapolated to accommodate to the human genome. Recruiting a team to write a new version became the next big job.

Because we had to start from scratch, Granger knew just the person to approach to help build the next generation of software: Eugene W. Myers, a unique character who slightly resembled Richard Gere. Gene believed that when it came to reading genomes, size did not matter. In May 1997, Myers published a paper with the medical geneticist James L. Weber describing how the entire human genome could be assembled using the whole-genome shotgun method, which was greeted with the now standard criticisms by the public Human Genome Project. Because of this, but most of all because of his algorithms, I admired Gene and had previously tried, without success, to recruit him from the University of Arizona. This time around, I could count on the prestige of what we were trying to do as an incentive; Granger also told me that after Celera had been announced, Gene called him asking if there was any chance that he could "come and play" with us. This was an exciting prospect, and I soon found myself talking to Gene about what it would take to actually assemble the human genome. I told him that I had read his paper on whole-genome shotgun assembly and had been impressed, but added that if he wanted us to try his approach, he had to be part of the team. The matter seemed to be settled when I agreed to match his university salary if he came out within the week to get going. To be honest, I felt that his salary was low for what I was asking of him, and after sleeping on it, he clearly felt the same, so by the end of the next day we had tripled the money. The following day he called again: Friends and colleagues all mentioned something called stock options; what were they, and was he going to get any? Yes, I replied, although I had not yet determined what the amount would be.

When he eventually arrived, Gene quickly began to appreciate the sheer scale of the problem. In essence, to ensure coverage of the entire 3 billion–letter-long human genome, the software would have to be capable of handling 30 million fragments. This was going to be the mother—and father—of all jigsaw puzzles, and we soon assembled a crack team to develop the necessary algorithm. As well as being led by Gene Myers and Granger Sutton, the two best mathematicians and computer scientists, I had Anne Deslattes Mays to turn their mathematics into software. All that we needed now was the facility to sequence the genome.

We had started out by wanting to house the two hundred new sequencers we had ordered in a space on the TIGR campus and had the go-ahead from PerkinElmer to adapt 20,000 square feet on the floor of a new building to cope with the weight and power demands. As time went by and construction delays set in, we all became convinced that the TIGR facility would not be up to the task, and we began a search for other quarters. Even then I had no idea that within a year we would be filling two buildings, each five times the originally planned TIGR space. We found them only about a mile from TIGR and negotiated a lease for the first with an option on the second. Robert Thomson was sent down from PerkinElmer to give me the benefit of a lifetime's experience in building and rebuilding facilities for the company. He had a can-do attitude and, in one memorable "all-hands meeting" to get the venture going, announced: "Someday I'm going to tell my grandchildren I was part of this."

We started what was a total transformation of the existing space, but it soon became clear that we had once again underestimated our requirement and would need at least part of the second building for our computer center. I like the building phase of projects, and this was turning into the biggest one of all. There were many memorable milestones. When we contacted Pepco, the local power company, it was clear that they did not have sufficient resources and would need to put in a new transformer and power poles to feed enough electrons to our computers and sequencing machines.

We started construction on the fourth floor of the building, where bacteria would be grown to produce the human clones and where both the robots for producing the DNA as well as the hundreds of PCR (DNA amplification) machines to prepare the DNA for the sequencing machines would be housed. A smaller sequencing laboratory was also built on the fourth floor so we could get up and running while the main lab on the third floor was being prepared. In order to get a head start on sequencing, we put two crude laboratories in the basement. As soon as robots and sequencers arrived on our loading dock, they were packed into these small, dark rooms.

By August 1998 things were beginning to take shape. We had opened a cafeteria in the basement. We were testing the new robots in the temporary laboratories. We had conference rooms, and since I did not have much time for sailing, I named each one after a body of water. Mark Adams had done a great job and had a team ready to work out new protocols and establish standard operating procedures (SOPs), complex and detailed documents that would help ensure quality control and consistency throughout the DNA laboratory—one of those apparently boring issues that is crucial. The public genome people assumed the effort to sequence any genome was too sizable for a single laboratory. As mentioned earlier, the yeast genome, which was only about three times the size of *Haemophilus influenzae,* had required some one thousand "monks" working for almost ten years in laboratories spread all over the world. The problem with that approach was that while a few centers did high-quality work, many others were of only average quality or even worse. The first yeast chromosome sequence to be published had to be redone, underscoring the management issues in having each diverse lab trying to read code in its own way, with varying degrees of success. In this system, quantity of sequence, not quality, was what mattered most.

The great pioneer of the field showed that it did not have to be this way. In 1977, when Fred Sanger and colleagues sequenced the first viral genome, that of the bacteriophage phi X 174, both accuracy and strategy were valued. After Sanger used restriction enzymes to cut the viral genome into smaller sections, the fragments were distributed among various lab members, each of whom sequenced his particular section numerous times to ensure a degree of quality control. When we sequenced the virus again a quarter century later, we found only three differences in the DNA code over five thousand base pairs.

Henry Ford understood the efficiency differences among distributed efforts, where many cars were built simultaneously in one factory by independent teams, so the end result depended on the quality of each team; therefore, for his assembly line, where each one of dozens of different teams built one car at a time, specialization and standardization determined overall quality. We had to adopt an assembly-line mentality at Celera and make a large number of incremental improvements across the entire process, as well as strategy changes, which would end up cutting both the time and cost of sequencing. The *Haemophilus* project had reduced the time to sequence a microbial genome from ten years to four months. At the time, recreating its sequence from twenty-five thousand individual fragments of DNA seemed like a massive undertaking and required the entire TIGR facility to work around the clock. When it came to the human genome we would need at least 26 million sequences, equivalent to one thousand *Haemophilus* genome projects.

The three major costs associated with genome sequencing are people, reagents, and equipment. The $300,000-plus cost of every 3700 sequencing machine could be amortized over all the millions of sequences to be processed, thereby contributing approximately ten to fifteen cents for every sequence read out of a total of around $1 to $2 per read. The way to reduce staffing costs was to automate more DNA processing steps, and Mark Adams carefully reviewed the robots that were available. One type, used for pipetting small but accurate amounts of liquids, relied on disposable plastic tips that were replaced in each cycle to prevent contamination of the next sample. This system worked on a small scale, but at Celera we were facing the consumption of a mind-boggling $14,000 a day in plastic disposable tips alone—close to $10 million over the course of two years. Instead, we found an obscure company that had developed a robotic pipetter with a self-cleaning metal tip. By examining every detail of every step with this degree of fiscal prudence, we could boost speed and cut costs.

Other examples include changes accepted by Ham Smith's team. For the previous two decades molecular biologists had relied on a "blue-white selection," a simple dye test that used a white color to indicate whether a bacterial colony contained a human DNA clone. Because we needed at least 26 million clones, we would have to make more than 50 million, at double the expense, if we used conventional methods. Ham was confident that he could construct a new cloning vector that would have essentially 100 percent efficiency and got started on it right away.

We were doing a great job, but because we still lacked the DNA sequencers, it was as if Henry Ford were trying to perfect his assembly line without having any tools. Based on ABI's promise that delivery of the first machine was perhaps several months away, I had set a very aggressive schedule, including having the entire fruit fly genome sequenced in less than a year—by June 1999. Now my ambitious plan was falling apart before it had even had a chance to begin.

Every day I would check in with Mike Hunkapiller's team, which had problems of every shape and size and at every level. They had problems getting all the parts delivered so quickly. They had problems ramping up production of the machines. They had problems with the consistency and reliability of the operation of the machines that they had managed to put together. They had problems with the robot arms within them, which would run amok at times. And those were just the problems they were happy to tell us about; we were to discover hundreds more on our own.

What eventually made the difference during this difficult start-up phase is that I led from the front. My team believed in me, and I believed in them. The extraordinary circumstances in which we found ourselves had turned my good people into extraordinary people.

12. *MAD* MAGAZINE AND DESTRUCTIVE BUSINESSMEN

A man cannot be too careful in the choice of his enemies.

—Oscar Wilde, *The Picture of Dorian Gray*

Only saint-like minds can watch someone in the next lab race them for an experimental result and not get violently upset.

—James Watson, *A Passion for DNA: Genes, Genomes, and Society*

The first tangible evidence that I had further offended the genome scientific establishment came soon after I announced that I was going to sequence the human genome with unprecedented speed. Once the publicly funded scientists had decided that they wanted nothing more to do with me or my teams, TIGR's application for government funding was turned down and the existing grant immediately withdrawn by Francis Collins. In hindsight, it was inevitable that my attempt to simultaneously jump-start the human genome project and join hands with the establishment was doomed to fail.

In the United States, the National Institutes of Health dominates the landscape of genome research funding. Proposals for research grants are drafted and refined over the course of many months, then submitted to the NIH, where they undergo "peer review" by a dozen or so scientists who are highly knowledgeable in that particular field. But, of course, in a newly emerging discipline the reservoir of existing expertise can be shallow, and even in an established field, a leading figure may be too busy to carry out reviews or do them conscientiously. The experts then form groups—study sections—in a process that in the case of American biomedical research often culminates with a stay in a shabby hotel in Bethesda. There the peer reviewers assign the grant a priority score between 1.0 (perfect) and 5.0 (nonfundable). Such is the demand for money that in reality a score higher than 1.5 is rarely successful.

All it takes for a grant proposal to fail is for one or more of the dozen reviewers not to like the field, the researcher, the institution, or the approach. A reviewer may admire the scientist who drew up the grant and respect the research, but in a highly competitive field, blocking a rival grant could boost the chance that the reviewer's own lab would receive funding. Similarly, it could help minimize the use of a new approach that has passed by the reviewer's own laboratory, one that had been written off as unworkable. Outright hostility and vitriol are not necessary to successfully kill a rival's grant application; one merely has to be lukewarm or offer faint praise.

The proposal that Mark Adams and I submitted on sequencing human DNA was, of course, doomed. Our score was greater than 1.5, and if we had been part of an ordinary venture, our proposal would have been beyond resuscitating—it would have been stillborn. The reason there are no second chances is that it takes nine months to complete the grant process, so even if your breakthrough idea was resubmitted, the resulting one- to two-year delay would guarantee that it could not stay at the front of a field.

What was different in my case was that I was able to find alternate sources of backing, thanks to my new arrangement with PerkinElmer, but perhaps what was even more important was that my new sponsors protected my project from the conservatism of peer review. There is a fundamental catch-22 cyclical conundrum in this process: One has to persuade one's peers to back new science and to back "good" science (that is, science that works). But where new ideas are concerned, there is no way to know if they indeed constitute "good" science until the experiments have been done. Anyone with an untested method, novel ideas, and unique insights has a battle on his hands. In my case I was promoting a new method that had no guarantee of working on the scale of the human genome and, even if it did, would leave many small gaps in the sequence, both of which I publicly acknowledged.

I had heard on the grapevine how senior figures such as Francis Collins, Eric Lander, Bob Waterston, John Sulston, and Sulston's bosses at the Wellcome Trust had discussed how to stop my program. They were initially divided over whether to speed up and change their own strategy or to stick to their existing plan. The American camp feared that they would lose everything if Congress decided that the public effort was a waste of money. The Wellcome reaffirmed its commitment to the existing plan. Watson said it was "absolutely critical, psychologically," to garner the backing of the NIH for the public effort to proceed.[1]

The problem was that much of their analysis of my work was not taking place behind closed doors in Bethesda but in the national media, in newspapers, on television, and in leading science journals. As the weeks went by, the attacks became harsher. Celera was going to produce the "draft genome," or the "Swiss cheese" or "CliffsNotes" version of the human genome.

That June a congressional subcommittee held a hearing on how my effort was going to affect the federally funded project. Collins, who had arrived dressed in a sports jacket, tie, and trousers just like mine, made a point of emphasizing our matching wardrobes to symbolize how "we are intending to be partners in every possible way."[2] Maynard Olson of the University of Washington complained that all he knew about my plans had come from a press release. Despite that, and despite Olson's own track record—his lab's yeast artificial chromosomes had actually held up the genome effort—he confidently predicted "catastrophic problems" with Celera's approach, and he warned that I would end up with 100,000 "serious gaps" in the human genome.[3] In my address I reminded everyone that sequencing the genome was not about a race but about research to understand and treat diseases. The public effort should be judged by how it worked with new initiatives, not by how it competed with them. Even my detractors conceded that I won this particular round of the public brawl. As Sulston later admitted, I emerged with my credibility intact while Olson's criticisms came over as sour grapes.

Perhaps one of the lowest moments in this public battle came in June 1998 with a quote that Francis Collins added after he had been interviewed by Tim Friend, the science reporter for *USA Today*. Their conversation had ended, but Collins apparently had a brainwave and thought of an even catchier way to describe the end result of all our toil at Celera. Originally he argued that we would come up with the "*Reader's Digest*" version of the genome. He called Tim Friend back and asked if he could call my genome the "*Mad* magazine version." Was he sure? Yes, Collins assured Friend. Later, when Collins came under attack for the remark, he denied that he had ever said it, in the traditional shoot-the-messenger style favored by politicians for millennia.

What disheartened me most about the guerrilla theater, the public ridicule and denigration by some of the icons of the field, was the possibility that it would demoralize my team and my backers. I was most concerned about its impact on the not inconsiderable figure of Tony White, who was indeed upset about the press attention but mainly for the reason that it was not directed at him. White had hired a press person in an attempt to get more of the limelight and complained bitterly about the copies of press stories on the walls of Celera that featured his upstart protégé, Craig Venter.

As well as resenting all those column inches, White continued to seem to show a total lack of comprehension of the Celera business plan. Although he had accepted my one absolute precondition of having the right to publish my data and release the sequence to the public, he often tried to renege on the agreement. He remained a devotee of the old-fashioned genomics business strategy of secrecy and/or patents. After all my dealings with HGS, I knew precisely the source of White's thinking, an attitude I still joke about in my lectures: the biotech industry mantra of "one gene, one protein, one billion dollars." Because a few human genes were, in fact, worth billions, it was widely assumed that there were hundreds or thousands more genes that would be equally lucrative. The logic was simple and simpleminded. The biotech companies HGS and Incyte took the lead in human gene patents, but today their stocks trade below their cash value despite their vast human gene patent portfolios. Most now understand what I have always believed: that human gene patents usually have less value than the cost to pursue them. Of the twenty-three thousand or so human genes, fewer than a dozen have generated real value for the businessman or the patent.

The conflicts with Tony White were intense from the first day and only grew worse. Whenever he visited Rockville, he would always resort to intimidation by yelling at employees or simply by being rude. In truth, Tony was not so difficult to endure because he seemed disengaged from the day-to-day business of the Applera Corporation, which now owned Celera and Applied Biosystems. Tony had just purchased (with company money) a $30 million jet (he gave the old one to Hunkapiller, who rarely used it) and would visit Rockville only once a month or so, and I got the impression that he spent most of his time flying between his various houses. (He was in the process of building a new one in Atlanta.) Tony had the Applera chief financial officer send him a daily note calculating his net worth based on that day's stock price, and he would call at least once a week to complain about the price being too low. Although Celera was a start-up company that aimed to have a long-term impact, he could not see beyond the quarterly reports and stock valuation.

Genes Are Not the Whole Story

Look through my DNA and compare it with that of all kinds of other creatures, such as my dog Shadow or a fruit fly, and you find a strong hint that huge tracts of what were once written off as meaningless junk contain a hitherto unrecognized "genetic grammar," making the language of our genes much more complex than previously thought. Evolutionary forces tend to retain important DNA sequences, while allowing unimportant sequences to change, which accounts for why genes are similar in all mammalian species. In the journal *Science,* Stylianos Antonarakis of the University of Geneva Medical School, Ewen Kirkness of The Institute of Genomic Research, Maryland, and colleagues compared my DNA to that of dogs and species as distant as elephants and wallabies and revealed large sections of supposed "junk" that are nearly identical. In all, some 3 percent of mammalian genome sequence does not code for protein, yet it is highly conserved and thus must be significant. These regions of what were once called junk have been redubbed "conserved non-genic sequences," or CNGs, a reference to how they are not conventional genes. Another study from an international team codirected by researchers at the Wellcome Trust Sanger Institute and the Broad Institute found additional evidence that these regions perform important functions. Although CNGs are not very variable, they may still carry mutations that can be detrimental to humans and cause multifactorial disease. But we are a long way from understanding their influence.

Perhaps they contain binding sites for proteins that regulate how genes are turned on by the ease with which they stick to so-called transcription factors (such as my apelike one, mentioned earlier), of which there are around eighteen hundred in the genome. Perhaps they are unidentified exons—that is, parts of genes that we have not recognized. Perhaps they help maintain the structural integrity of the genome, ensuring it is the right shape for the cellular machinery to interpret the code, or represent some other yet to be defined functional unit. Some symptoms of Down syndrome, caused by an extra copy of a chromosome, might be linked to the presence of additional CNGs, for example.

He would often return to his favorite theme—could I explain to him once more how I was going to give the human genome sequence away and still have Celera make money?—and when even he felt too embarrassed to pose this question, he would call to say that an old friend was asking him the same thing. He knew the answer, of course, but did not know how to express it convincingly. Once again I would explain that the raw genomic sequence was of little value to scientists, biotech companies, pharmaceutical companies, or the public. Tony found it hard to grasp that the human genetic code was 3 billion base pairs of gibberish, just endless strings of As, Cs, Gs, and Ts, and would be worthless to anyone who did not have the ability to identify the tiny fraction that coded for proteins, for example.

It was often overlooked that Francis Collins and his friends could portray the public genome effort as "pure" and free of patents, because most of the data it produced were placed without any understanding or context into GenBank, the public repository of DNA sequences. What was of real value was the complex analysis of the genetic code to discuss what it meant. To accomplish precisely this, Celera was building new software tools for the world's most advanced computer, which was dedicated to this one job. After sequencing the human genome we would sequence the mouse genome, providing a crucial tool—comparative genomics—to differentiate the really important parts, the so-called evolutionary conserved regions common to both genomes, and to determine their function. We would also look for spelling mistakes in the genome, single-letter nucleotide polymorphisms, or SNPs (pronounced snips), that are linked with the risk of disease or drug side effects or that govern the efficacy of a treatment.

So what is it we sell, exactly? Tony would usually ask after this explanation. We were trying to come up with the equivalent of Windows software for the genomic world, I would say, although I would joke that I was no "Bill Gates of the genome," as I had often been called in the press. Another way I liked to put it was that we wanted to become the "Bloomberg of biology": We wanted to sell access to the information that was gathered, packed, and organized in a comprehensive, user-friendly database. We wanted to wed leading-edge molecular biology with heavyweight computation to reveal the logic of biology to our paying customers. Tony never liked that answer either.

Peter Barrett, who had been sent from PerkinElmer to be the chief business officer of Celera, understood the database business model. He was smart (he held a Ph.D. in chemistry), an affable guy, and had managed to survive and prosper in the PerkinElmer corporation over a twenty-year period. Because he had a high level of distrust of others in management and was something of a loner, he might not have been the best person to help Celera's case. That being said, he clearly put 110 percent effort into building the business and helped make us millions in revenue before we had sequenced even a single letter of code.

I staunchly believed, as I still do today, that the database business was a viable model to make Celera a profitable company, and I think that I fully lived up to the bargain I made. Peter did remain sensitive to the patent issues, believing in the mantra of the biotech and pharmaceutical companies. While I was trying to defend the concept of not patenting our sequences unless they were clearly of value to new diagnostics or pharmaceutical development, Celera's patent attorney, Robert Millman, was by now complaining behind my back to White. Millman against my wishes wanted to patent each and every sequence, to carpet bomb our sequence with legalese.

Millman's actions should not have come as a surprise to me: After all, he had worked with Bill Haseltine on drafting the documents to block the publication of the first genome, *Haemophilus influenzae,* in 1995. Now he was living in what he thought was patent paradise—or, as he liked to put it, a "patent attorney's wet dream."[4] His vision of a great gene grab had also gripped the imagination of William B. Sawch, the Applera chief legal counsel, who encouraged Millman to go over my head, and Millman began to report to Tony White. He would complain that I was throwing money out the window by not filing patents on every gene, whether fruit fly, human, or mouse. Using computer-generated guesses as to whether a sequence contained a gene, Millman wanted to lay claim to patents before our paying customers could, let alone our competitors.

The battles went all the way to the Applera board, where I had to argue my case for filing patents only when there was clear value. I was fighting for my integrity and that of my team because we had promised to make the human genome publicly available. The stress was more than I could handle at times. By then I had gotten to know Bill Clinton a little and was inspired by the way he dealt with the endless pressures from the job, the media, and his political opponents. Not letting your opponents see you cringe and sweat can be more damaging to your attackers than a good counterpunch (though the latter can be very satisfying).

Despite these obstacles, the morale at Celera was not only high, it was electric. Everyone who visited us in Rockville would comment on it. People were happy, excited, and energized in a way I had never experienced before. We loved giving tours of Celera, especially me, for I was immensely proud of my team and what I had created, which felt like a scientific Camelot. New ideas, new approaches, and new techniques were instantly rewarded and put into action. Each and every person there knew that he or she was contributing to the whole effort and that we were making history.

My specialized teams in each area had the same basic instruction: Accomplish the impossible. I had also stressed that if any one team failed, the entire process would fail, for in shotgun sequencing every step rests on the success of the previous one. Each of my knights cultivated his own subcultures of merry men and women. In honor of their responsibility for working up new algorithms and computational approaches to genomics, the team headed by Gene Myers and Granger Sutton fostered a true geek culture, complete with high-octane espresso makers, foosball, and ping-pong tables. Myers even called his band of followers "the geek group." Each Monday a battle would commence, when geeks, clad in plastic Viking helmets and armed with Nerf guns that shot foam balls and even the occasional inflatable mace, waged war on the bioinformatics group, who used Nerf crossbows as their weapons of choice. During combat, Wagner's "Ride of the Valkyries" often blasted out of speakers. Ham Smith's team included a group of young, attractive technicians who helped to assemble the library of clones and whom I liked to call Ham's harem. We did not have a Round Table, but we did have an excellent cafeteria in the basement with good food cooked by a great chef, Paul. The cafeteria became the central meeting point where almost everyone ate, bonded, and brainstormed daily. Whenever I felt low, seeing with my own eyes what we had built and talking to the Celera team always renewed my energy and my ability to struggle on.

We were now housed in our two 100,000-square-foot buildings, which included four floors and a basement. On the first floor of Building 1 were the administrative offices, my office, and those of the senior scientists; the second floor became the proteomics facility; the third floor would be devoted to the ABI 3700 DNA sequencers and cubicles for the sequencer technicians; and on the top floor DNA was processed.

The genome effort would start on the top floor with Ham Smith and his team, who would construct the sequencing libraries by using the shearing forces in a nebulizer—which sprays the solution of DNA through a tiny nozzle to gently break the DNA from chromosomes into much smaller fragments. After the appropriate pass through the nebulizer and then through gels, the resulting DNA fragments could be sorted according to size. Ham could isolate pieces that were 2 kb (two thousand base pairs), 10 kb, and 50 kb, in length. The random pieces of DNA would then be inserted into plasmid vectors that would enable the DNA fragments to be inserted into E. coli and grown millions of times. In this way we wanted to create tens of millions of fragments to make three distinct libraries: a 2 kb library, a 10 kb library, and a 50 kb library.

The libraries would then be taken down the hall to the bacterial facility, where each would be "plated." The bacteria would be diluted so that when the resulting bug gruel was smeared across an agar growth plate (a waxy substance containing key nutrients to feed the bacteria), each individual bacterium would be separated from its neighbor by about one millimeter. As the bacterial cells divided over and over again, colonies of E. coli containing a single fragment of human DNA would grow in each spot, becoming visible after a day.

Not so long ago, scientists would use sterile toothpicks to transfer bacteria from such colonies to a growth tube in order to make substantially more DNA. In what we called the Picking Room, we replaced the toothpicks and technicians with large robots that used a very accurate mechanical arm with a small TV camera attached to study the colonies. If the bacteria were too close, the robot ignored them, since different clones could be intermingling. But if the camera clearly revealed a single colony, then the robot arm stabbed it with a metal probe and transferred its precious DNA cargo to a growth plate (a plastic dish with 384 small wells containing growth media), where the bacteria would be multiplied millions of times. The probes were automatically cleaned each time, and in one day our four robots could process more than 100,000 clones. They were fascinating to watch and would become a favorite with visiting camera crews.

Removing the human DNA from the bacteria proved to be one of the more difficult steps to scale up. The DNA itself grows in the plasmid, which is separate from the bacterial chromosome. In a typical molecular biology laboratory a good technician can perform a hundred plasmid preparations every day. To cope with the clones pouring out of the Picking Room, we would have needed a thousand technicians.

The wells in the 384-well plates were about 1.5 inches deep and very narrow, which created some unique problems. Early tests revealed that the bottom of the wells did not get sufficient oxygen, limiting the growth of the bacteria there. The team solved this problem in a clever fashion by placing stainless steel balls (the size of shotgun pellets) in the wells. A large number of the plates were placed on a circular platform and then spun slowly past a series of magnets placed at different heights so the balls would rise and fall through the bacterial media. By mixing the contents this way, we achieved uniform bacterial growth.

At the same time our chemistry team devised a new way to burst open the bacterial cells to release the plasmid containing the human DNA. By placing the 384-well plates in a centrifuge, we could drive the remains of the bacteria and their DNA to the bottom of the wells, leaving the plasmid and its precious cargo of human DNA in solution. The plasmid DNA was then readily purified by a new method that saved more than $1 for every plate of 384 wells that we processed.

The next step was to attach the four color dyes to the four bases of the genetic code. We would do this using the DNA amplifying workhorse of molecular biology, PCR, which could copy DNA while adding the dyes at the same time. At Celera we had 300 PCR machines working simultaneously. Now, at last, we were ready to read some DNA.

Approximately 520 384-well plates containing now purified and reacted DNA were passed through the 3700 DNA sequencers, which would read out the order of the base pairs of the genetic code. Inside the sequencer, the DNA was separated one molecule at a time in a fine capillary tube. As the DNA reached the end of the capillary, a laser would activate the dyes that were attached. The activated dyes were then detected by a small TV camera and the data sent into the computer. This was the critical step as the biological information molecule, DNA, was transformed from the analog signal of biology into a digital code. Its four chemical base pairs were translated into four colors, which in turn were translated into a series of ones and zeros representing the four genetic bases. Reading each piece of DNA from end to end provided five hundred to six hundred base pairs of code.

We had a process that worked. Now we had to repeat it 26 million times and then put the sequence together again.

This chain of steps would hold the key to sequencing the human genome and future advances in my applications of this approach to the environment. I had known that DNA automation coupled with new computational analysis was the key to understanding our genome since developing the EST method. When I began using the very first DNA sequencing machine in 1987 with Jeannine Gocayne, we both became proficient at seeing patterns in the four-color readout. Attempting to match these patterns in thousands of sequences, let alone millions, was clearly beyond human cognition. And yet computers are, relative to the human mind, only primitive pattern seekers. We had always pushed at the limits of whatever computer hardware, software, and analytical methods had to offer, and now we were attempting to go where no one had been before.

Building 2 at Celera would handle this massive computational challenge. The basement was dedicated to powering this awesome task and housed tons of lead-acid batteries to provide a stable and continuous supply of electricity to the computer system on the floor above. The computer room had cost us more than $5 million before the first hardware had even arrived, simply to provide air-conditioning, fire suppression systems, and security measures, the latter at the insistence of our insurance carrier. The room had no outside walls in case of bombings by Luddite groups (not an uncommon occurrence at data centers) and had to be entered by passing through security guards and using a palmprint reader.

Some in my organization became a little paranoid, for we were subjected to not infrequent mail and telephone threats. The FBI would visit occasionally to remind me that I was a likely target for the Unabomber and instructed the mailroom to check my correspondence and parcels with a metal detector. (I had one at home as well.) Marshall Peterson insisted we cut down a nearby strip of trees (Mad Dog feared they could become a nest for snipers) and that we move the executive offices to one of the higher floors. Internet security was even more critical, as we faced daily hacking attacks and had a world-class team working around the clock to thwart them.

With its Compaq Alpha chip the Celera computer could top around 1.2 teraflops (teraops), equivalent to 1.2 trillion calculations per second. The computer also had 4 gigabytes of RAM and around 10 terabytes, or ten thousand gigabytes, of hard disk storage space. Peterson's team would come to be incredibly proud of what they had built and the speed with which they accomplished it, an unprecedented achievement in the computer world of the day.

In 1999 our computer was rated by Compaq engineers as the third largest in the world and the largest computer in civilian hands. (Today it would not be in the top few hundred, and even PCs can accommodate 64 gigabytes of RAM.) Overseeing all this would be a *Star Trek*–style control room where giant wall screens and dozens of smaller computer displays tracked the CPU utilization, the computer room temperature, who was in the facility, CNN, the weather, the state of Internet traffic, the 300 ABI 3700 DNA sequencers, the power grid, the extent of database use by each subscriber, and, for the sake of Tony White, the Celera stock price.

Because the business plan approved by the board and by Tony did not call for any customers in the first year, or really until we had succeeded in sequencing the human genome, the only index of success that our masters really understood was the stock price. Initially it hovered around $15 a share, which meant that the company was worth on the order of $300 million, roughly equivalent to the cash investment. But when this price persisted, Tony became increasingly anguished, threatening to sell or close down Celera. By then the great day that the PerkinElmer board had approved the formation of Celera and had convinced themselves they could be doing something great for humanity was a distant memory.

Peter Barrett and I had paid visits to the major pharmaceutical companies and found, unsurprisingly, that the ones that were most forward-thinking in terms of genomics were the most interested in what we could offer. To subscribe to our database meant paying from $5 to $9 million each year over a five-year term. The companies were free to pursue drug development without royalties to Celera (a concession granted only after a major internal battle with White). Amgen became the first to sign up.

The pharmaceutical companies all had major secrecy concerns. Some worried that spies perched in the trees outside could photograph my office computer screen. We were also subject to computer security issues that arose from the very structure of our business: Each company worried that it rivals—and even the Celera team itself—would monitor how it was using our genome data. And, of course, most companies wanted to block or limit the publication of the human genome data to deny their competitors access to the sequence. This poured more gasoline on the already combustible relations between the scientific and business teams at Celera.

The Celera program had been devised on the assumption that for years we would be the principal source of human genomic data. Because of this we had won the concession that we would release our data every three months on the Internet, despite the fact that the board of directors of PerkinElmer, Tony White, Peter Barrett, and the pharmaceutical companies had all hated that provision. But when Collins and the Wellcome Trust announced that instead of working with us they would compete and rapidly do a "rough draft" of the genome, I was immediately blocked from releasing the data as I had promised. I did not fight the order, given all these developments, since I was not unhappy with having to wait to publish the whole human genome in one go, with one dramatic announcement and a unique scientific paper.

Despite the many adjustments to the public program, its supporters used the change in the Celera data release plan as evidence that we could not be trusted and had dishonorable intentions. As Maynard Olson argued: "Clearly, one possibility is that Celera's game plan, from the beginning, was a classic 'bait-and-switch' scam. In this scenario, the company's strategy was to use the promise of free, unrestricted access to the data to undercut support for the public project and thereby set the stage for a lucrative monopoly in selling the sequence."[5]

The reality was more straightforward. The changes were crucial if Celera was to survive and thrive. My biggest challenge at this point was coming up with terms for the pharmaceutical companies that I could live with. In other words, I had to be able to satisfy their obsession with secrecy and still be able to release the human genome sequence to the world. I worked with the scientific team to project our best estimate of when we would have the first analysis done and a paper ready to submit to a scientific journal. To encourage the pharmaceutical companies to sign up for the database, I agreed that we would not publish our data prior to its appearance in the journal, but we were at liberty to release any data that matched sequences already available to scientists. One of the major companies went as far as insisting on a financial penalty for each exon coding sequence that I released that was not found already in public databases.

All the pieces were now in place. Money was pouring into Celera, because companies were paying millions to see our data and analyses. My scientific teams could now publish and release data. As had been the case with the EST sequences at TIGR and HGS, the more my competitors in the public effort tried to damage my efforts, the more they aided my goals. We had the equipment, the ideas, and the strategy. There was just one hurdle left: to get the robots, sequencers, and people working efficiently and in harmony on sequencing the human genome.

13. FLYING FORWARD

> *That the fundamental aspects of heredity should have turned out to be so extraordinarily simple supports us in the hope that nature may, after all, be entirely approachable. Her much-advertised inscrutability has once more been found to be an illusion due to our ignorance. This is encouraging, for if the world in which we live were as complicated as some of our friends would have us believe, we might well despair that biology could ever become an exact science.*
>
> —Thomas Hunt Morgan, *The Physical Basis of Heredity*

Many have asked me why, of all the creatures on this planet, I chose *Drosophila,* just as many have asked why I did not go straight for the human genome instead. The truth was, we needed a test bed; we needed a proof of concept. We needed some degree of reassurance before we spent nearly $100 million on sequencing the human with my untested method. And as any biologist knows, work on this little fly has driven the frontiers of biology, particularly genetics.

The genus *Drosophila* includes vinegar flies, wine flies, pomace flies, grape flies, as well as fruit flies, some twenty-six hundred species in all. But say the word *Drosophila* to any scientist, and he will immediately think of only one species, *Drosophila melanogaster.* Because it breeds quickly and easily, this airborne speck is a model organism for developmental biologists; they use it to shed light on the miracle that occurs between fertilization and the growth of an adult. Among many insights they have provided, fly studies have helped lay bare the workings of the homeobox genes that control the basic body plan of all organisms.

Any student of genetics is familiar with the fly research of Thomas Hunt

Morgan, the father of American genetics. In 1910 he noticed a white-eyed mutant male among red-eyed wild types. He bred this white-eyed male with a red-eyed female and found that their progeny were all red-eyed: The trait was recessive, and we know now that two copies of the gene, one from each parent, are required for fly eyes to be white. As Morgan continued to cross-breed the mutants back to one another, he noticed that only males displayed the white-eyed trait, and concluded that it was probably carried on the sex chromosome (the Y chromosome). He and his students studied the characteristics and inheritance of thousands of fruit flies in an enterprise that still goes on today in molecular biology labs around the world. By one estimate more than five thousand people worldwide study this little insect.

I knew firsthand the value of this research when I used cDNA libraries of *Drosophila* genes to help my studies of adrenaline receptors, revealing their equivalent in the fly—octopamine receptors—and shedding some light on the common evolutionary inheritance of the nervous systems of fly and man. When I was trying to make sense of the human brain cDNA libraries, the most revealing insights came from computer matches to *Drosophila* genes, when well-understood fly genes suggested the possible functions of similar human ones.

The *Drosophila* sequencing project was launched in 1991 when Gerry Rubin at the University of California, Berkeley, and Allen Spradling at the Carnegie Institution decided that the time was right to begin a fly genome project. It was in May 1998, when the Berkeley *Drosophila* Genome Project was one year into a three-year NIH grant and had finished 25 percent of the sequencing, that I made the offer that Rubin admitted "was too good to turn down" at the notorious Cold Spring Harbor meeting. By the same token, however, my strategy was risky: Every letter of code we produced would be studied by close to ten thousand fly scientists around the world, and Gerry's own high-quality genome data would serve as a benchmark to judge whether our own were any good. The original plan called for the completion of the sequencing of the fly genome within six months—by April 1999—in order for us to be ready to mount an attack on the human genome. I could not imagine a more dramatic and public way to demonstrate that our new strategy worked. I gave myself the coldly comforting reassurance that if we failed, at least it was better to do so quickly with the fly than draw it out by tackling the human. But the truth was that to fail at all would be the most spectacular burnout in biology. Now that Gerry had put his reputation on the line as well, all of us at Celera were determined not to let him down. I asked Mark Adams to take the lead on our end of the project, and because Gerry had a first-rate team at Berkeley, the collaboration went smoothly.

As we did for all genomic projects, we started by thinking hard about the DNA we were to sequence. Like humans, flies are varied at the genetic level. If the genetic variation in a population was more than 2 percent and we had fifty different individuals as a sample group, the reconstruction would be tough. The first job was for Gerry to inbreed the flies as much as possible to give us a homogenous set of fly DNA. But that alone was not sufficient to ensure genetic purity: If we extracted DNA from the whole fly, we would still have extensive contamination from the bacteria in the diet and gut. Gerry chose to isolate the DNA from fly embryos to avoid these problems. But even cells from the embryos had to be taken apart, in order to isolate the nuclei, where the DNA we wanted resided, to prevent it from being contaminated by the DNA in the mitochondrial power packs, which resided outside the nuclei. The result was a vial containing a wispy solution of fly DNA.

Once Ham's team received the pure fly DNA in the summer of 1998, they went to work constructing the libraries of DNA fragments. Ham himself liked nothing more than to cut and splice DNA, with his hearing aid turned low so nothing would distract him from his bench work. The libraries were supposed to launch an industrial-style sequencing operation, but all around us were the sounds of drills, hammers, and saws. With a small army of construction workers very much in evidence, we were still struggling with major problems, including debugging sequencers, robots, and other equipment, in our attempt to create a sequencing factory from scratch in months instead of years.

The first 3700 DNA sequencer did not arrive at Celera until December 8, 1998, and it was greeted with much fanfare and a communal sigh of relief. As soon as it was out of its wooden crate, we placed it in a windowless room in the basement—its temporary home—and started test runs as soon as we could. When it worked, we obtained very high quality DNA sequence data, but those first machines were very erratic. Some were dead on arrival. New problems continued to appear in machines that did work, and often on a daily basis. The software controlling the robotic arm had a major bug: On occasion the arm would fly at high speed across the instrument, smashing into the wall and halting that sequencer until a repair team came to fix it. Some machines suffered from wandering laser beams. Tinfoil and Scotch tape were used to stop an overheating problem that made the yellow Gs fade away from the sequence due to evaporation.

Even though the machines were now being installed on a regular basis, up to 90 percent were out of commission at the outset. The ABI service team was too small to cope and kept falling further behind. Some days we had no sequencers working at all. I had faith in Mike Hunkapiller, but it was shaken when he started to blame the failures on my team, the dust from construction work, the minor temperature variations across each floor, the phase of the moon, and so on. Some of us were beginning to go gray with the stress.

Dead 3700 machines, waiting to be shipped back to ABI, stood in the cafeteria as silent witnesses to the crisis, and we eventually ended up eating in a sequencer mortuary. My frustration was giving way to near panic, for I needed a certain number of operable instruments every day—230, to be exact. For the price tag of around $70 million, ABI either could provide us with 230 machines that worked 100 percent of the time or 460 that worked half the time. Mike was also going to have to double the number of trained technicians so that the instruments could be repaired as fast as they failed.

Mike, however, had no interest in doing any of this without more money. He had another customer by now, the public genome effort, which had started buying hundreds of machines without even testing them. While the future of Celera depended on these machines, what Mike did not seem to appreciate was that the future of ABI depended on them as well. As the bickering became intense, this issue promised to be the first real test of the mettle of the Applera board and of Tony White. The inevitable showdown came at a major meeting with ABI engineers and my staff at the Celera facility.

After we cited the incredible failure rates, quantified as the mean time between failures and the repair times, Mike once again tried to blame my team, but even his own engineers did not buy it. Finally Tony White weighed in: "I don't care what it takes or who we have to kill." This was the one and only time that he really came through for me. He ordered Mike to provide new instruments ASAP, even if he had to divert them from other buyers and even though the cost of doing so was yet to be determined.

He also told Mike to provide twenty more repair staff to fix the machines faster and to figure out the underlying causes of the problems. This was easier said than done, because there was a shortage of trained people. Eric Lander had lured away two of his top engineers, for a start, and according to Mike, this was our fault, too. He turned to Mark Adams and said, "You should have hired them before someone else did." That comment marked a low point in my esteem for him. The fact was, I was blocked from hiring ABI staff by the terms of our agreement, while Lander and others in the public genome effort were free to recruit them, and soon the company's best engineers were working for our rivals. By the end of the meeting I still felt under pressure but saw a glimmer of hope that the situation was going to improve.

And so it did, albeit gradually. Our inventory of machines rose from 230 to 300, so that when 20 to 25 percent of them were down with problems, we would still have 200 or so functioning and could manage our targets. The technicians did heroic work and steadily improved the repair rate, cutting the down time. The engineers in Foster City battled on to fix the more fundamental problems. Throughout that period I held on to one thought: What we were undertaking was possible. There were thousands of reasons we could and should fail, particularly given the deadlines, but failure was not a possibility I would accept.

We started sequencing the fly genome in earnest on April 8, around the time we were supposed to have finished. Even though I knew White wanted me out, I did my best to work with him so that I could accomplish my goals. The stress and worry followed me home, of course, but my closest confidante was the one person with whom I could not share these problems. Claire had clearly shown her contempt for my total absorption in Celera and for how I seemed to be repeating the mistakes of TIGR/HGS. By July I was beginning to feel the kind of profound low I had experienced only once before, in Vietnam.

Because the production line method was not up and running, we faced a grueling slog to put the pieces of the genome back together. To find overlaps and not be distracted by repeats, Gene Myers had come up with an algorithm that used a key principle of my shotgun strategy: sequencing both ends of all the clones produced. Because Ham had produced clones of three precise lengths, we knew the two end sequences were a precise distance apart. As before, this "mate-pair strategy" would give us a great organizing method to put the genome together again.

But since each end had been sequenced separately, for this assembly strategy to work we had to carry out careful accounting to make sure we could reunite each pair of end sequences: If we failed to associate a sequence with its proper mate only once in every hundred attempts, the strategy would fail. One way to prevent this would have been to use bar codes and readers to track every stage of the process. But at the outset we lacked the necessary software and equipment for the sequencing team, and so we had to do it by hand until bar coding could be implemented. This would be no obstacle for an old-fashioned sequencing lab, but at Celera a small team of fewer than twenty people was handling a peak flow of 200,000 clones every day. We could anticipate some errors, such as when a 384-well plate was read the wrong way, and then use software to spot the telltale error pattern and correct for it. There were, of course, still residual glitches, but it is a testimony to the team's skill and dedication that we could cope with the errors we did find.

Despite all the problems, we managed in four months to produce 3.156 million high-quality sequence reads, around 1.76 billion base pairs of sequence in all, contained between the ends of 1.51 million DNA clones. Now it was up to Gene Myers, his team, and our computer to put all the pieces together into the *Drosophila* chromosomes. The sequencing accuracy dropped as the pieces became longer. For *Drosophila* the sequences averaged 551 base pairs, and the mean accuracy was 99.5 percent. If we had just two 500-letter-long sequences that overlapped by 50 percent, most of us could find the point of overlap by sliding the two sequences past each other until the base pairs agreed. This is the way a monk would do it, but for shotgun sequencing there were not enough monasteries in the world to finish the job.

For the *Haemophilus influenzae* we had 26,000 sequences. To compare each one to all the others required 26,000 squared, or 676 million comparisons—the equivalent of one million monk years (the number of manual comparisons a monk could do in a year). The *Drosophila* genome, with its 3.156 million reads, would require about 9,900,000,000,000 or 9.9 trillion comparisons. For human and mouse where we produced 26 million sequence reads, about 680 trillion comparisons were required. Perhaps it is no wonder that most scientists were skeptical about the chances of this approach working.

Although he had vowed not to fail, Myers also had his doubts. He was by now working all hours and looked gray and exhausted. His marriage was in trouble, and he began hanging out with James Shreeve, the journalist and author who was shadowing us at the time. In an attempt to distract Gene, I took him to the Caribbean to relax and sail on the *Sorcerer,* but he spent most of the time hunched over a laptop, his black brow knitted over his black eyes in the bright sunshine. Despite the incredible pressure, Gene and his team would come to generate more than a half-million lines of computer code in about six months for the new assembler.

If sequence data were 100 percent accurate and there were no repetitive DNA, then genome assembly would be a relatively simple task. But in reality genomes are loaded with repetitive DNA of various types, lengths, and frequencies—the equivalent of a big stretch of blue sky on a jigsaw. The short repeats, consisting of fewer than five hundred base pairs, are relatively easy to deal with: They are shorter than a single sequence read, so the surrounding unique sequences would allow us to figure out where they went. But longer repeats were challenging. Our way past this was, as mentioned before, the mate pair strategy, sequencing both ends of each clone and having clones of varying lengths so as to provide the most overlaps.

The algorithms encoded in those half-million lines of computer code generated by Gene's team used a stagewise process, which began with the safest moves, such as the simple overlapping of two sequences, and then progressed to more complicated strategies, such as using the mate pairs to link up islands of overlapped sequences. This is like solving a complicated jigsaw puzzle by putting smaller islands of assembled pieces together into larger islands and then repeating the process—only our jigsaw puzzle had 27 million pieces. And it was crucial that the pieces were of a high-quality sequence: Imagine trying to do a jigsaw if the color or image on a given piece was blurred. For the long-range ordering of the genome sequence, a substantial fraction of the reads needed to be in mate pairs. Given that all the data were still being tracked manually, we were very relieved to find that we had more than 70 percent of the sequence data in mate pairs: The computer modelers had told us that any less would mean Humpty Dumpty could not be put back together again.

Now we could turn the Celera Assembler on the sequence data: In stage one, the data was trimmed to the highest accuracy; in stage two, the "Screener" removed contaminating sequences from the DNA of the plasmid vector or for E. coli. As few as 10 base pairs of contaminating sequence would block any assembly. Stage three was the "Screener," where each fragment was checked for matches to known repetitive sequences from the fruit fly genome, courtesy of the hard slog of Gerry Rubin. The locations of repeats with partial overlaps were recorded. In stage four, the "Overlapper" compared each fragment with all other fragments, a huge number-crunching exercise that we had successfully tested by smashing up the public *C. elegans* code and seeing if the Overlapper could reassemble it correctly. (We made several requests to publicly funded *C. elegans* genome scientists [Waterston and Sulston] for the sequence data they used to reconstruct the genome but were rebuffed every time.) Making 32 million comparisons every second, our algorithm was hunting for at least forty base-pair matches with less than 6 percent differences. When two pieces overlapped, they were assembled into a larger piece, a contig (contiguous fragment).

In an ideal world, that should have been enough to put the genome back together. But we had to contend with stutters and repetitions in the DNA code, which meant that a single DNA fragment could overlap with several different pieces, creating false connections. To simplify the problem we kept only the uniquely joined pieces, called "unitigs." The software that did this, the "Unitigger," in effect threw away all the DNA that we could not assign with confidence, leaving just the unitigs, the correct subassemblies of fragments. In effect, this step not only gave us some room to change our minds about how to put the fragments back together, but also reduced the complexity of the problem substantially by winnowing down 3.158 million fragments to 54,000 unitigs containing two or more fragments, a forty-eight-fold reduction. The 212 million overlaps were reduced to only 3.1 million, a sixty-eight-fold reduction in the problem size. The pieces of the jigsaw were gradually but systematically falling into place.

At that point we could use our knowledge of the way sequences were paired from the same clone, using the Scaffolder algorithm. All possible unitigs with mutually confirming pairs of mates were linked into scaffolds, imposing larger-scale order on all these small pieces of code. An analogy I use in lectures to describe this step compares it to building with Tinkertoys, which consist of sticks of different lengths that can be inserted into holes on wooden nodes (balls and disks) to build a bigger structure. In this case the nodes represent the unitigs. By knowing that paired sequences lie at the end of clones that are two thousand, ten thousand, or fifty thousand base pairs long—the equivalent of being a given number of holes apart—they can be lined up.

Using Gerry Rubin's sequence, which accounted for about one-fifth of the fly genome, a test of this approach produced only five hundred gaps. When it came to applying our own data, a test we carried out in August, we ended up with more than 800,000 little fragments. Having so much more data to work with indicated the strategy had done badly, it had failed—exactly the opposite of what we expected. Over the next few days, the sense of panic grew and the list of possible errors lengthened. Adrenaline was pumping through the top floor of Building 2, in what was jokingly called the Tranquillity Room, a lunatic reference to how I had named the meeting rooms in the main building after the Earth's oceans. But the room was anything but peaceful as the team went around in circles for at least two weeks in search of a solution.

In the end the problem was solved by Arthur L. Delcher, who had worked on the Overlapper. He spotted something odd about line 678 of the 150,000 lines of code, where a trivial slip meant that a significant fraction of matches was being thrown out. When it was fixed and the computer finished its next run, on September 7, we had 134 scaffolds that covered the working (euchromatic) genome of the fruit fly. We were ecstatic and relieved, and it was time to announce our success to the world.

The genome sequencing conference that I had founded years earlier provided the perfect opportunity, and I knew that we could expect a record crowd that was eager to see if we could deliver on my promise. I decided that Mark Adams, Gene Myers, and Gerry Rubin should describe our achievement: the sequencing, the assembly, and the impact on science, respectively. Eventually, I had to move the conference from Hilton Head to the bigger Fontainebleau Hotel in Miami because of the demand. Representatives from every major pharmaceutical and biotech company, genome scientists from around the world, and a substantial number of analysts, reporters, and others from the investment community were all in attendance. Our competitors, Incyte, had invested heavily in the send-off party, an in-house video, the works, to convince the delegates that it offered the "Best View of the Human Genome."

We all gathered in the main ballroom, which, as is traditional for conference venues, was a vast hangarlike room decked out in neutral colors and hung with chandeliers. It was supposed to accommodate two thousand people, and as the crowd grew larger, there was soon no longer any standing room. On September 17, 1999, as part of the opening session, Gerry, Mark, and Gene gave an update on the *Drosophila* genome effort. After a brief introduction Gerry Rubin announced how the delegates were about to hear about the best collaborative effort in which he had ever participated. The atmosphere became electric. The audience realized that he would not have spoken so glowingly unless we had something exciting to say.

An expectant hush fell as Mark Adams began to describe our factory-style operation at Celera and the new approaches we had established for genome sequencing. This was all a big tease; he stopped short of any mention of an assembled genome. Then came Gene, who talked the audience through the background to whole-genome shotgun assembly, the *Haemophilus* effort, and the major steps in our genome assembler, complete with an animated computer simulation of how a genome would be put back together. As the allotted time began running out, many were probably beginning to think that our presentation was going to be all PowerPoint and no data. But with a wicked smile Gene finally got around to mentioning that the audience might be interested in seeing some real data, not just simulations.

The data could not have been more clearly or more dramatically delivered as Gene Myers unveiled the genome. He knew that, on its own, sequence data was not enough, and so for good measure he showed a comparison with the sequences that Gerry had painstakingly put together the old-fashioned way: They were identical. He compared our assembly to all the preexisting markers that had been mapped to locations on the fly genome over decades. Out of thousands, only six did not agree with our assembly; after examining each we had ascertained that Celera was correct and that the errors were in the earlier work by other laboratories using older methods. Oh, and by the way, we had just started sequencing human DNA, and it looked as if repeats were going to be less a problem than they had been with the fly.

The applause was loud, long, and sincere. The buzz that erupted during the break after the talks showed that we had made our point. One of the public team was seen by a journalist to shake his head and remark: "Those fuckers are actually going to do it."[1] We left the conference feeling totally reenergized.

There remained two outstanding issues, both familiar. First was how to release the data. Despite my memorandum of understanding with Gerry Rubin, my business team was not happy with the idea of uploading the valuable fly data to GenBank. They came up with a proposal: Put the fly genome in a separate database run by the National Center for Biotechnology Information, where anyone could use it who agreed not to resell it for commercial purposes. Michael Ashburner, an excitable chain-smoker from the European Bioinformatics Institute, was unimpressed by how the sequence had been posted on a special server along with a protective notice, and felt that Celera "had screwed us."[2] (The subject line in one of his e-mails to Rubin read: "What the fuck is going on with Celera?").[3] Collins was also dissatisfied, but most important of all, so was Gerry Rubin. In the end, to the dismay of the likes of Millman and White, I sent the data to GenBank.

The second issue was that we had a genome sequence for the fly, but what did it all mean? We had to go further and analyze it, as we had done with *Haemophilus* four years earlier, if we were to publish in a journal. Annotating and describing the fly genome could take well over a year, a year that I did not feel I had because our focus now had to be on the human genome. In discussions with Gerry and Mark, we came up with a novel way to solve the problem, one that would involve the *Drosophila* scientific community, be exciting science, and move things forward rapidly. We decided to hold what we termed an "annotation jamboree," inviting top scientists from around the world to come to Rockville to analyze the fly genome over the course of a week to ten days. We would then write up our results and publish a series of papers on the genome.

Everyone loved the idea. Gerry began inviting leading groups to the jamboree, while the Celera bioinformatics team worked out what computers and software we needed to make their trip worthwhile. We agreed that Celera would cover the costs of travel and accommodation. And we crossed our fingers that the excitement of doing science this way would win the day. Among those who were to be included were some of my harshest critics, and we hoped that their political posturing was not going to undermine the event.

About forty *Drosophila* scientists turned up that November, and even our critics had found it too tempting an offer to ignore. The first meetings were a little rough, as expectations faced the reality of analyzing more than 100 million base pairs of genetic code in a few days. While the visiting scientists slept, my team worked around the clock to develop software tools to meet some of the demands we had not anticipated. By the third day the thawing began as the new tools enabled scientists to, as one told me, "make more exciting discoveries in a few hours than they had in their entire careers." At the sound of a Chinese gong the team met every afternoon to summarize their latest findings, discuss any problems, and make plans for the following session.

These meetings became more fun as the sheer excitement of pure scientific discovery intoxicated everyone. We were providing the first look into a new world, and what was glimpsed greatly exceeded everyone's imagination. There soon became too few hours to discuss all that we wanted to and absorb what it all meant. Mark held a dinner party, but it did not last very long, as everyone quickly drifted back to the lab. Lunches and dinners were soon being eaten before computer screens full of *Drosophila* data. Long-sought families of receptor genes were revealed for the first time, along with a surprising number of fly counterparts of human disease genes. High-fives, whistles, and whoops came as each discovery was made. Amazingly, one couple even found the time to become engaged.

But there was one significant concern: The team had discovered about 13,000 genes in contrast to the more than 20,000 they had expected. Since the humble worm *C. elegans* had on the order of 20,000 genes, many had reasoned that the fly must have more, since it had a nervous system and ten times more cells. There was an easy way to check that we were not making any mistakes in our calculation: We could take the 2,500 known fly genes and see how many turned up in our sequence. After extensive analysis, Michael Cherry of Stanford University reported that he had found all but 6 genes. After some discussion, all 6 were dismissed as artifacts. The fact that all the fly genes were accounted for and correct gave us a boost of confidence. The community of thousands of scientists devoted to *Drosophila* research had spent decades hunting down every one of the 2,500 known genes, one study at a time, and now they had all 13,600 laid out before them on a computer. After eleven days we had found out more than enough to provide an initial analysis of the genome.

A memorable moment came during the inevitable photo session during the final bout of backslapping and hand-pumping. Mike Ashburner got down on his hands and knees so that I could put my foot on his back for a photo; this was his way of acknowledging what we had accomplished in the face of his doubts and skepticism. The diminutive *Drosophila* geneticist even thought of a caption: "Standing on the Shoulders of a Giant." "Let's give credit where credit is due," he later wrote.[4] "Celera really pulled out the stops for the jamboree." Despite their attempts to portray the glitch in releasing fly data to public databases as reneging on our promises, my opponents were forced to concede the jamboree was "hugely valuable to the fly community."[5] We all parted friends after experiencing scientific nirvana.

We decided to publish three major papers: one on the whole genome, for which Mark would be first author; one on the assembly, with Gene as first author; and one on comparative genomics to worm, yeast, and human genomes, with Gerry as first author. The papers were eventually submitted to *Science* in February 2000 and published in a special issue dated March 24, 2000,[6] less than a year after my discussions with Gerry Rubin at Cold Spring Harbor. Just prior to the publication, Gerry arranged for me to deliver the keynote lecture at the annual *Drosophila* conference in Pittsburgh, attended by hundreds of top fly scientists. My team placed a copy of the CD-ROM containing the entire *Drosophila* genome sequence as well as the *Science* papers on every seat in the auditorium. Many in the room had been troubled when our collaboration was first announced, but after giving me a very warm introduction, Gerry assured the crowd I had delivered on every promise that had been made and that it had been a wonderful partnership. My talk concluded with some of the early findings that the jamboree team had made and a summary of what was on the CD-ROM. When I was rewarded with a prolonged standing ovation, I felt the same surprise and pleasure as I had when Ham and I first presented the *Haemophilus* genome at a major microbiology meeting five years earlier. The *Drosophila* genome papers went on to be some of the most cited scientific papers in history.

While thousands of fly researchers around the world were thrilled with my data, my critics quickly went on the offensive. John Sulston attacked the genome as flawed and a failure even though it was more complete and accurate than the one that resulted from his decade-long slog sequencing the worm, one that would require four more years to complete after a draft was published in *Science.* Sulston's colleague Maynard Olson called the *Drosophila* genome sequence a mess that Celera had left for the public genome effort to clean up. In fact, Gerry Rubin's team worked rapidly to close the remaining gaps in the sequence with the publication and comparative analysis of the finished genome less than two years later. These data confirmed that we had one to two errors per ten thousand base pairs across the genome and fewer than one error in fifty thousand base pairs of the working (euchromatic) genome. In the repetitive sequences, however, the data made it clear that even better algorithms could have a big impact.

Despite the public acclaim for the *Drosophila* project, however, during the summer of 1999 the tensions with Tony White had reached breaking point. White was by now becoming obsessed with my press coverage. Every time he visited, he had to pass by the framed copies of the articles about Celera's exploits in the hallway next to my office. We had blown up one piece, the cover of *USA Today*'s weekend magazine, which featured me sitting cross-legged in a blue-checked shirt with the headline: "Will This MAVERICK Unlock the Greatest Scientific Discovery of His Age?"[7] Copernicus, Galileo, Newton, and Einstein floated around me but there was no sign of White.

On a daily basis his press person would call to see if he could take part in the seemingly endless stream of interviews being concluded at Celera. He would only be satisfied—and then just for a while—when the following year, she managed to get him on the cover of *Forbes* magazine as the man who took PerkinElmer from a $1.5 billion market cap to the combined $24 billion that ABI and Celera represented.[8] ("Tony White has transformed a sad sack named PerkinElmer into a high-tech gene hunter.")

Tony also became obsessed with my public engagements. About once a week I would deliver a talk—a fraction of those I was invited to do—because the world wanted to know about our ongoing work. He even complained to the board of PerkinElmer, by now renamed PE Corporation, that my travel and speeches were breaking corporate rules. While I was on a two-week working vacation at my house on Cape Cod, Tony flew down to Celera with his CFO, Dennis Winger, and William Sawch, general counsel of Applera, to interview my key employees to reassure themselves that my leadership was sound. In other words, they hoped to get enough dirt to justify firing me. White was stunned when each person told him that if I left, he (or she) would, too. This created huge tensions, but it bonded my team more closely than ever. We would celebrate each victory as if it were our last.

When we published the fly genome sequence—the largest in history to that date—Gene, Ham, Mark, and I had a private toast because we knew that we had survived Tony White long enough to have our science validated. We had proved that the whole-genome shotgun method could work on a large genome. We now knew that turning the shotgun approach on the human genome would work, too. Even if Tony White pulled the plug on our operation the next day, we knew that we had this key accomplishment under our belts. I wanted more than almost anything to walk away from Celera and the world that he represented, but because I wanted to sequence *Homo sapiens* even more, I had to compromise. I massaged his ego so I could survive long enough to get on with the job and finish what I had set out to accomplish.

14. THE FIRST HUMAN GENOME

Generally, the first reaction to the prospect of being scooped is a combination of despair and hope that your opponent, "X," will fall dead. You may consider giving up, but this could leave you without any tangible results to show for years of toil. . . . So it is hard not to think about retooling your effort to try the same approach as your competition. Even though you are behind, by being a little more clever you might overtake him. He, of course, might then become hellishly mad.

—James D. Watson, *A Passion for DNA: Genes, Genomes, and Society*

Long before we started sequencing the first human genome, long before we were even sure that we would be able to do it, we had fun speculating about whose DNA would have the unique honor of being the first to be read from beginning to end. Who would possess the scientific curiosity, the self-confidence, and the peace and security of mind to want to have his or her own genome sequenced? Who would have sufficient understanding of the profound interplay of nature and nurture to want to see his own personal genetic programming published on the Internet, particularly when most people had been frightened by genetic determinists into thinking that this would lay bare all their biological secrets?

Technical issues were involved, too, many of which boiled down to how sex introduces more genetic variety into humans than the old-fashioned, asexual form of reproduction used by microbes. When it came to sequencing the genomes of bacteria, we had selected reference clones—all of which were identical, as the name suggests—that would provide a homogeneous DNA sample. For the fruit fly genome we had used highly inbred strains to ensure that we had DNA with as little variation as possible. But when it came to the human genome, there were as many genetic variations as there were people on the planet.

Because the structure of human DNA is a double helix with two complementary strands, it did not matter which strand of DNA we sequenced. But there is a complication with humans. Each of our twenty-three chromosomes exists in pairs: twenty-three from our mothers, including the X chromosome, and twenty-three from our fathers, including either an X chromosome for daughters or a Y chromosome for sons. (Females therefore have two X chromosomes, while males have one X and one Y.)

An obvious issue was whether to select a man or a woman. With a male we would have the advantage of there being both an X and a Y chromosome but the disadvantage of only half the amount of the DNA for the X and Y with the full amount for the other twenty-two paired chromosomes. If we chose a female, then we would have two doses of the X chromosome but no Y chromosome. If we studied only one person, should we pick an average subject or a President Clinton? What were the liabilities and risks to that person? Would such a person ever agree?

What became clear early on was that our decision did not really matter that much, for given the extent of human genomic variation, there would undoubtedly be a huge effort put into sequencing many more human genomes when the technology was ready. However, it did make sense for us to obtain as much genetic diversity as possible, given the scientific and commercial interests in genetic tests to find sequence differences linked with disease. This implied that we might sequence a pool of DNA from several people and create a consensus genome sequence, one not representative of any individual but an amalgamation of humanity, a reference genome.

Gene Myers and the team did several calculations to see how many subjects we could have in the pool without introducing so much variation that it would jeopardize our ability to assemble the reference genome using available algorithms and computers. Five or perhaps six was the limit if we allowed for substantial coverage of one person's genome to aid in the assembly. We decided that we would try to have a mixture of DNA from both males and females, and include some ethnic diversity.

At TIGR, before Celera had even been formed, Ham Smith had fretted about how to establish the perfect human sequencing libraries, not least because he did not have much experience working with human DNA. Ham and I had several discussions about how to procure human DNA, including from commercial sources. But to create the libraries, and be completely confident that we knew what they contained, Ham would have to start from scratch.

Human samples also meant long and involved procedures associated with obtaining informed consent, which would prevent us from even beginning the process during the six months that we would spend building Celera. Because Ham and I were keen to get moving, the next step was obvious: When it came to finding human DNA donors, we felt that there could not be two better informed individuals on the planet—individuals who had a deep understanding of the risks associated with having their genome sequenced and made public—than the two of us. Neither Ham nor I bought into simplistic notions of genetic determinism, which supported the idea that we were only what our genes made us, and that the trajectory of our lives could be accurately predicted from our genetic code. At the same time we both possessed a natural curiosity about our own genomes. There was never a question in our minds that we were undertaking any medical risk other than, perhaps, to our state of mind: We could expect political attacks from our detractors if it became known that we used our own DNA.

Once we agreed on this approach, we each decided that we would deliver a rich and easily produced source of DNA for the library: in the form of sperm (and were soon joking about who would need the larger tube). In the end we settled on using standard 50-milliliter sterile tubes and freezing the contents. While Ham could easily bring his sample directly into the lab without his technicians knowing the source, we thought it would be too obvious if I strolled in with the frozen tubes and handed them over. Because FedEx boxes with frozen reagents arrived at TIGR almost every day from Applied Biosystems, I simply took one of the opened boxes containing dry ice, added my sample, and delivered it to the lab, where most assumed that the sample had come from Mike Hunkapiller or Tony White. The subterfuge had to be repeated several times because the initial experiments consumed a great deal of DNA.

Once Celera was up and running, the issue of which additional DNA to sequence grew as complicated as we had feared. Lawyers became involved and offered divergent views of what should or could be done. To oversee the process I called on Sam Broder, the former National Cancer Institute director and now chief medical officer of Celera, and set up a stellar committee of outside experts. I informed Sam at the outset that we already had two DNA samples that had been turned into libraries by Ham at TIGR and that these were being used for all the initial sequencing to jump-start the Celera program. I revealed to him that Ham and I were the donors and explained that the remainder should include females and as much ethno-geographic diversity as possible. I left it up to Sam whether he should notify the committee that we were already generating sequences from two human donors; he thought it was better not to, but instead to establish procedures that would not be inconsistent with what Ham and I had already done.

The Eyes Have It

Read any popular account of genetics, and you will often see it said that DNA determines everything from susceptibility to disease to IQ (whatever that is) and eye color. In classrooms around the world children are taught that brown is dominant—that is, all it takes is for one parent to give you a gene responsible for a dominant characteristic, and you will have that characteristic, too. Thus, if one parent has brown eyes, so do his children, while two blue-eyed parents almost always have blue-eyed children.

Let's pretend that you have not met me or studied the dust jacket of this book and have decided to find out my eye color by studying my genetic code—the biological equivalent of chartering the *Queen Mary II* to cross the Hudson River. One volume of my code, chromosome 15, is a good place to start. There you will find a gene called OAC2, which is the major determinant of brown and blue eye color. The gene acts in specialized cells called melanocytes, which produce the pigment melanin, which is responsible for eye color. The physical basis of my eye color, like anyone else's, is determined by the distribution and content of melanocyte cells, though the process is more complicated than is commonly thought.[1]

Eyes are less likely to be blue or gray based on the exact genetic spelling of this particular gene, according to a study of more than six hundred normally pigmented people. (The non-blue/grays had the base pairs A/T or T/T in one variant, and A/G or G/G in another, or the combination of both variants.) According to these data my genome indicates that I have a higher possibility of having blue/gray eyes. Instead of possessing what the scientists call a "non-blue/gray" variant, I have two variants—one with C/C and A/A and the second with G/G and A/A, and I do, indeed, have blue eyes. While my genome tells a straightforward story, eye color actually depends on several genes, and although it is not very common, two blue-eyed parents can produce children with brown eyes. And although blue and brown eyes dominate among Caucasians, we also have gray/green/hazel and every shade in between. The simple and simplistic textbook account of the genetics of eye color does not do Nature justice.

The committee expressed two major concerns. One was that if it was possible to identify which individuals had given their DNA, there were risks that they might lose their death and/or life insurance if any disease genes could be identified in their genome. Similarly, if they had mutations associated with various undesirable social traits or personality disorders, it could cause those individuals problems if their identity was revealed along with their genetic code. The policy we ultimately established was that, for reasons of liability and to protect our donors, Celera would never disclose their identities. However, the committee did accept that the individuals themselves could disclose that they were donors if they so wished.

The second major issue concerned sequencing genomes of various ethnic backgrounds. I met with the committee only once, and it was on the "race" issue, for there was real anxiety that the data might be used by some to justify racism. It seemed to me to be fundamentally wrong to sequence the genomes of five white males to represent humanity, particularly when at the genetic level we all look pretty much the same. When the committee heard the argument expressed that way, it quickly agreed on diversity. We wanted to pool about twenty potential donors and advertised in *The Washington Post* and around Celera and Applied Biosystems. Unsurprisingly at least two reporters offered their DNA, one of whom would write an article on the Celera donor process.[2]

Each and every donor, including Ham and me, was required to take a class on the risks and the informed consent process, and to sign the necessary agreements. When Broder provided us with a consent document that he had drafted with the advisory board, I joked that we didn't want DNA from any man who could read all thirty pages of this complex legal document and still manage to ejaculate, because he had to be a lawyer.

Each donor was paid $100 for his or her sample, which for women was blood, taken from the arm. Men gave both sperm and blood (although a few refused to provide sperm). (When this process was described to a well-known public figure, she complained, "This makes complete sense: Men get paid to have an orgasm, and women get stuck with a needle.") Once we had each sample, its donor was assigned a code number, and only Broder had the identifying key.

From each donor we attempted to grow cell lines and create sequencing libraries, which we followed up with test sequencing. The five finalists were selected by my senior staff, including Broder, on the basis of anonymized information, which included the code number, the sex, and the self-identified race, and of having high-quality libraries that provided good sequence data as well as the permanent cell lines: These were Ham and me, and three females who described themselves as African American, Chinese, and Hispanic. I still do not know the identity of any of the women. Although several donors, including the reporters, later revealed their identities, there was no way to establish a link between them and the sequenced DNA without resequencing another sample from them. The sequence that we ultimately published was a composite of the five donors, and the fact that our genome sequences could even be assembled as a composite is a testament to the similarity of humans across the planet at the level of DNA.

The government program faced even greater problems over the decision of whose genome was to be sequenced. Collins and his colleagues trumpeted the fact that they had blended the DNA of fifteen to twenty individuals so that the resulting genome sequence would be from an anonymous everyperson. Over the years many DNA BAC libraries had been produced from samples contributed by helpful postdocs, lab donors, and so on, long before anyone had thought much about thorny issues such as ethical and informed consent. After one or more of the donors self-identified, all the libraries were thrown out, marking another setback for the public program and another instant policy change. Then almost all the public genome ended up coming from only one or two donors (a fact that was kept quiet for as long as possible).

As soon as the last fragment of the *Drosophila* genome had been sequenced, I turned the Celera facility over to a full assault on the human genome. By then—the morning of September 8, 1999—all the hard work by our technical teams and by ABI had cut the sequencer failure rate, which had been as high as 90 percent, to as low as 10 percent. This meant that we still had at least thirty $300,000 sequencing machines down for repair on any given day, but even at this failure rate, we had sufficient capacity on the three hundred functioning machines to be able to sequence the human genome in less than one year.

But by then the pressure was on us: The public program had announced that it had already sequenced about a quarter of the genome. In another change of direction, my rivals announced they would produce just a crude version of the genome and finish this "first draft" by the following spring, no doubt accompanied by a media event. The key differences in what we were doing at Celera and the altered publicly funded approach came down to standards and strategies: the whole-genome shotgun technique versus the clone-by-clone traditional approach. I knew that we had the winning strategy, and that even with the same or even greater sequencing capacity, the government-funded labs could not compete unless they abandoned their standards and changed their plan to match ours.

The year before we started our effort, in September 1998, the official government line had morphed into: We are going to do a draft genome before Celera can and finish the job by 2003, the fiftieth anniversary of Watson's joint discovery of the double helix. In place of this original plan of publishing high-quality data over the course of a decade, they were now making an effort to dump as much raw sequence into the public databases as quickly as possible. My self-proclaimed rivals—the five surviving genome centers, which had nicknamed themselves the G5 (the group had started out as the G18)—had convinced themselves that by doing so they were blocking me from both patenting the genome and getting credit for finishing first. I was baffled by the silliness and immaturity of their thinking. While my many critics were obsessed with the release of the Celera data, the public-funded labs were heedlessly dumping sequences into the public databases that the pharmaceutical companies were gleefully downloading nightly so they could file patents on them. This naïve policy by all those opposed to patenting of the human genome therefore had precisely the opposite effect: Gene patents were filed sooner and faster, and almost all were based on the government data, not Celera's.

The downgrading of the government objectives was met with little comment or analysis, thanks to a masterful job in public relations. No one seemed to appreciate that by changing its objectives the public effort had also, in effect, exchanged the aim of a highly accurate and complete chromosome-by-chromosome effort to sequence the human genome to a quick and dirty "rough draft" that made what we were undertaking at Celera appear thorough and comprehensive. So much for the original mantra of "quality first"; so much for the fears that a draft genome would sap the motivation to finish the job properly.

And so much for the shortcomings of the 3700 that lay at the heart of Celera's effort. Sulston's team had published a review of the machine in *Science*,[3] claiming that because it produced shorter sequenced fragments—"reads"—it offered no advantage over the existing instruments. ("There is no immediate gain in throughput in terms of capital investment.") The review triggered a drop in both ABI and Celera stock prices. Ironically, the assessment verdict was essentially ignored by the government-funded teams after my decision to use the 3700, as were the usual cumbersome evaluations and analyses required of them before purchasing expensive equipment, and our usually conservative rivals scrambled to buy the relatively untested 3700s as fast as they could. ABI reported about $1 billion in sales in the year after Celera was launched. The Wellcome Trust alone spent more on the 3700s than Celera so that its Sanger Institute could sequence only 25 to 30 percent of the human genome. Meanwhile, MIT lent money to Eric Lander to buy even more instruments than the government was prepared to fund, based on future returns from overhead on grants he had received from Francis Collins (grants that are still worth more than $40 million a year), which gave him the biggest operation in the public consortium.

Thanks to the change in the strategy of the G5, my bosses stood to make much more money. Hunkapiller and White loved feeding the public program, which now wanted to buy millions of dollars' worth of 3700 sequencers and reagents, and were like arms merchants who had started a war so that they could sell weapons to both sides. It was frustrating having to work so hard to build team spirit among the Celera sequencing group as they watched our business "partner" equip our rivals at a faster pace.

Using the same instruments for reading the genetic code meant that—aside from the not inconsiderable matter of the government-funded program's having more than tenfold greater resources of money and manpower—the key difference between the Celera and public programs lay in our respective scientific strategies. To most the word *sequence* implies that the base pairs of the genetic code are actually assembled in their proper order; no one would think that he had assembled a jigsaw puzzle simply by throwing the pieces on a table. However, because the government-funded labs were doing thousands of mini genome projects by sequencing BAC clones one at a time, they had thousands of mini jigsaw puzzles to solve, order, and orient, whereas we had only one big one to do. I never imagined that they would claim to have assembled BAC clones or chromosomes when all they had were the puzzle pieces. I was betting on the integrity of our science to prevail and on our programmers, our algorithms, and our massive computer to outcompete the much bigger public effort.

In assembling a DNA sequence from shotgun sequencing, a minimal amount of coverage of sequences is required. For example, a 1X, or onefold coverage of a 100,000-base-pair BAC clone, means that you have generated 100,000 base pairs of DNA sequence. But this does not mean you have sequenced every letter of the clone one time. The catch is that these are *randomly* generated DNA fragments. (You would be highly unlikely to end up with a complete newspaper, for example, if you tore up fifty newspapers into fifty separate pages, mixed them in a box, and pulled out fifty pages at random.) Put these random fragments back together, and, as is predicted by statistical methods, you will find that 1X coverage actually represents only about 66 percent of the DNA sequence of the clone. (Some parts will be duplicated; others will be missed). Threefold, or 3X, coverage is required to cover 96 percent of the sequence. Using the sequence assembler method of the government program, it would take around 8X to 10X coverage to order and orient the pieces to reconstruct a BAC clone. We had thought that we would need that level as well. But after our success with the fly, I knew we could get by with much less to achieve better than 99.6 percent coverage of the human chromosomes. Thanks to our paired end strategy, in which we sequenced DNA at each end of clones that were two thousand, ten thousand, or fifty thousand base pairs long, only a fivefold, or 5X, coverage was required to obtain a DNA sequence with the correct order and orientation.

Competition with Celera had also helped focus the efforts of our rivals. When the public project had first carved up the genome, some labs were very territorial and gambled that they would be able to sequence chromosomes and parts of chromosomes even before they had the money, equipment, and capability to do so. By September 1998 the entire genome was spoken for, but not everyone who had staked a claim was capable of mapping it quickly and was ready for high throughput sequencing. Overall the public effort was in danger of falling apart because of a limited supply of mapped BAC clones, despite its far greater total sequencing capacity than Celera.

Understandably, Eric Lander had been unhappy with this state of affairs, and in October 1998 proposed abandoning the agreement to divide up the genome and instead sequence clones chosen at random from a library covering the entire genome, an idea that threatened to destroy the fragile consensus of the public project.That December, however, he accepted a compromise: Sulston and Waterston would guarantee the supply of mapped clones for the public effort. By March 1999, endorsed by a "thrilled" Vice President Al Gore, the consortium announced it would produce at least 90 percent of the human genome sequence in "working draft" form by the spring of 2000, "considerably earlier than expected."[4] But because the accelerated public program was dominated by the four major labs, it created bitterness and resentment among minor players, and Collins had even coolly referred to phasing out the centers that could not keep up, "much to the dismay of their leaders."[5] Bruce Roe at the University of Oklahoma, an early DNA sequencer (and, it turns out, the one responsible for some of the more colorful quotes about me in the press), put it more pithily: He was "the guy being treated with K-Y jelly by the NIH."[6]

Although Lander had clearly understood that the government effort would not be able to assemble a human genome sequence without adopting my approach, rather than acknowledging that our method was better or even useful, he proceeded to quietly adopt it while still publicly attacking it. Worse still, Collins and others attempted to use taxpayers' monies to fund a commercial competitor of Celera, Incyte Genomics of Palo Alto, California. In this deal Incyte would provide paired end DNA sequences to help the government-funded laboratories compete with us. They would do this with the help of the SNP Consortium, backed by the Wellcome Trust and a group of drug companies, by hunting for single-letter nucleotide polymorphisms (SNPs, or snips). As well as boosting the capacity of the government effort to assemble its genome data, a by-product would be the generation of a "windfall" of SNPs, effectively doubling the number available to the pharmaceutical companies in the consortium, so they did not need to deal with Celera. By working through the SNP Consortium Collins could also deny that "he" (the NIH) was funding Incyte to help compete with Celera. Another Collins rationale for using the SNP Consortium was that they would not have to publish the data. (The Consortium was not subject to government/Wellcome Trust rules.) They could both deny they were using our paired end strategy and ensure that Celera could not benefit from this data. It was one of the consortium members, Allen Roses of Glaxo Wellcome, who, having felt a sense of outrage at these tactics, had told me what Collins was trying to do. According to Francis Collins, "We could not justify even a single day passing where researchers around the world, aiming to understand important medical problems, would not have free and open access to the data being produced."[7] But he and the genome centers generated millions of paired-end sequences (which have never been released to this day, except as part of the assembly).

Tim Friend of *USA Today* eventually did a full exposé of the Incyte scheme with a story headlined "Feds May Have Tried to Bend Law for Gene Map."[8] Collins was furious, and his policy director supposedly vowed to punch Friend in the nose.[9] But the point was and still remains that taxpayer funds had gone via a third party to purchase from Celera's main competitor precisely the sort of data I would have given them for free if they had been serious about collaborating.

By now the race to read the genome had captured the imagination of many, and the public perception of who was winning became an important issue for both sides. The government-backed labs wanted to impress the politicians that they still deserved funding. As for Celera, we were a public company, which relied on the support of its investors. During peak periods, when major announcements were made by Celera or the government-funded effort, as many as five hundred or so press stories appeared every month, and sometimes even thousands.

To deal with the media, Collins had a press team, while Varmus had his own person, and every government-funded lab had one or more press officers. The media, however, returned again and again to one theme: Craig Venter as underdog, a sole crusader and outsider who was pitted against the collective might of the establishment. Under his leadership Celera was taking on the official Human Genome Project, a $3 billion to $5 billion, government-backed international effort with major centers in Britain, France, Germany, Japan, and the United States.

While my team liked this David and Goliath spin, an annoyed Collins and his colleagues would grumble about how they were overwhelmed by my "huge public relations advantage."[10] I had a "clever press campaign"[11] orchestrated by a "PR machine,"[12] which was, of course, "well oiled."[13] Collins griped about the unseemly articles depicting the race to the genome, where I was standing on the helm of my yacht as he crouched on his motorcycle. ("What drivel!")[14] Sulston moaned that "trying to get reporters to print the admittedly more complex analyses that we felt were being ignored was going to be an uphill struggle."[15] My "good advisors" had used "ruthless manipulation"[16] and "the penetrative, unremitting power of Celera's public relations"[17] to sway the world's media. However, at the end of the day, as Sulston himself admitted, the public program "did poorly on the PR front."[18]

The whining of the public project about my massed ranks of spin merchants and opinion formers was a constant source of amusement at Celera, for my "PR army" was in fact a young woman named Heather Kowalski, who had left her job as press officer at George Washington University in November 1999 to help Celera cope with the constant demands of the media. While she did not have as much experience as many of the other applicants, there was something about her attitude that I really liked. The demands of the media were so relentless, particularly when I went on trips, that she became a constant traveling companion and advisor, and worked so hard that she deserved her success.

Heather knew there was only one way to cultivate the media: by being honest and straightforward, to build their trust. By adopting this simple approach she was able to deal with even the most awkward journalist. Heather had an unusual dose of common sense, which proved helpful both in enabling me to field questions and in dealing with the near-daily attacks and counterattacks. Unlike some members of the army of press officers deployed by my rivals, she did not reprimand reporters if they did not follow the Celera line. Most important of all, she was frank and did not shrink from telling me when she thought I was saying or doing something stupid or misguided (a not infrequent occurrence). Others came to rely on her to give me bad news when they were afraid to do so directly.

Work did continue despite the increasing public scrutiny, and the sequencing of the 3 billion base pairs of human DNA went even better than for *Drosophila.* We were now producing between 50 million and 100 million base pairs of DNA every twenty-four hours, and the sequence was of extremely high quality. The new software was now complete and the barcode reading system was functional, so that the tracking of the paired end sequences became routine. The big unknown was still how well the assembler would work with close to ten times as much data as we had for *Drosophila.*

If we wanted to, we could also draw on the data released by the government effort every day into the public database, GenBank. Like other taxpayers, we had helped to fund this effort, after all. Pharmaceutical companies downloaded the data nightly, and Incyte was very open about its use of GenBank to create a database that they used to compete with us. Francis Collins did not complain about these outright commercial uses and, indeed, used them as a further justification of the value of federal effort. Yet after I indicated that we would do a de facto collaboration with the taxpayer-funded effort by including its data in our assembly, there were anguished howls of protest. The G5 had discussed whether they could withhold their data from Celera even though their mantra had always been that they were providing the sequence for free to all. There were even suggestions of scientific fraud, such as when Sulston told the BBC that the Celera effort was a "con job."[19]

To make use of the GenBank data I asked Gene Myers and his team to come up with a second version of the algorithm that we used to assemble the genome. He called it the "Grande" in honor of the industrial quantities of coffee he would have to consume to get it working. We had had a similar backup plan for the *Drosophila* genome based on the mapping data that Gerry Rubin and his colleagues had produced over several years. We never used it, but it was a security blanket for the combined public-private *Drosophila* genome effort; it helped reassure us that we were producing a quality sequence. Ultimately, Celera's goal and commitment to its shareholders and database subscribers was to produce a very high quality human genome sequence that could be used to drive the development of new pharmaceuticals and disease treatments. Our working assumption was the more sequence data, the better the genome assembly. Our goal was to provide the best, most complete version of the human genome to aid scientific discovery, disease gene discovery, and new therapies. Patients with cancer or other diseases did not care who sequenced the genome; they just wanted new hope for a cure or a treatment for their disease.

Once again this pragmatism proved tough for my team, with Myers fretful over having to depend on public data of variable quality compared with the uniformly high quality data we produced in-house. When we assembled our genome, the data from the public program had not yet accounted for the entire genome but offered only patchwork coverage, with a lot more data on some areas than others. And there were other issues that had to do with mislabeling and chimeric BACs (muddled-up sequences). The public program would fix most of these problems over the next six years, but the low-quality data would interfere with our strategy and degrade the quality of our assembly,[20] something we did not learn until we had sequenced the mouse genome.

As the grandstanding continued, the press battles became more and more wearing. By now the G5 meetings had been marred, as one of the participants put it, by "adolescent, locker room attitudes" where I was routinely vilified. Then one day a new opportunity came for a truce when Tony White telephoned me to say that he and Mike Hunkapiller were talking to Eric Lander (he was by now their number one customer) and that Eric was interested in exploring my idea of a cooperative effort once again.

On the face of it, the public program had much more to gain from collaboration than I did, given that I could already access their data on GenBank. I was skeptical and suspected that the priority in these discussions would not be the human genome but Eric Lander and how he could gain an advantage. Eric's peers in the public program were skeptical, too, but Eric believed that collaboration was the only way to avoid my being declared the clear winner in the race to finish the genome.

Despite my misgivings, I decided to give collaboration another shot. My scientific advisory board was keen on the proposal, too, since any failure by the public program might trigger all kinds of fallout, including decreased funding for NIH. Mike and I were going to be in Boston for a meeting and agreed to see Lander, who was based in the Whitehead Institute nearby in Cambridge. We met in a private room at a Boston hotel; Lander had wanted to keep the encounter secret.

I told him that a full collaboration of the kind I had had with Gerry Rubin and the *Drosophila* community was probably not possible, given all the animosity, but I was still open to a cooperative effort in which we would exchange data and publish the analysis of the genome in either one joint paper or in two simultaneous ones. Eric, who had been involved with the birth of many biotech companies, clearly understood our database business. We would offer a DVD of the code to any scientist for his own use as long as it would not be sold in any way or form. Although we wanted to make all our sequence data freely available to the scientific community, we did not want rival database companies such as Incyte to be able to download and resell Celera data, as they were doing with the federal data.

The only issue that seemed important to Eric was that he be a coauthor on the Celera genome publication, regardless if there were one or two papers, because we were planning to use data that he and others had published in GenBank. We would include proper attribution for any data that we used, as was scientific custom. But I pointed out that by his terms I should be an author on any paper from the federal funded program since over the years I had deposited substantial human data into GenBank that was now being used by the public genome effort. Ditto any successful human gene hunter.

We agreed to talk again soon, and Eric stressed that even if we did come to an agreement, he was speaking only for himself and was not certain that he could bring the other public labs on board. I doubted the talks would go anywhere, but I hoped they might at least help cool the rhetoric and calm the press. When I reported back to my senior team, they were even more dubious. After all the attacks and abuse, they wanted to "kick the butts" of the federally funded labs, and for me to give in now was "just wrong."

On October 7, 1999, I received a three-page memo from Rich Roberts, chairman of my scientific advisory board, most of which consisted of a draft document sent to Rich by Eric Lander. The document outlined the various points that had arisen in our earlier conversations, and when it came to sequence data that had been created using both Celera and public data, Eric's understanding was perfectly clear: "As Craig described, Celera's business plan entails attracting customers by virtue of its value-added database, rather than by exclusive access to otherwise unavailable sequence data. A key issue is that Celera wants to prevent competitors from using Celera's data to rapidly create competing value-added databases; this justifies a delay (12 months) in depositing the Joint Analysis to GenBank."[21]

Discussions went back and forth with Lander through November, and he even took part in a Celera scientific advisory board meeting held on October 10, 1999, in the same Wye River compound in Maryland where, the year before, President Clinton had tried to advance the peace process between the Israelis and the Palestinians. After the teleconference with Lander, Celera senior scientific staff and the board felt that we had made real progress toward cooperating with the federal genome program under terms with which Celera could live. However, we were soon to learn that, as he had warned, Lander had represented only one person in these discussions—himself.

On November 12, 1999, Eric decided to tell Francis Collins about the proposal. Eric said he would summarize what had been agreed upon, adding that "there is enough common ground about the spirit of an acceptable agreement to think we can do it and the time is right to have a discussion with them (Craig, Arnie [Levine], maybe Mike and Tony White) and us (you, Harold, me and anyone else essential)." Even then, however, he sensed trouble. When it came to the key issue of the parties that would attend the core discussion between the two sides, Eric was pushed out of the meeting by Collins and replaced by strong detractors of Celera: Sulston, Waterston, and Martin Bobrow from the Wellcome Trust's board of governors. With Varmus, Collins would of course be present. One of the few figures in the federal establishment with whom I had a forged a good relationship, Ari Patrinos from the

Department of Energy, was not invited. Tony White insisted on going, as did Hunkapiller, since Collins et al. were now their major customers and had in the past few weeks made much of depositing their billionth letter of code into GenBank and of completing chromosome 22, equivalent to 2 percent of the genome. I sensed that we were approaching a disaster of the first order; that much was apparent to Ari, too.

In preparation for the gathering, Collins prepared a document in which he substituted the terms that had been negotiated in all the discussions with Lander. The document was labeled "shared principles," although they were only shared by the members of his team. We agreed to meet at a hotel near Dulles Airport where Francis and I had made an earlier failed attempt at living in harmony. Reluctantly, and knowing that I was bound to end up worse off than before, I went to Dulles on December 29.

Collins and Bobrow began with the usual political rhetoric, presenting themselves as saints for submitting their raw data to GenBank each night, while we were sinners because ultimately we were helping drug companies that wanted to protect their investments of hundreds of millions of dollars to turn these data into treatments. When the discussion finally turned to combining data sets and a possible joint publication, those "shared principles" quickly proved to be nothing of the kind. The problem was that our critics neither understood commercial realities nor were open to reason, discussion, or compromise. We were expected to concede data, methods, and credit for our work; they had to concede nothing in return. Waterston declared that the Celera data should be made available immediately, and he did not care if Incyte or any other company used it to compete with Celera. Once Tony White's explosive outburst in response to that last suggestion had died down, it was clear that the meeting was over, but its repercussions would be felt for years.

Despite its own unwillingness to negotiate or compromise, Collins's group could now say it had met with us in the true spirit of brotherhood and cooperation but that Tony White had responded with outrageous and unreasonable demands. (White had indeed insisted that the merged data set be protected from use by others for three to five years, but that demand, made out of sheer frustration with the intransigence of the public side, was a gift to anyone who wanted to blame the breakdown of negotiations on Celera.)

While insisting that the data had to be made public, free for commercial use with no restrictions, Collins's group was also deeply frustrated and embittered by the thought that we would use their data to compete with them.

James Shreeve likened us to 1950s sci-fi B movie monsters "that absorb the energy of the bazooka shells and missiles launched at their flanks, growing ever larger from the attempts to destroy them."[22] Even as the Dulles meeting was breaking up, Collins sidled up to me to ask once again for coauthorship on Celera papers.

The government/Wellcome Trust attendees subsequently summarized their view of Celera's position in a four-page letter to me, dated February 28, 2000, which included the ultimatum (at the bottom of page three) that unless they received a response by March 6, they would assume there was no further interest on my part in discussions about working together. The memo arrived in my office while I was out of the country, and my assistant, Lynn Holland, informed Collins that I was away for two weeks and that I would respond when I returned. The Wellcome Trust now resorted to a crude tactic to put more pressure on me. On Sunday, March 5, the Trust released a copy of the February 28 letter, which was marked "confidential," to the *Los Angeles Times.* Collins denied having anything to do with the leak. By that time he was being criticized by his own management since the affair was casting a shadow over a potential doubling in funding for the NIH. As one member of the public project admitted, "It would have been politically disastrous for the National Institutes of Health to be involved in such a leak."[23]

Tim Hubbard of the Wellcome Trust Sanger Center circulated a memo to Sulston, Morgan, and others dated March 5 that explained why it was leaking the document:

> *Celera officials have made numerous statements about their honorable intentions which have been very widely reported. It is clear from this and other published documents* (Forbes) *that their attitude is actually quite cynical* [sic] *and focused on obtaining the maximum monetary gain regardless of the effect on worldwide medical research There is only one human genome and it is clear that they wish to lock up as much as they possibly can, regardless of the inhibiting effects this will have on both academic and commercial research and development.*

Isn't this sour grapes after Celera's success with *Drosophila*? the memo perceptively asked. Of course not: "Celera staff has been making unsubstantiated and greatly exaggerated claims."

Collins and his group had often given tips to two reporters on the *Los Angeles Times,* Paul Jacobs and Peter G. Gosselin. In this case they had had enough notice with the Wellcome leak to turn the letter into a front-page story, one that ran to 1,348 words.[24] They quoted Tony White as saying that it was a breach of trust that would probably doom prospects for any further discussion of a joint effort: "Sending that letter to the press is slimy." I made the same point when I was asked to comment the next day, confirming that the release of the memo, which was one-sided at best and, at worst, inaccurate, had been timed so that I didn't have an opportunity to respond to it before the deadline.

The following day *The Washington Post* reported: "If release of the letter was indeed meant to pressure Celera, it didn't work. . . . Celera's efforts have gone remarkably well, and by combining its data with those produced by publicly funded researchers, the company expects to publish a completed human gene sequence this year, three years earlier than the deadline set by the Human Genome Project. That means Celera, not the academic researchers who have spent years of their lives on the genome project, could reap the credit for one of the great achievements of modern science."[25] That same day I reiterated our position that Celera was still interested in collaboration and that the institute's letter had "dramatically misstated" our position on intellectual property protection.

Within a few days Gerry Rubin also came to our support by telling *The New York Times* that NIH officials might have pushed us too hard: "I don't think the director of a publicly traded company can afford to give away everything they have." Celera and I had "completely honored the letter and spirit of their agreements" when it came to the fruit fly genome, and Gerry said he was puzzled at the pressure being put on Celera to release its data when it had already made public far more than any of its competitors.[26] One reporter observed: "More and more, the Human Genome Project, supposedly one of mankind's noblest undertakings, is resembling a mud-wrestling match."[27] As the gloves came off, this reality also dawned on John Sulston at long last: "Many accused me of mud-slinging, jealousy protecting my turf. . . . I had entered the world of politics."[28] The leak, he admitted, was a debacle.

A few days later Collins and Morgan revealed they still had one more trick up their sleeves, one that would also backfire. By this time, thanks to the bull market, and perhaps even to the unveiling of chromosome 22, Celera shares were soaring. Acting through Neal Lane and Sir Bob May (now Lord May of Oxford), the respective science advisors of President Clinton and Prime Minister Tony Blair, Morgan had been lobbying hard to have the two leaders make a joint statement concerning the intellectual property of human genes. After extensive editing and a number of postponements, during a slow news time the White House decided to go ahead with a press conference that included Collins and Lane. Prompted by the concern that Celera might announce its first draft, the statement was also made to allay British fears that the White House smiled more on its efforts than on the public program. At the National Medals of Science and Technology award ceremony President Clinton stated,

> *This agreement says in the strongest possible terms our genome, the book in which all human life is written, belongs to every member of the human race. Already the Human Genome Project, funded by the United States and the United Kingdom, requires its grant recipients to make the sequences they discover publicly available within twenty-four hours. I urge all other nations, scientists, and corporations to adopt this policy and honor its spirit. We must ensure that the profits of the human genome research are measured not in dollars but in the betterment of human life.*

Sir Bob thought it "a bland statement of principle" that "might tilt the moral landscape" and one that "ought to have strengthened, not weakened, the market" by clarifying issues of patents and ownership. Lane agreed that it simply reaffirmed existing policy: "Nobody raised the flag, nobody said this is likely to send a signal that you are changing policy, or in the wrong direction, or could scare people."

But that was not the impression that had been given to the press by White House spokesman Joseph (Joe) Lockhart. When he briefed reporters earlier that morning, he suggested that the president planned to restrict genetic patents; this occurred both in a CBS Radio News interview and during the press "gaggle"—the off-camera huddle of reporters who crammed into Lockhart's office. This was seen as a major blow to the biotech community and to Celera in particular.

When the stock market, which was then at the peak of "irrational exuberance," responded dramatically with a sickening slump, the White House quickly acted to try to put this destructive genie back in the bottle. Unusually, Lane found himself summoned to give a briefing that lunchtime. He was still ignorant of the bloodbath in the markets, and in the James S. Brady Briefing Room he said: "I want to make it absolutely clear that this statement has nothing to do with any ongoing discussions between the public and the private sector."[29] With him was Francis Collins, who seized the opportunity to tell the press how the public program was ahead of schedule and under budget. He seemed to gloat over the developments, sending out a very different message: "I am happy to be here on what I think is a rather significant day, where a very important principle about access to the human genome sequence—our common, shared heritage as human beings—is being endorsed by the leaders of the free world."[30]

At the press conference "the squabble with the Celera folks" was raised and one reporter asked if the statement was designed "to encourage Venter and Celera to get back into talks and to formulate a formal agreement on how they will share their information." Lane replied that the joint statement "applies to everyone," and Collins repeated the familiar mantra: "It is not just a matter of patenting that's being referred to; it's also the immediate release of the data."

When another journalist pointed out that "stocks are going down today, rapidly," Lane replied that he "saw no reason" to link the slump with any statement that had been made. "Our understanding is that Celera is supportive of this statement," he added, but moments later he admitted: "I'm not aware that they were told in advance what this statement would say." He ended by saying what would have done little to comfort shareholders that day: "We want to make people's lives better, and this statement lays out principles we think will do that."

Lane and Collins's reassurance had little effect, for the market continued its fall the next day. Celera alone lost close to $6 billion in valuation over the two-day period, and the market in biotech in total around half a trillion dollars. Today, Sir Bob maintains that the statement "said sensible things" and that the way the market reacted was "perverse. . . . [I]t was looking for an excuse to correct itself." Neal Lane cited the market reaction as an example of how "a relatively small thing, when it occurs in the White House, can have a really big impact."

An editor of *The Wall Street Journal* called and asked me how it felt to be the most powerful person in the U.S. economy. "Poorer" was my answer. I had been on track to be the first biotech dollar billionaire, but my stock had plummeted by more than $300 million in the first few hours of the decline alone. The year before I had sold the *Sorcerer,* being too busy with the genome to make much use of it, but I was now negotiating to buy a beautiful 135-foot schooner in the south of France in anticipation that I would soon be able to sail once again. I had even had the masts repainted. This boat required a crew of twelve, and it was going to cost me around $15 million, plus $2 million or $3 million annually to run. The schooner's German owner followed the markets and was decent about it when I told him I didn't think I could afford the purchase any longer. I lost another $30,000 breaking the contract.

While I also lost paper money that I never had or counted on, hundreds of billions of dollars that would have gone into research for developing new treatments had vanished overnight. Investors who believed in me and my vision likewise suffered the consequences. There was legal fallout, as well: A shareholder suit was filed by one of the law firms that specializes in class action suits each time there is a dramatic fall in a company's stock price. Its argument went, in essence, that the government was punishing Celera for the failed negotiations with Collins et al., and Celera had failed to disclose that these critical discussions were under way with the government. Such is the strange parallel universe that lawyers inhabit: We were being sued for failing to mention a collaboration that never took place.

With hundreds of billions of dollars draining from the market, the pressure on the White House was intense. The day after his speech, President Clinton issued a correction explaining that his statement had not been intended to have any impact on the patentability of genes or on the biotech industry. But the damage was done; Collins and Morgan had succeeded in causing a major embarrassment to the White House. After the market crash and the letter release to the *L.A. Times,* the president ordered Neal Lane to end the genome war. "Fix it. . . . Make these guys work together."[31]

Lane, who had been taken aback by the genome conflict, was happy to comply and passed the message to Collins. The first noticeable effect was a decrease in the attacks on Celera, which John Sulston referred to as a "muzzling": "Celera's success in silencing Francis was very valuable to the company."[32] For me it was simply extraordinary and a sad day for science that it had required the intervention of the president of the United States to put an end to the endless vilification. However, the one lesson I learned from all my encounters with government was that only a fool would try to keep aloof and apart from the politics. I had been worried that Collins would maneuver to use the White House to give the impression that the government and Wellcome Trust were the only parties involved in sequencing the human genome, whether they finished first or not. A triumphal broadcast by Collins from the Rose Garden, backed by the president and the British prime minister, was going to carry more weight than anything my one-woman PR army could put together.

At one point it had even seemed as though it was I who had the ear of the president. As well as dining with the Clintons after the Second Millennium Evening in March 1998, Claire and I visited their private quarters in the White House. As we drank Diet Coke, wine, and beer into the early hours, we discussed everything from Stephen Hawking's lecture to how the president made the bed too quickly (when Hillary believed it should be left to breathe) and even a question session, during which Hillary explained that she felt Collins had been wrong to mix science and religion, given his position as a federal official.

But as we entered the last lap of the genome race, it was Collins who was seated with the First Lady during the State of the Union address, while I seemed to have fallen out of favor. When the Eighth Millennium Evening came around, it was suggested that I talk about the genome, but two days after he had taken part in our Wye River meeting, Lander stepped in instead and I was uninvited.

At the White House in October 1999, Lander discussed "Informatics Meets Genomics"[33] and gave the usual line on how the data had to be made public. ("It's unambiguously the case that information about the human genome has to be freely available to everyone in the world.") Clinton was particularly struck by Lander's statement that all human beings are more than 99.9 percent alike genetically and thought "of all the blood that had been shed . . . by people keeping us divided over that one-tenth of a percent."[34]

"Craig is now persona non grata" is what my assistant, who had once worked in the White House, was told by her contacts there. It took us a while to find out why this was now the case. At the Eleventh Genome Sequencing and Analysis Conference in Miami Beach the month before, we held a special plenary session on the application of DNA technologies. Someone from Applied Biosystems had recommended inviting a woman from the FBI to discuss forensic DNA sequencing, which was an expanding market. Just before the start of the session I had met her and listened with surprise as she described how just a year earlier she had been at the White House to obtain a DNA sample from the president for the ongoing investigation into his then-alleged affair with Monica Lewinsky.

When I introduced her at the meeting, I made a gauche reference to the Lewinsky affair, which by then had been in the public domain for a year:[35] I reminded the attendees, most of whom had come to the previous conference, that the last time we met, the president had sent an apology for having to withdraw at the last minute from being a special keynote speaker. Now I knew the reason, I explained: He had had an appointment with our next speaker. If the matter had ended there and then, I doubt the White House would have been concerned, but the FBI agent not only began her talk by discussing Monicagate but proceeded to show a slide of the notorious blue dress with circles drawn around the three most famous DNA samples on the planet. She discussed how much DNA she was able to obtain from each spot, and how each contained a substantial amount of semen. She went on to describe taking a blood sample from the president on August 3, 1998, and then performing DNA analysis on his blood and on the sample that she had isolated from the blue dress. She showed slide after slide demonstrating the matches between the blood sample and the DNA in sample K39, and explained that the chance of a random match in the case of Caucasians was one in 7.87 trillion. Her final slide was a view through a microscope of the presidential seal composed of sperm.

To say that I was shocked that an active-duty FBI agent went into this much detail about so highly charged a case is an understatement. How my role in all this was presented when it was reported back to the White House is unclear. However, I eventually found out that overprotective staffers had decided to ban me as a result and blamed me for the FBI agent's talk. I did not want to be a victim of a butterfly effect, where the flap of an insect's wings—the FBI agent—would cause a hurricane of bad feeling in the form of a one-sided presidential proclamation on behalf of Collins and the government program. But I could see the storm clouds gathering.

Then on May 4 came a call at home from my friend in the Department of Energy, Ari Patrinos. This was in itself nothing out of the ordinary, since we often talked, usually each Sunday. But what he casually mentioned to me that night certainly gave me pause for thought. He invited me to come to his town house in Rockville for a drink, and by the way, he added, Francis Collins might just stop by. Francis lived in the same town house complex, almost across the street from Ari. Given what I had been through already, I was understandably reluctant. The Celera sequencing program could not be going better, and the end was now in sight. Celera would sequence the genome first, and it would be of much higher quality than the work being produced by the government. And besides, the last time I had a cozy chat with Francis, I ended up with a class action lawsuit and lost hundreds of millions of dollars.

In Francis's later account of our meeting, it was he who made this important and gracious gesture, assuming the role of the peacemaker who had decided to end hostilities, bury the hatchet, and offer me the olive branch. ("I approached a mutual friend of Venter and myself [Ari Patrinos]), and asked him to set up a secret meeting.")[36] But the idea had in fact been Ari's, and, as he recalls, "I had a lot less trouble convincing Craig to attend the secret meeting than Francis. Francis balked for several months, claiming he needed to secure permission from his NIH boss."

I have since learned that Ari had been motivated not by the order of the president to "make these guys work together" but rather by his realization that we were both so insulated from reality by our respective entourages, peer groups, and advisors that any attempt at reconciliation would be doomed unless it could be carved out away from the charged atmosphere of a formal meeting. (As Ari himself put it, "They had to be extracted from those cocoons and be brought to an environment where they could be themselves.") At the time I reasoned that Ari had made the call because Collins and the government-funded lab workers were concerned that Celera would make a preemptive announcement that we had finished the genome long before they could mend the situation with the White House.

They were right to fret, of course. The House Subcommittee on Energy and the Environment had called a hearing on the human genome sequencing project and invited me to testify on April 6. With the encouragement of my team, I took a candid, blunt approach. I carefully explained the difference between a sequenced genome, in which the order of the base pairs of the genetic code are known across the genome (which Celera was producing), versus having the sequences from BAC clones, in which most are not assembled or ordered (the public genome effort). Heather and I thought that the hearing presented the perfect opportunity to announce our progress. We issued a press release—"Celera Genomics Completes Sequencing Phase of the Genome from One Human Being"—for the benefit of the government genome workers, so they knew that we were very close to the big announcement, which could come at any time after the assembly process was completed. Collins grumbled that the press had misconstrued what we meant by this, but we were not bluffing, and I knew he knew that as well. Equally, I knew that to have presidential endorsement for what we had done would elevate our achievement beyond the reach of academic squabbles and ensure its place in the history books.

Ari reassured me that he would make certain any discussions in his house between him, Collins, and me would be off the record and that I could deny any serious conversations had ever taken place. With strong pressure from Ari and driven by my feeling that I needed to do something about the White House, I agreed to stop by one evening.

Ari lived in a typical three-level town house with a recreation room/family room in the basement, where Francis was already waiting. Ari started plying us both with beer, and the atmosphere was tense. The conversation started slowly with current trivia, as though it was really a chance encounter. After a few drinks the discussion turned more serious, dealing with a possible joint announcement at the White House or at least involving the president. The issue of a joint or simultaneous publication in the journal *Science* was also covered. No commitments were made on anyone's part other than to keep our discussions confidential. After a few more drinks Francis and I left together, joking that there might be a photographer crouching in the bushes. As he walked to his house, I got into my car and drove home.

I told Claire what had transpired, and her response was blunt: I was insane to be taking part in such discussions. But they continued nonetheless, and Claire was not alone in voicing her disapproval of them. Heather, who had become my key advisor and friend, was so angry that she either yelled at me or just refused to talk about the subject. But I kept going because, as was the case in our previous discussions, I felt Collins had no choice now but to take my position seriously.

Although Collins had not discussed his plans with the Wellcome Trust or with Waterston and Lander, we set ourselves a simple goal: to make a joint announcement with President Clinton at the White House when Celera finished its first assembly of the human genome. At that time, the government effort would give a progress report on its own efforts, and we would announce that we would work together toward a joint publication in *Science*. We had gone as far as to contact Donald Kennedy, the *Science* editor, to see how he would handle the resulting papers and, of course, Celera's requirement to somehow limit its commercial competitors from simply downloading and reselling its data. This would not have been an issue had we been in Europe, where new laws had been enacted that permitted the copyrighting of unique databases. The U.S. Congress was still wrestling with the issue, and no resolution was in sight. Kennedy encouraged all of us to work with him, excited by the prospect that *Science* would publish this historic achievement.

Secrecy soon became even more important, because the White House did not want to be upstaged. I had to bring the senior members of my team into the discussions, and, of course, they were unhappy that we were giving away our advantage, our chance to embarrass our attackers, and our opportunity to single-handedly take the prize. Heather was still angry and thought I had lost my mind. Ham and Gene looked uncomfortable and remained mistrustful of the public program, which was hardly surprising given all we had endured. They were very concerned, as was I, that the public relations battle with the government human genome effort would rob us of the time that we wanted and needed to sort through our data to check their quality and analyze their meaning. Ironically, we now needed more time to cope with the mixed quality of the public data.

My colleagues were not the only ones unimpressed by the possibility of a compromise; I was chastised by others, sometimes brutally. At one dinner in La Jolla with Richard Lerner, the head of the Scripps Research Institute, the endgame of the genome race came up. When I told him that I had at least contemplated a joint announcement with the government, Lerner was livid. He likened it to leading the marathon in the Olympics only to stop just before the finish line to let the next runner show up so we could hold hands and cross together. His view was that NIH and the government needed to be taught a lesson for their arrogance, and that if I accepted any less, then I would be letting down the thousands who were rooting for me to inflict revenge for all the wrongs that the government had perpetrated on the scientific community.

I was asked by the White House to provide a date when I felt we would have the first assembly completed, and I worked with Gene Myers and the bioinformatics group to figure out how long it would take our computers to put the sequence together. I gave them an estimate and added what I thought would be a generous buffer, and they in turn informed me that that particular date was not available but offered instead one a few weeks earlier, which panicked the team even more. As the discussions continued in Ari's basement (his wife and kids were becoming discontent with being evicted every time Francis and I rolled up), June 26 became the date set in stone. Having a deadline spurred my team to work even harder, although they were essentially going all out as it was. There was a real danger that we would burn out before finishing.

In my conversations with Francis and Ari we worked out over meals of pizza and beer that President Clinton would preside over the event and that Tony Blair would take part via a live video link from London. First the president would speak, followed by Blair, then Collins, and finally me. Each of us would have about ten minutes to make a statement, and we agreed to share the texts of them ahead of time. While I was pleased with the arrangement, I was still cautious about what would be said and how my team's work at Celera would be portrayed. The last time Clinton and Blair had spoken publicly on the issue, Celera stock had dropped more than $100 per share. If that happened again, it would cost me my job for having agreed to these discussions.

A week before the meeting, "senior administration officials" informed the press that Collins and I would be making an announcement on the completion of the genome sequence and that President Clinton might be involved.[37] *The Wall Street Journal* said, "A joint announcement at the White House would send a powerful symbolic signal that the government and private sector can collaborate on projects of such significance, the administration official said, adding that the agreement to share a podium at the ceremony is the result of weeks of negotiations."[38] The article went on to indicate that "a joint ceremony would coincide with other talks aimed at coordinating scientific publication of the findings of each group, perhaps by the fall, though there isn't any assurance that such an agreement will be reached."

As the date drew closer, the tension increased, but by then the momentum was unstoppable. Once I had agreed to a draw, we wanted to announce it at the White House as quickly as possible before hostilities could flare up again. I received a daily call from either Ari or Neal Lane asking if the assembly had been completed yet. It had not, but I assured them it would be soon and checked almost hourly on the progress of the computation. We were using "bleeding edge" chips from Compaq that were nevertheless struggling with the amount of number crunching. The massive computation needed to be restarted many times due to computer crashes. Gene and his team were not sleeping at all in order to check the progress of each stage of the assembly process. One of the advantages of writing the algorithm in sections was that we could evaluate the success of each step before proceeding to the next. This also helped a great deal on the psychological front because we knew we were making real progress.

The initial assembly was completed a few weeks before the announcement but was based on the so-called compartmentalized methodology rather than on the whole genome. In essence the genome was done by assembling clusters of data into hundreds of smaller chunks, which could be assigned to the right place in the genome using public mapping data, rather than determining the entire sequence in one go. But the public data did not cover the whole genome, and we still wanted to go through with the whole-genome approach. The moment that the assembly algorithm, Grande, looked as if it would really deliver the genome came only a day or two before the big announcement.

When it was clear that we would indeed succeed in the first assembly of the human genome, I turned my attention to my White House speech. I knew that the president, the prime minister, and Collins were using professional speechwriters, but after all I had put into this project, I wanted to pen every word I was going to deliver. We were told that there would be extensive press coverage (an understatement) and that our comments would be carried live around the world on several channels, including CNN and the BBC. The pressure increased when we were informed that this would be the first time in history that a key scientific advance was announced from the White House, and I found myself struggling over the decision of what to say and how to say it. Meanwhile, I received the draft of President Clinton's comments, which were so generous and inspiring that I began to feel even more inadequate for the task. The next day a copy of Collins's speech arrived, and I have to admit that I was impressed. Now I was beginning to regret not having used a speechwriter. I plugged away late every night, sometimes producing only a few sentences or a paragraph. The White House was pressing me for the copy of what I was going to say on live television from the East Room, and the calls became more frantic and insistent. Then I received a copy of Tony Blair's speech, and my blood began to boil. Drafted by his chief scientist, Sir Bob May, it was so partisan that I suspected the Wellcome Trust had had a major influence on its drafting.

I was so upset that I called Ari and told him that if Blair went ahead with those remarks, I would boycott the White House ceremony and hold my own press conference. Ari tried to calm me and promised to call Neal Lane immediately. He told me not to do anything rash—in fact, not to do anything or call anybody until I heard back from him or Neal. Neal finally called and wanted to go through Blair's speech with me, line by line, to clarify what had offended me. When I finished, he clearly understood my position and was sympathetic, but he said there was little he could do. "I can change anything you want in Collins's speech and even the president's, but you are asking me to change a major international address by a foreign head of state. I just can't do that."

The lack of reassurance reminded me of Collins's prevarication in our basement discussions when he told me, as Lander had done before him, that he could not speak for his colleagues. Collins had also said he had nothing to do with the release of the ultimatum letter to the *L.A. Times* and had blamed the Wellcome Trust. If someone deceives me once, then shame on him; if it happens twice, then shame on me. I was not going to let it happen a second time on live television from the White House. I was firm: If the speech went out as it was, I would not show up. Neal pleaded with me to wait until he had at least tried to get it changed.

Being an optimist I kept working on my text and was sitting at the computer in my home office after midnight when the phone rang. It was a relieved-sounding Neal Lane, who assured me that everyone had gotten my message and that Tony Blair's speech would be rewritten. Could I see a copy first? I had his assurances that the speech would be changed and that I would be pleased. Now would I agree to participate? I had never known Neal to be anything less than honorable and straightforward, so I accepted his word. The conversation quickly turned to my speech. I promised him a copy by 6:00 A.M. but walked him through what I wanted to say. Neal seemed pleased. The next time we would see each other would be at the White House in the morning. We were about to unveil the book of humankind to the world.

15. THE WHITE HOUSE, JUNE 26, 2000

In the long history of humankind (and animal kind, too) those who learned to collaborate and improvise most effectively have prevailed.

—Charles Darwin

Some duplication of effort may be the only way to reach a goal within a certain time span. No two minds ever take exactly the same path, and placing all your marbles on one person's intuition never makes sense if you have the money and want to move quickly.

—James Watson, *A Passion for DNA: Genes, Genomes and Society*

I did not sleep the night before not one but two heads of state were to unveil the results of the greatest concerted undertaking in biology. The coming celebration would be hailed by some as the most notable intellectual moment in history itself.[1] Despite my threats to boycott this momentous announcement, I was now determined that this was going to be my day, perhaps one of the most important in my life. After speaking with Neal Lane, I had continued to polish my speech; changing a word here and there. I crossed out a sentence; I moved paragraphs around. Again and again I called or e-mailed Heather and other friends for feedback, keeping them up through the hours of darkness. I had to get the speech just right.

By six that morning, as I had promised, I e-mailed the text of my speech to the White House. I took a hot shower before dressing in a dark blue suit with a red power tie. A long day stretched ahead of me in the sultry heat of Washington. Claire and I were to be picked up from our Potomac home by a driver for the twenty-five-minute ride to the White House. Our marriage

(top, left) The Venters, a typical American family, on a visit to Ocean Beach, California, in 1948. (Mother, Elizabeth; me, age two; Father, John; and older brother, Gary)

(top, right) Me, age three, outside Bayside Manor home in Millbrae, California, not far from San Francisco airport.

(above) Me, age five, in kindergarten class photo (second from left, front row).

(right) A happy seven-year-old enjoying life.

Eighth-grade report card, showing the result of my refusing to take spelling tests the year before. (Some parents may, perhaps, find some hope on seeing similar report cards from their children.)

Key to Marking:

A Excellent · B Good · C Average · D Barely Passing · F Failing

O Outstanding · S Satisfactory · U Unsatisfactory · Inc. Incomplete because of absence

Effort

This rating tells whether the child is working to the best of his ability. Unsatisfactory effort is a matter of serious concern.

U in effort or F in any subject should be investigated by the parent.

STUDENT'S NAME John C. Venter

		Reports 1	2	3	4	Teacher and Room
READING		D+	C	C	C-	
	Effort	U	S	S	S-	
	Conduct	S-	S	S	S	D J Richardson
ARITHMETIC		D	B	B-	C-	
	Effort	U	O	S	U	
	Conduct	S	S	S	S	J. York - 183
ENGLISH		C	B+	C	C-	
	Effort	S-	O	S+	S-	
	Conduct	S+	S	S	S-	D J Richardson
SOCIAL STUDIES (Incl. Hist. & Geog.)		C-	C+	C	D	
	Effort	S-	S-	S-	S-	Kathryn Jaeger
	Conduct	S	S	S	S-	181
~~HOMEMAKING~~ MANUAL ARTS		C+	B-	B	B+	
	Effort	O	O	O	O	R J Williams
	Conduct	O	O	O	O	
GENERAL SCIENCE		C+	B	B	B+	
	Effort	S	S	S	S+	David E. Ramos 182
	Conduct	O	O	O	O	
SPELLING		C-	C-	C-	D+	
	Effort	U	S-	S	U	
	Conduct	U	S	S	S	D J Richardson
PENMANSHIP		C-	C-	C-	C-	D J Richardson
STUDY	Effort	S-	S	S	S-	
	Conduct	S-	S	S	S-	D J Richardson
MUSIC, Vocal						
MUSIC, Instrumental						
ELECTIVE						
PHYSICAL EDUCATION		S	S	S	S+	
	Effort	S	O	O	O	
	Conduct	O	O	O	O	H. Weinberg
CONDUCT OUTSIDE OF CLASS		A	A	A	A	D J Richardson
Number of days absent					2	
Number of times tardy						

ASSIGNMENT FOR NEXT YEAR All room assignments are final. Please do not request changes.

GRADE 8 ROOM 128 DATE 6/12/59 Present Teacher's Signature D J Richardson

Mills High School swim team, 1963 (me, front row, fourth from the left).

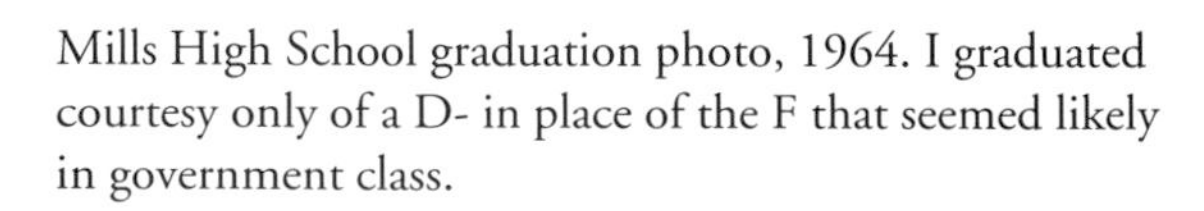

Mills High School graduation photo, 1964. I graduated courtesy only of a D- in place of the F that seemed likely in government class.

A toughened version of me, with friends in Counterinsurgency School in the swamps of Virginia as preparation for Vietnam, 1967.

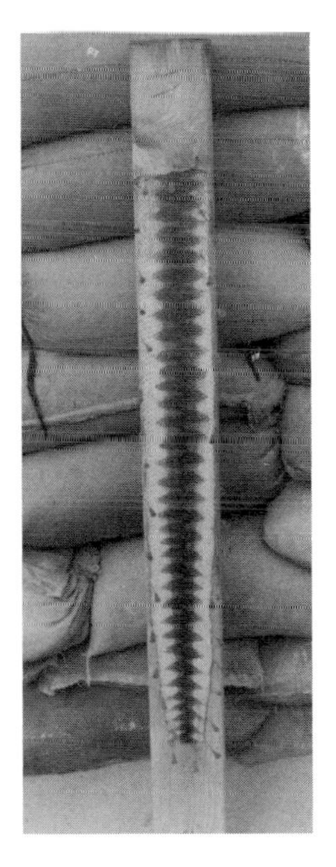

hina Beach, Da Nang, Vietnam; me with a venomous sea snake at bumped into my leg while I was swimming in the surf.

a-snake skin pinned to a board with hypodermic needles while ying in the hot sun outside of a bunker. This memento now ngs on the wall of my office.

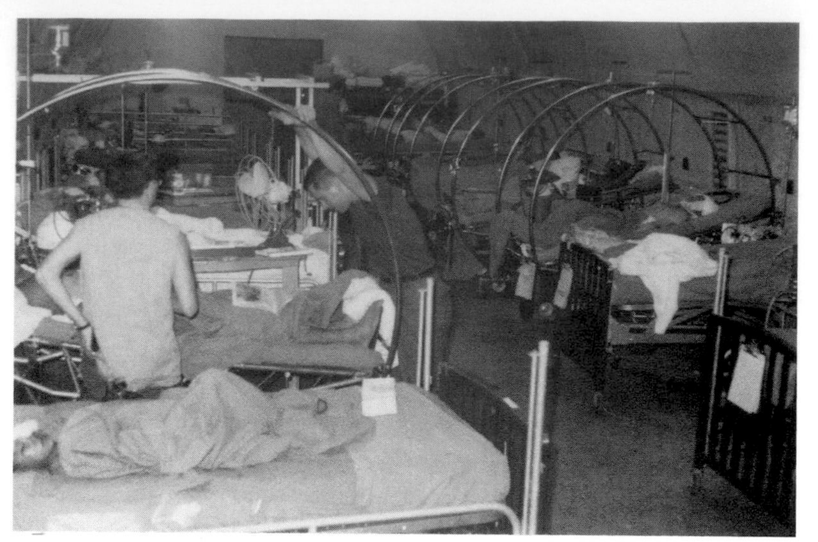

Intensive care ward with Striker frames in a Quonset hut hospital, Da Nang, Vietnam, where I spent the first six months of my tour. This photo shows the striking diversity of patients we treated at any given time, from wounded and burned children to Koreans to POWs.

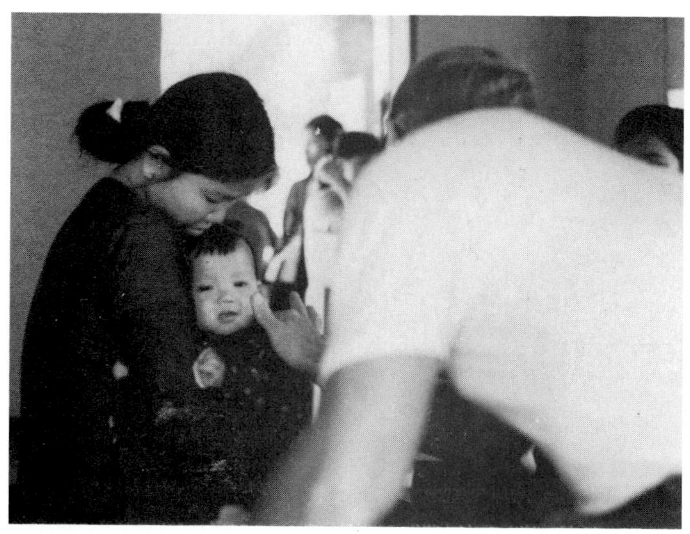

Treating patients at an orphanage outside of Da Nang in 1968. Skin infections were common.

Vedding day, November 1968, Geneva, Switzerland. I met Barbara Rae, of New Zealand, while on R & R from Da Nang n Sydney, Australia.

Vorking in a San Diego boatyard on my nineteen-oot folk boat, *PAIX*, soon after starting at UCSD as a nior-college transfer student.

ngle-handed sailing on *PAIX* out the channel from Iission Bay into the Pacific Ocean.

Me with my mentor, the late Nathan (Nate) O. Kaplan, and Barbara after we graduated from UCSD with new Ph.D.s in 1976.

In Buffalo, New York, enjoying the afternoon with son, Christopher Emrys Rae Venter, born at the end of the blizzard of 1977.

A winter break at Falling Leaf Lake, near Lake Tahoe, with Christopher, Barbara, and de friends and College of San Mateo mentor Bruce Cameron and his wife, Pat Cameron.

My early Buffalo Medical School lab team, which focused on the adrenaline receptor isolation.

As a single parent celebrating Christopher's third birthday in an old row house near the Buffalo Medical School, 1980.

Racing my eighteen-foot Hobie Cat off a Canadian beach was important escape fro work for me.

My second wedding ceremony, with my new wife, former graduate stude Claire M. Fraser, in Centerville, Massachusetts, in October 1981.

My father after a good day of golf, shortly before he died in his sleep on June 10, 1982, from sudden cardiac death.

My NIH pre-genomics lab team after the transition from receptor biochemistry into molecular biology.

Sirius, my thirty-three-foot Cape Dory in Bermuda shortly after surviving a major gale while sailing from Annapolis to Bermuda.

In my NIH lab, trying to boost the morale of my team after *The Washington Post* reported that Jim Watson had declared at a Senate hearing that monkeys could do my EST method of gene discovery. Claire bravely wore the monkey suit to mock Watson.

At the helm of my Frers-designed eighty-one-foot sloop, *Sorcerer,* at the start of the great ocean race from New York City to Falmouth, England, in May 1997.
(Photo by Rolex.)

Receiving the New York Yacht Club trophy for winning the contemporary division of the transatlantic race from Commodore Bob James in Falmouth, England, in 1997. The last time the trophy was presented to an American was in 1905, to the skipper of the schooner *Atlantic.*

February 17, 2001. The *Drosophila* genome sequencing team leaders (me, Gerald [Gerry] Rubin, Gene Myers, Susan E. Celniker and Mark Adams) at the American Association for the Advancement of Science (AAAS) meeting to receive the Newcomb Cleveland Prize for 2000. *Science* also highlighted the sequencing of the fruit fly genome as the "Breakthrough of the Year" in 2000.

eting with Ari Patrinos and Francis Collins at the White House prior to the announcement he sequencing of the human genome, and seeing the *Time* magazine cover for the first time. *oto by Marty Katz.)*

The human genome announcement in the East Room of the White House, June 26, 2000.

Warm congratulations from President Clinton following my presentation.

Taking questions from the White House press corps with Ari Patrinos, Francis Collins, and the White House science adviser Neil Lane. *(Photo by Marty Katz.)*

ıe top Celera genomic team in the Green Room at the Washington Hilton Hotel following
e White House announcement: Mark Adams, Nobel laureate Ham Smith, Gene Meyers,
d me. *(Photo by Marty Katz.)*

ıe Myers, Mark Adams, and me waiting for the onslaught of questions from a packed
room at the Washington Hilton genome press conference. *(Photo by Marty Katz.)*

An intensity and scale of attention that is highly unusual for a scientist from hundreds at th human genome press conference at the Washington Hilton. *(Photo by Marty Katz.)*

The issue of *Science* featuri my team's description of th human genome. Seeing thi cover still generates strong emotions in me. *(From SCIENCE Volume 291 Number 5507 (16 Feb 2001). Reprinted with permission fror AAAS.)*

Celebrating with siblings Keith, Susanne, and Gary at our mother's eightieth birthday in San Francisco.

The ninety-five-foot *Sorcerer II* in the Galapagos Islands at the *Master and Commander* movie site.

At peace at sea, near Cocos Island, Costa Rica, during the *Sorcerer II* Expedition.

Me with fiancée Heather E. Kowalski.

had been under a severe strain as a result of my immersion in the genome project, and we did not talk much on the way. Once again I read and reread my speech.

On arriving at the White House we were quickly taken through security to meet Francis, his wife, and Neal Lane. Soon after, the dapper, compact figure of Ari Patrinos arrived, and the White House photographer took photos. Several featured Ari, Francis Collins, and me holding up a copy of *Time* magazine. Francis and I were shoulder to shoulder on the cover, but I was pleased to see that I was slightly out in front. The effort to get us both in that photo was more than the record of the completion of a monumental scientific task, because a great deal of wrangling behind the scenes was involved.

Time had been following the genome story almost from the beginning, mostly thanks to the science writer Dick Thompson (who now works with the World Health Organization, in Geneva). The White House, Francis, Ari, and I had agreed to give the magazine an exclusive on the story of the discussions leading up to the White House event. This involved secret interviews with all the key players and late-night photo sessions, with the main one taking place after midnight in the Natcher Auditorium at the National Institutes of Health. Prior to the White House event I was informed by *Time* that I had been selected by the editor to be on the cover by myself but they were pressured by a senior White House official to include Francis as well. I was reassured it was "my cover," and they would not change their plans if I was unhappy. The next day Collins phoned to plead with me, saying it would send the wrong message to have only one of us appear on the cover, and I reluctantly agreed. When I told Dick Thompson, he asked if I was really sure. I said I was feeling magnanimous and it was the right thing to do.

On the great day, happily, all the rivalries were swept aside by everyone's feeling of being part of an historic achievement. At the White House, Francis and his wife welcomed us graciously, and there was an electric atmosphere and a mood of high anticipation. When President Clinton came in to greet the four of us, he was very upbeat. He later pulled me aside to tell me that we had a good mutual friend in Thomas J. Schneider, whose wife, Cynthia, was the ambassador to Holland, and who was clearly a fan of what I had done and had made an effort to bring it to the president's attention. As Clinton himself wrote later of the encounter between Francis and me, "Craig was an old friend, and I had done my best to bring them together."[2] As the president, Francis, and I walked together from the hall into the East Room of the White House, a band struck up "Hail to the Chief," and we entered to face a standing ovation. I sat on the right and Francis on the left of the presidential podium. At the back of the room were banked row upon row of TV camera crews on a platform. Two large plasma screens carried a live video linkup to Downing Street and the British prime minister, Tony Blair. In London the audience in Number 10 was giggling at the stark contrast between the pomp and ceremony in the White House and Blair standing stiffly and alone at a podium before a camera.

I gazed around and noticed that the audience contained members of the president's Cabinet; senators; members of Congress; ambassadors from the United Kingdom, Japan, Germany, and France; and the great and the good of the genome community, including Jim Watson, in a white suit, who was sitting in the front row with my advisory board members Norton Zinder and Rich Roberts. There was Claire sitting with Francis's wife, and, of course, my team from Celera, including Ham Smith, Granger Sutton, Mark Adams, Heather Kowalski, and, in a surprisingly sharp outfit, Gene Myers. Perhaps the only person scowling in that ebullient group was the well-dressed Tony White, though across the Atlantic, John Sulston was cringing: It was not clear that the public project had reached its magic 90 percent mark, and he was worried that the event was dishonest[3] in spirit: "We just put together what we did have and wrapped it up in a nice way and said it was done. . . . Yes, we were just a bunch of phoneys!"[4] But even he had come to feel that on that great day it felt anything but political.

I was proud and elated when the president began his remarks on the great genome enterprise, likening what we had achieved to drawing a great map of humanity.

> *Nearly two centuries ago, in this room, on this floor, Thomas Jefferson and a trusted aide spread out a magnificent map—a map Jefferson had long prayed he would get to see in his lifetime. The aide was Meriwether Lewis and the map was the product of his courageous expedition across the American frontier, all the way to the Pacific. It was a map that defined the contours and forever expanded the frontiers of our continent and our imagination.*

I grinned and wondered what the president would think if he knew that Lewis might have been my distant relative. But my smile soon faded as again and again I found myself thinking back to how far I had come since my period in Vietnam, and I began to feel very emotional. In my speech I was going to mention how my time in service to my country had given me so much drive and determination. I began to worry that I would lose my composure, which I had the tendency to do when discussing my wartime experiences.

The president moved on from Meriwether Lewis to genetic cartography.

> *Today, the world is joining us here in the East Room to behold a map of even greater significance. We are here to celebrate the completion of the first survey of the entire human genome. Without a doubt, this is the most important, most wondrous map ever produced by humankind.*

After paying tribute to the one thousand researchers across six nations who drew up this remarkable map, and of course to Crick and Watson and the forthcoming fiftieth anniversary of their discovery of the double helix, the president made a reference to God. The successful completion of the genome was

> *more than just an epic-making triumph of science and reason. After all, when Galileo discovered he could use the tools of mathematics and mechanics to understand the motion of celestial bodies, he felt, in the words of one eminent researcher, "that he had learned the language in which God created the universe." Today, we are learning the language in which God created life. We are gaining ever more awe for the complexity, the beauty, the wonder of God's most divine and sacred gift.*

The religious gloss was familiar: Francis had been working with the president's speechwriter.[5] This gave me pause. I realize that in America such references are a political necessity, but it detracted from all my hard work and that of an army of genome scientists to have this huge advance in the rational pursuit of the secrets of life linked to a particular belief system.

I certainly believe, like the president, that science amplifies and reveals the wonder of the world, but the thought of being a self-replicating bag of chemicals that resulted from 4 billion years of evolution is far more awesome to me than the notion that a cosmic clockmaker snapped his fingers to put me together. By the time these heretical thoughts had passed, the president had returned to reality by reminding us what the effort was really about—not revealing the mind of God but the genetic roots of diseases such as Alzheimer's, Parkinson's, diabetes, and cancer.

And then came the inevitable mention of the many battles I had fought over the years. The president referred to "the robust and healthy competition that has led us to this day" and how the public and private efforts were now committed to publishing their genomic data simultaneously "for the benefit of researchers in every corner of the globe." At this point I became nervous because I knew that it was now Tony Blair's turn to address the journalists. Although I had won some kind of concession that he would change his speech and remove the snipes about Celera, I still did not know precisely what he was going to say.

The prime minister's face loomed large on the plasma screens. After a homely reference to his little boy and the usual platitudes, I was pleased and relieved when he singled me out. "And I would like, too, to mention the imaginative work of Celera and Dr. Craig Venter, who in the best spirit of scientific competition has helped accelerate today's achievement." It did not dawn on me at the time, but Blair's speech triggered understandable fury in Britain because of what he did not say: How could he mention Celera and not the Sanger?

The torch for the public effort was carried by Francis Collins (no doubt to the irritation of Eric Lander), who began with a moving reference to the funeral of his sister-in-law the day before. She had died of breast cancer, much too soon to benefit from the new insights that were being achieved. Francis, who regarded the very act of reading the genome as an occasion of worship,[6] described how it was "humbling for me and awe-inspiring to realize that we have caught the first glimpse of our own instruction book, previously known only to God" and then paid a generous compliment to me and the Celera effort. "I congratulate him and his team on the work done at Celera, which uses an elegant and innovative strategy that is highly complementary to the approach taken by the public project. Much will be learned from a comparison of the two. I'm happy that today the only race we are talking about is the human race."

To top it all, Francis gave me a kind introduction, describing me as someone who is "never satisfied with the status quo, always seeking new technology, inventing new approaches when the old ones wouldn't do," and an individual who is "articulate, provocative, and never complacent." I had made "profound contributions to the field of genomics." As gratifying as his words were, I could not help but feel some regrets at this point. How different the genome effort would have been had his attitude today been representative of the government-led effort toward Celera from the beginning!

Now it was my turn. As a White House staffer moved a small step into place behind the podium to raise me above the lectern, I began with a self-effacing comment that I was shorter than the two previous speakers and then began to speak all those words that I had invested so much effort in over the past few weeks.

> Mr. President, Mr. Prime Minister, members of the Cabinet, honorable members of Congress, ambassadors, and distinguished guests. Today, June 26, in the year 2000, marks a historic point in the 100,000-year record of humanity. We're announcing today, for the first time, our species can read the chemical letters of its genetic code. At 12:30 p.m. today, at a joint press conference with the public genome effort, Celera Genomics will describe the first assembly of the human genetic code from the whole-genome shotgun method. Starting only nine months ago, on September 8, 1999, eighteen miles from the White House, a small team of scientists headed by myself, Hamilton Smith, Mark Adams, Gene Myers, and Granger Sutton began sequencing the DNA of the human genome using a novel method pioneered by essentially the same team five years earlier at The Institute for Genomic Research.
>
> The method used by Celera has determined the genetic code of five individuals. We have sequenced the genome of three females and two males, who have identified themselves as Hispanic, Asian, Caucasian, or African American. We did this sampling not in an exclusionary way but out of respect for the diversity that is America, and to help illustrate that the concept of race has no genetic or scientific basis. In the five Celera genomes there is no way to tell one ethnicity from another. Society and medicine treat us all as members of populations, where as individuals we are all unique, and population statistics do not apply.
>
> I would like to acknowledge and congratulate Francis Collins and our colleagues in the public genome effort in the U.S., Europe,

and Asia for their tremendous effort in generating a working draft of the human genome. I'd also like to personally thank Francis for his direct actions in working with me to foster cooperation in the genome community, and to shift our collective focus to this historic moment and its future impact on humanity. I would also like to thank the president for his commitment to public-private cooperation and for making this day even more a historic event. Obviously, our achievements would not have been possible without the efforts of the thousands of scientists around the world who have gone before us in the quest to better understand life at its most basic level. The beauty of science is that all important discoveries are made by building on the discoveries of others. I continue to be inspired by the work of the pioneering men and women in the broad array of disciplines that has been brought together to enable this great accomplishment. I would like to particularly acknowledge Charles DeLisi from the Department of Energy and Jim Watson from Cold Spring Harbor, both here, for their vision in helping to initiate the Genome Project. The completion of the human genetic blueprint would not have been possible without the continued investment of the U.S. government and basic research. I applaud the president's efforts and the work of Congress during the last several years in producing the largest funding increases to fuel the engines of basic science.

At the same time we could not overlook the investment of the private sector in research in America. There would be no announcement today if it were not for the more than $1 billion that PE Biosystems invested in Celera and into the development of the automated DNA sequencer that both Celera and the public effort used to sequence the genome. In turn, some of the investment was driven by the public investment in science.

Thirty-three years ago, as a young man serving in the medical corps in Vietnam, I learned firsthand how tenuous our hold on life can be. That experience inspired my interest in learning how the trillions of cells in our bodies interact to create and sustain life. When I witnessed firsthand that some men lived through devastating trauma to their bodies, while others died after giving up from seemingly small wounds, I realized that the human spirit was at least as important as our physiology. We're clearly much, much more than the sum total of our genes, just as our society is greater than the sum total of each of us. Our physiology is based on complex and seemingly infinite interactions among all our genes and the environment, just as our civilization is based on the interactions among all of us. One of the wonderful discoveries that my colleagues and I have made while decoding the DNA of over two dozen species, from viruses to bacteria to plants to insects, and now human beings, is that we're all connected to the commonality of the genetic code in evolution. When life is reduced to its very essence, we find that we have many genes in common with every species on Earth and that we're not so different from one another. You may be surprised to learn that your sequences are greater than 90 percent identical to proteins in other animals. It's my belief that the basic knowledge we're providing the world will have a profound impact on the human condition and the treatments for disease, and our view on our place in the biological continuum.

The genome sequence represents a new starting point for science and medicine, with potential impact on every disease. Taking the example of cancer, each day approximately two thousand die in America from cancer. As a consequence of the genome efforts that you've heard described by Dr. Collins and myself this morning, and the research that will be catalyzed by this information, there's at least the potential to reduce the number of cancer deaths to zero during our lifetimes. The development of new therapeutics will require continued public investment in basic science, and the translations of discoveries into new medicine by the biotech and pharmaceutical industry.

However, I am concerned, as many of you are, that there are some who will want to use this new knowledge as a basis of discrimination. A CNN-*Time* poll this morning reported that 46 percent of Americans polled believe that the impact of the Human Genome Project will be negative. We must work together toward higher science literacy and the wise use of our common heritage.

I know from personal discussions with the president over the past several years, and his comments here this morning, that

> genetic discrimination has been one of his major concerns about the impact of the genomic revolution. While those who base social decisions on genetic reductionism will be ultimately defeated by science, new laws to protect us from genetic discrimination are critical in order to maximize the medical benefits from genome discoveries.
>
> Some have said to me that sequencing the human genome will diminish humanity by taking the mystery out of life. Poets have argued that genome sequencing is an example of sterilizing reductionism that will rob them of their inspiration. Nothing could be further from the truth. The complexities and wonder of how the inanimate chemicals that are our genetic code give rise to the imponderables of the human spirit should keep poets and philosophers inspired for millennia.

The president thanked me for "those remarkable statements" and added, "When we get this all worked out and we're all living to be 150, young people will still fall in love, old people will still fight about things that should have been resolved fifty years ago—we will all, on occasion, do stupid things, and we will all see the unbelievable capacity of humanity to be noble. This is a great day."[7] It was indeed.

Immediately after the East Room announcement we gave a briefing in the White House press room and then moved on to the Washington Hilton for the major press conference. Each genome team had its own green room to rest and prepare in the hour beforehand. By this time the atmosphere had grown festive, and everyone in the Celera group was floating on air—almost everyone, that is. Heather took me aside and told me that Tony White was very unhappy; in fact, he was furious, fractious, and pouting. He had not been met at the White House gates. He had not been invited to address the throng on how he had signed the check that launched Celera. Indeed, he had not even had a chance to meet the president. Although I could have lobbied for access on his behalf, this was really a job for his press person; in any case, who gets to talk to the president is a matter for the White House, and it did not help his case that Tony had spent a great deal of time bad-mouthing Clinton. Things went from bad to worse for White as Collins and I made our way to the press conference. A *60 Minutes* camera crew was filming us, walking backward as the reporter asked us questions. The three of them backed into Tony, knocking him to the floor, and then stumbled over him. You could almost see the smoke steaming out of his ears as he awkwardly got up.

The press conference was unlike any I have been to before or since. Held in the ballroom, it included close to six hundred people and an unbelievable collection of television cameras and photographers. Flash after flash strobed across our faces. Gene Myers and Mark Adams, who deserved so much of the credit, joined me on stage with the government-funded scientists for the questions and answers. To everyone's surprise the tone remained positive, cooperative, and cordial. We were all one big happy genome family. But there was, of course, one question at the back of our minds. Would the truce last when we were back in our laboratories? Would the spirit of generosity and cooperation prevail when we returned to real life to write up our achievement for publication in a scientific journal?

16. PUBLISH AND BE DAMNED

A scientific man ought to have no wishes, no affections—a mere heart of stone.

—Charles Darwin

The truce with our rivals did not, in fact, last long. The greatest prize was to publish the full details of what we had done, to reveal to our peers and our critics the human genome, which we had sequenced in all its glory. Most important of all, we wanted to show the world, after having taken the first detailed look at the human instruction book, our analysis of what it meant. All this had been planned for the prestigious journal *Science.*

But, of course, the animosity was by now so bitter and so deep that this was wishful thinking. Within weeks of the White House announcement the public effort was lobbying behind the scenes against the publication of our paper with a series of incendiary missives. One sent to *Science* declared: "You have lowered a proud journal to the level of a newspaper Sunday supplement, accepting paid advertisement in the guise of a scientific paper."[1] An e-mail circulated among researchers calling on them to boycott *Science.* Once again the motivation was data release despite our efforts to make our data freely available to the scientific community.

While I understood the arguments and even the principles about publication, I believed there was a vendetta against Celera and me for having stolen a prize that others had been intent on claiming for themselves. The constraints on me and my team at Celera were truly mild; the only real one on the data came from the Celera shareholders, who had put up millions of dollars and rightly wanted to ensure that their investment wasn't undercut by enabling the data to be used by Celera competitors.

We were working with the *Science* editor Don Kennedy and his team on drafting the best form for the agreements that would permit researchers free, open, and unrestricted use of the Celera human genome sequence, while at the same time limiting the ability of commercial companies to repackage and exploit our data. But my critics wanted to block our publication unless there were no restrictions. To my amazement, they argued that if our business competitors could not use our data, the scientific community could not, either. And to my amusement this twisted philosophy was endorsed by the same Francis Collins who would write a touching folksong about how the genome belongs to everyone.

Don kept me fully informed of the criticisms. The irony was that the person doing much of this lobbying was Eric Lander, or "Eric Slander," as my team came to call him. From his experience with the companies he had started and his consulting work in the biotech industry, Eric certainly knew the score, and he had even agreed to the same terms that we now offered to *Science* before that disastrous meeting with the public program in Dulles. Eric was well aware that Celera could not let other companies take its data for free so they could undercut its own flourishing database business. In November 2000, Lander nonetheless persuaded Varmus and others from the MIT mafia, as some called them, to sign a letter to Kennedy urging him not to publish the Celera paper. Don Kennedy held firm, while Lander and Collins did all they could behind the scenes to block its release. Meanwhile, Don and I received high-level support for the data release agreement, including from Bruce Alberts, president of the National Academy of Sciences, and David Baltimore, Nobelist and president of the California Institute of Technology in Pasadena.

As a final gambit Lander and Collins threatened not to publish their own genome paper in *Science* but to take it instead to the journal's rival, *Nature,* if Kennedy did not reject my paper. Perhaps they had forgotten how Watson had told *Nature* that no U.S. genome scientists would ever publish in *Nature* again when he tried to block publication of my human genome directory paper in 1994. Once again it was the case that the end justifies the means.

We reached an agreement with Don Kennedy and *Science* to provide a DVD of the sequence to any scientist who wanted a complete copy, as well as access to a free website where anyone could do extensive data searches against the human genome for sequence matches. Celera would also provide a subscription service to academic institutions and biotech and pharmaceutical companies that provided extensive software and all of the genomes that had been sequenced (including the mouse genome), as well as massive computational facilities to enable comprehensive analysis of the genomes.

By January 2002 the Celera database business was generating more than $150 million in annual revenue, making it profitable in less than three years. The subscribers included most major universities and academic institutions, from the University of California to Harvard to the Karolinska in Stockholm to none other than the National Institutes of Health. While the scientific world (except, of course, the Wellcome Trust, which blocked its grantees from subscribing) was using Celera sequence data, Collins, Lander, and Sulston insistently repeated that Celera data were not available.

While this new round of battles raged, my team at Celera was working around the clock to analyze the human genome sequence that we had worked so hard to generate. The pressure on us was enormous, and it was all self-inflicted. The effort was led by Mark Adams, Gene Myers, Richard J. Mural, Granger Sutton, Ham Smith, Mark Yandell, Robert A. Holt, and me. The *Science* paper went through more than one hundred drafts, but the result was a comprehensive first look at the genome and its genes. We all knew that history would judge us by the quality of this document, and that meant the quality of our analysis and what it said about being human. I wanted a detailed, rigorous examination of what we had found, presented with confidence.

Our paper was published on February 16, 2001. This was no ordinary paper. It had 283 coauthors,[2] was forty-seven pages long (ten times the usual length), featured a five-foot-high color-coded fold-out map of the genome, and referred the reader to a massive amount of supplementary data on the *Science* Web site. One major surprise was how few genes we had actually found. Incyte and HGS had both claimed to have isolated and patented more than 200,000 genes using my EST method. At times they claimed there were more than 300,000 human genes. I had published a paper a few years earlier indicating that the number was far smaller, in the range of 50,000 to 80,000. The reality was that at most there were only 26,000 human genes.

The larger estimates we made had resulted from the simple assumption that genes would be evenly distributed across the genome, an assumption that turned out to be untrue. In fact, there were regions (we named them deserts) that contained millions of bases of genetic code but few or no genes, such as on chromosomes 13, 18, and the X. In contrast, some regions or chromosomes were densely packed, such as chromosome 19. To me this distribution was fascinating and immediately raised the question of what it said about human evolution. A comparison with the code of the fruit fly gave strong hints.

We share large sets of our genes with the fly, but the genes that we do not share speak volumes about the 600 million years of evolution since we had a common ancestor. These are the genes that made us uniquely human, and include a huge expansion in the numbers of genes associated with acquired immunity, intracellular and intercellular signaling pathways, and, most of all, the central nervous system. Such expansions took place in the gene-dense regions of our chromosomes more than elsewhere, notably by duplicating genes of a given category. For example, the genes associated with cell-to-cell communication were repeated over and over and were able to mutate and evolve new functions in these gene-rich areas. The deserts tended to be linked with the more ancient regions and function of our code, those associated with the basic process of staying alive.

Seeing the *Science* paper in print gave me one of the most intense feelings of satisfaction I have ever experienced. Despite the battles, the nonsense, the pettiness, and the constant complaints that what we were doing was hopeless, impossible, and unworkable, we had succeeded. It was one very, very sweet moment: I had successfully sequenced the human genome, not in fifteen years but in nine months, and made history with one of the best scientific teams ever assembled. No reward, award, or praise could ever substitute for that wonderful feeling.

When it had become clear that our publication in *Science* was going to proceed as planned, the public program carried out its threat to boycott the journal and publish in the British journal *Nature* instead. That was fine by me, because we would now have the cover of *Science* to ourselves. Even today, some of the *Science* staff remain upset by how Lander and Collins et al. still publish their follow-up genome papers only in *Nature.*

Unfortunately, my rivals continued their effort to undermine Celera's achievement. Their next attack, an article in the *Proceedings of the National Academy of Sciences,*[3] claimed that our sequence was not superior to the government/Wellcome Trust sequence, on which they said it heavily depended. According to Sulston, the whole-genome shotgun had not worked as claimed, even though Richard Durbin of the Sanger Center had admitted Celera's product was "in some respects better than ours."[4] Lander dismissed it as an "utter failure"[5] and a "genome tossed salad."[6] They deployed mathematical arguments that even they could not have believed seriously undermined our work. Gene Myers, in particular, was bitter, wounded, and indignant. Although we considered taking legal action, we decided it was better to rebut the attacks with data and facts in the scientific journals, remembering my mentor Nate Kaplan's words, "The truth will out."

The biggest general science gathering on the planet, the annual meeting of the American Association for the Advancement of Science, was about to be held in San Francisco. Since the AAAS is the publisher of *Science,* this was the natural venue for the release of the *Science* paper. The government-funded group published at the same time in *Nature* after having spent the previous Christmas dealing with a fourteen-page list of criticisms that they had been sloppy. Collins and I were invited to deliver the keynote lectures. We were both greeted with standing ovations. Afterward, I was overwhelmed by autograph seekers and fans.

That night Celera sponsored a celebration party at the San Francisco Design Center that was fueled by oysters, caviar, and chilled vodka. Mike Hunkapiller was there to share in the celebration, having missed the White House event because he had come down with chicken pox. The party was the most enjoyable of my career, but it was marred by the absence of Claire, who had said that she was too tired to come to California. I danced with several women, but with Heather most of all. She had never stopped believing in me during all the battles of the past few years.

We followed up the human genome with the mouse genome sequence. When I had suggested that the public program do mouse while we do human, it was met with outrage, but my rationale was sound. Having the mouse genome, which we completed in only six months, gave Celera great advantages in comparative genomics.[7] This time it was easy to ignore the public program because there were essentially no mouse data available and we had used only our shotgun data from unique mouse strains. By not contaminating our genome with poorer-quality public mouse data, we ended up with a much better assembly result than we had for the human. The mouse/human comparisons were the first to reveal that mammals share more than 90 percent of their genes and that they occur in largely the same order on our chromosomes establishing unequivocally the evolutionary relationship at the genome level With grants from the NIH we went on from there to sequence the rat genome and the genome of the Anopheles mosquito, which carries the malaria parasite.

By now I had built new programs at Celera, including the largest proteomics unit in the world (to help figure out what all those genes do) as well as for research on cancer vaccines. I purchased a pharmaceutical company in South San Francisco that focused on small molecules. I worked with Mike Hunkapiller to form Celera Diagnostics. I felt that I was moving Celera in the right direction, from reading the genome to using its code to discover new tests and treatments.

Recognition from the establishment for our work came thick and fast. I went to Saudi Arabia to receive the King Faisal International Prize for Science, along with the great entomologist, biologist, and writer E. O. Wilson, and to Vienna to receive the World Health Award from former Soviet president Mikhail Gorbachev. I received a multitude of honorary degrees from top universities around the world and was given the Paul Ehrlich and Ludwig Darmstaedter Prize, the top science prize from Germany. From Japan came the Takeda Award, and from Canada the Gairdner Award. However, politics intruded even here, when government-funded scientists tried to prevent me and my team from sharing the Gairdner Award, protesting that I was the sole recipient of the Paul Ehrlich prize, and complained that the American Society for Human Genetics had given me an award for "research that was unpublished and unseen."[8]

By now my team was spent. Gene, Mark, Ham, and I knew that we had reached a high point rarely achieved in life and that it would be somewhere between tough and impossible to ever get there again. Gene and Mark began to cast around for new opportunities. I started to think about a return to TIGR and all the other science I wanted to do. I would have stayed at Celera if it had been an independent company, but it was increasingly clear to me that I could not and would not survive continuing to work with Tony White, and so I began to plan my exit. White, in fact, had been planning it, too, still smarting from the way he had been sidelined at the White House.

However, a tragedy would intervene in all our lives. On September 11, 2001, I was headed for the airport, after having given a lecture in San Francisco the night before, when the news came about a plane hitting one of the World Trade Center buildings. Like everyone else, I watched, transfixed in horror, as the second plane struck. At that time there were fears that the attacks would be nationwide, so I decided to drive out of the city to Millbrae, where my mother and stepfather live, and checked into the Hyatt, where I would be stranded for several days.

As the extent of the tragedy unfolded, I wanted to do something to help, as did every American. Given that thousands of people had lost their lives in the inferno and collapse, I thought that the Celera sequencing facility could help with the important job of DNA analysis to identify the dead, which would have proven to be a vast task with conventional labs. Because of the state of the remains, the best approach was to study the more plentiful mitochondrial DNA. I called Mike Hunkapiller, not least because Applied Biosystems makes a forensic sequencing kit that is widely used by law enforcement. Mike suggested that we call Tony, and he in turn was supportive and gave me the go-ahead to offer help.

I had already met the head of the forensic lab in New York City, Robert Shaler, several months earlier, after a lecture I gave at the New York Museum of Natural History. I called and told him what I was thinking. At that time all air traffic was still grounded, but with the help of Tony White's corporate jet and special clearance, I set off with Mike's head of forensics, Rhonda Roby. We were the first nonmilitary flight allowed to fly across the country and had to check in with the North American Aerospace Defense Command (NORAD) every thirty minutes or risk being shot down.

Rhonda and I were met by the New York State Police on landing and given an escort to downtown New York. The atmosphere in the examiner's office was frantic, almost chaotic. We met with Bob and with the head of the New York State forensic office and figured out what Celera had to do to become a certified forensic laboratory. We were asked if we wanted to visit the site. I did, not least because my brother's office had been on the third floor of the first building. (I was relieved to find out that everyone in his office escaped without injury.) When Rhonda and I climbed out of the squad car, along with the head of the state police lab, I was overwhelmed by the sight, smell, and extent of the disaster. In the forensic tents we were shown the remains of the victims, such as they were. A heart. Body parts. Small bones and other flecks and fragments. In all there were about twenty thousand individual samples to analyze, identify, and return to the victims' families. The sights and smells transported me back to my time in Vietnam, to the Tet Offensive and having to decide which ones of the seriously injured I would help to live and whom I would leave to die. But in these tents there was no one left to save.

When we returned to Celera, it became clear that we had a long road ahead of us to become a forensic lab, for in making identifications there is no room for error. Rhonda Roby and Yu-Hui Rogers (now the head of the Venter Institute genome sequencing center) set up what we called the "Soaring Eagles Team" and worked incredibly hard to obtain certification from the FBI and from New York State. However, as time went by, the enthusiasm initially shown by Tony White evaporated, and he began to insist on a return to commercial operation. A pullout by Celera was orchestrated by Applied Biosystems, and Rhonda was called back to Foster City. I was frustrated and embarrassed, having given my word to Bob Shaler that in the wake of this awful tragedy we wanted to make a difference, not a profit.

The tensions between Tony White and me reached breaking point at an investor meeting in San Francisco. In a desultory way I delivered my standard business talk; there was seldom any real substance to what I was allowed to say. As the questions started, Tony and Dennis L. Winger, the Applera CFO, came to stand on either side of me, like dutiful apparatchiks keeping watch over a party official in the former Soviet Union. I left as soon as the session was over and took Heather, who had flown out with me, for a ride to see some of my old haunts. In a way I think I was taking stock of my life, the long journey from Millbrae to Celera. I drove to where my father was buried at the Presidio. We walked among the old ruins of the gun emplacements that were built to defend the city, one of my favorite high school hangouts. We then headed to Sausalito, where I used to sail and to spend weekends, and ended up having dinner with Dean Ornish, whom I had met at dinners with President Clinton and found to be a warm guy. Dean had developed a low-fat diet coupled with lifestyle advice and meditation as a means of changing the risk for heart disease and was now trying to do the same for cancer. The people around the table—his prostate cancer group—were as far from a Tony White or an Eric Lander as I could get. They were inspirational. Late in the evening, as we drove back to the city, I knew what I had to do.

Because I am not good at keeping things bottled up, I confided in two Applera board members that I wanted to go, but in a way that would not hurt Celera (in other words, a way that would not hurt my friends and colleagues). I was all for a graceful exit, but when it came to Tony White, I should have realized there was to be a final reckoning. If a relationship is to end, there is always one decision left, one that can show who holds the power, who is in control. In January 2002 the Applera board met just outside my office at Celera. Attendees included Mike Hunkapiller, the man who had opened this chapter of my life, and the president of Celera Diagnostics, Kathy Ordoñez, who had joined the company from Hoffmann–La Roche. At four in the afternoon the board sent the corporate lawyer, Bill Sawch, to see me. I was fired.

I was days away from being able to take advantage of one quarter (775,000) of my stock options; this windfall would now have to be forfeited. I would have thirty days to sell any other stock that I held. Sawch had two prepared press releases, one if I cooperated and a second one, as a threat, if I did not. The first was released on January 22. The Applera Corporation announced that I had stepped down as president of its Celera Genomics Group. Tony White, who would take over temporarily, praised my "incredible achievements" but said that Celera was now moving into drug discovery: "Our Board of Directors, Craig, and I all agreed that Celera's ongoing best interests would be served by making room for additional senior level management experienced in pharmaceutical discovery and development." And I found myself in the release agreeing that "I am confident that this move will leave Celera optimally positioned to continue to write history." I left with having sequenced the *Drosophila,* human, mouse, rat, and mosquito genomes, a profitable database business, a proteomics pipeline, a pharmaceutical business, a new diagnostic business, and $1 billion cash in Celera's coffers.

Even when you know the end is near, there is no way to really prepare yourself for the reality when it hits. Even though I wanted to leave Celera, I wanted to have some time to thank my team for their incredible determination and their motivation. They had given me 110 percent. Instead, I had to leave immediately and began to clear my office of memorabilia. My totem of good luck—the dried skin of the sea snake—would not let me down, but for now it was quickly packed away with the framed articles and the rest of my Venterabilia. I would not be allowed back. I would not be allowed to see or say good-bye to my senior team or any of the more than a thousand employees I had hired. Lynn Holland, who had been my personal assistant since my days in TIGR, and another member of my office team, Christine Wood, were in tears.

I was scheduled to deliver a major speech that night at the Economic Club in Washington, where I was guest of honor. I went home to change into a tuxedo and called Claire, but she was so devastated by events that she could not even talk to me. I gave an off-the-cuff speech at the club that Heather still claims was my best ever. The next day the reality began to sink in. Claire had been upset because she knew that I wanted to go back to TIGR and my job, as originally planned. By now she wanted more than anything else to remain as the head of TIGR. I, too, felt a real sense of loss, for only now did it dawn on me how important my team was to me after all the effort we had shared in building Celera.

Depression

The attacks and setbacks I have experienced over the years would have plunged some people into profound depression. That is not to say I have not been down from time to time, but I have been fortunate that I have been mostly able to escape deep clinical depression. Is this because of my genes? A team led by Kay Wilhelm of Sydney's St. Vincent's Hospital and the University of New South Wales in Australia found that the influence of adversity on the onset of depression was significantly greater for those who inherited on chromosome 17 a short version of the serotonin transporter gene, known as 5-HTTLPR, from both parents.

The difference in length is in a part of the gene called the "activation sequence" that controls how much of the protein is made. As a result of having a shorter version, around one-fifth of the population makes less of a protein responsible for transporting the brain chemical serotonin, which plays a key role in mood and pain regulation, appetite, and sleep, and is affected by Prozac. They have an 80 percent chance of becoming clinically depressed if they experience three or more negative events in five years. Once again we have a study that undermines simpleminded genetic determinism: Brain chemistry depends on both genes and circumstances, on both biology and society.

The work also showed that those with a long version that gave them "genetic resilience" against depression had only a 30 percent chance of developing the mental illness, given similar circumstances. The remainder—about half of all people—have a mix of the two genotypes. Many other studies have linked the short version to anxiety-related personality traits including harm avoidance and neuroticism and increased experimentation with illegal drugs. Fortunately for me, I have two copies of the long form and more serotonin.

I worked with my close friend and counselor Dave Kiernan on the final negotiations with the Applera attorneys. I was to agree not to say anything critical about Tony White for one year, and if I did not bad-mouth him or poach key people, they would give me the rest of my stock at the end of that time. Mark Adams was at the top of their list; he had to stay. But we managed to change the terms so that Ham, Heather, Lynn, and Chris could leave with me if they chose to. Once we had agreed on the terms, I stopped dwelling on the past and began to look to the future.

Within twenty-four hours a few of my closest colleagues at Celera would not take my calls, while others made it clear that as soon as I was established elsewhere, they would follow me if they could. My closest staff and associates wanted to walk out immediately but I urged them to stay. I think it was hardest on Lynn, Chris, and Heather, because they had to deal with my exit, the press, and Tony White strutting around in my place. My departure made the headlines, but in the spirit of our agreement, I was unavailable for comment. By April my job had been taken by Kathy Ordoñez, and 132 workers had been laid off, about 16 percent of Celera's staff.

I was so affected by leaving Celera that, according to a *Forbes* reporter, I had blurted out: "There is at least as good a chance that I'll commit suicide as die from some disease."[9] Even if I had said this (I don't remember having done so), it was probably a dramatic flourish to the end of an evening's drink and gossip. Anytime my life has gotten tough, I have found a simple and effective way to stop feeling sorry for myself: All I have to do is think of the incredible life, incredible fun, and incredible science that I would have missed if I had drowned myself in Vietnam.

I sought solace in the one thing I knew could cheer me: I headed for my boat and set sail for the turquoise seas of St. Barts, the little, *très chic* piece of France in the Caribbean. Whenever I have struggled to understand my life and my science, and whenever I have sought new challenges, I have looked to the open seas as a kind of haven. Sailing out of sight of the land and beyond the reach of mobile phones and television, I have found peace and time to think, refresh, and renew.

Each major transition in my life has been accompanied by a new sailing adventure with an expanded horizon. To keep my sanity when I was stationed in Vietnam, I would sail my 19-foot *Lightning* around Monkey Mountain and for a few miles up the coast from Da Nang. When writing my doctoral thesis I covered a distance of hundreds of miles from Catalina Island to Mexico in a little open boat. When I was experimenting with automated sequencing, I had the greatest sail of my life in my Cape Dory 33, *Sirius,* through mountainous seas in the Bermuda Triangle. Before I started to sequence the human genome, I had raced across the Atlantic Ocean in my 82-foot sloop *Sorcerer,* which I sold when I focused on this extraordinary challenge. Now, in the wake of leaving Celera, I found myself in a new yacht, sailing new seas, and seizing new scientific opportunities.

She first caught my eye when I saw her advertised in *Yachting* magazine. She was a sloop, 95 feet overall, only two years old, and hardly used. The boat had been designed by German Frers, the designer of *Sorcerer*, built in Auckland, and had been lying in New Zealand. Frers told me that *Sorcerer*'s sister was almost identical in design except that her hull was even bigger and faster. I flew to New Zealand and sailed her from Auckland to the volcanic Hauraki Gulf Islands. I fell in love with her immediately and made an offer. There was no time to sail her home, so I arranged for her to be shipped by freighter to Florida. Since she arrived in December 2000, I have had enormous fun and pleasure sailing her off the coasts of Cape Cod and Maine in the summers, and in the warm waters of the Caribbean in the winters.

While I was sailing in the turquoise waters of the Caribbean Sea in January 2002, I found myself thinking about what to do after sequencing the human genome. I had put so much of my energy and myself into Celera that the thought of building my life from almost nothing all over again was hard to contemplate. One option that I had always counted on was for me to return to TIGR, but given all the resentment that had built up with Claire and others because I had abandoned them to build Celera, it was no longer feasible.

I could have walked away from science to lie on a beach or to sail into my grave, but that reminded me of my patient in Vietnam who had given up and died because living was too tough and too painful. I was not done yet. All my life I have been a dreamer and a builder, and this was not the time to stop. I decided that it would be easier to start from scratch, just as I had done all those years ago when I worked at the NIH. I decided to move forward, to at least strive to do something new that could have an even bigger impact than sequencing the human genome.

I had kept most of my Celera stock, but after the announcement of my departure, its price fell to an all-time low. As a consequence I could not sell many of my shares, since the price was below my option price. But I was fortunate in that, though not as great as they could have been, my finances were more than sufficient to support what I now wanted to do. I had given 50 percent of my Celera holdings to my not-for-profit foundation, which could sell the stock when the price was higher, and as a result I now had more than $150 million to work with to do the science I wanted. It was the thought of the science, and the ideas I had not had time to pursue, that slowly brought me out of the doldrums. I could make the human genome more directly relevant to patients; I could look at what genomics could do for the environment; I could use sequencing to explore the incredible diversity of the sea or the city air. There was so much more left for us to understand. And I would pursue the ultimate challenge: to synthesize life itself. I decided to build a new research effort from scratch, as I had done before. Heather, Lynn, and Chris all told me they would leave Celera as soon as I needed them. I was refreshed, renewed, and ready to try to start over again.

Beyond the Genome

We are now beginning to realize that some genetic influences lie beyond our own DNA code. The lives of my grandparents—the air they breathed, the food they ate, and the stress they suffered—may affect me despite my never having directly experienced those things. These are "transgenerational" effects and have been observed, for example, in a study of a remote parish in northern Sweden. Thanks to Överkalix's registries of births and deaths and its detailed harvest records, it was found that people whose paternal grandparents had eaten less between the ages of nine and twelve seemed to live longer. The effect was sex specific: The diet of grandfathers was linked only to the lifespan of grandsons, and that of grandmothers only affected granddaughters. These effects may be due to an "epigenetic" mechanism, one that affects the way genes are turned on and off down the generations, rather than the mutations or changes in the genes themselves.[10] The Human Epigenome Project is now under way to figure out how the genetic code is put to use by the body.

17. BLUE PLANET AND NEW LIFE

Organic life beneath the shoreless waves
Was born and nurs'd in ocean's pearly caves;
First forms minute, unseen by spheric glass,
Move on the mud, or pierce the water mass;
These, as successive generations bloom,
New powers acquire, and larger limbs assume. . . .

—Erasmus Darwin, *The Temple of Nature*

From my very first studies of beating heart cells, I have been motivated by what interests me, not by my peers. In those dark days after I left Celera, I was once again inspired by my own experiences. Throughout my life I have been addicted to oil—I love cars, motorcycles, powerboats, sailboats, and airplanes—and as a result have burned tons of that product of ancient biology, adding more than my share of carbon dioxide to the environment. But over the years I have changed from being a superconsumer of fossil fuels who was ignorant of the consequences to someone who has become concerned enough about their effects on the environment to seek alternatives. Where better to start my new adventure than with the oceans, which sustain our planet and my sanity? I figured that we needed to work out exactly what is *in* the oceans if we are ever to assess accurately the effects of climate change, such as ocean acidification, and I could do that with a project that would possibly give us, as a bonus, new tools to help combat global warming.

Because we live on land, a terrestrial, even humancentric view of life has dominated our view of the impact of climate change. But when seen from space, our Earth is the blue planet. The first stages of life probably began here around 4 billion years ago in salty waters, where the line that divides inactive molecules and other chemistry from the biochemistry of life was crossed, and something emerged that we would today define as being alive. That vital something was a self-replicating cell, a complex mixture of proteins and genetic material swaddled in a lipid envelope. Today there is a fantastic diversity of ocean life, from whales to microbes, much of which is poorly understood, particularly at the microscopic end of the size spectrum. New solutions for the problem of climate change could arise from understanding that diversity of life and how it harnesses the sun's rays and absorbs carbon dioxide. I thought that I could exploit this understanding to go even further: I could try to mimic the events that took place billions of years ago in the oceans to come up with new kinds of life, a prospect that raised even more extraordinary possibilities.

When I returned from the Caribbean, I started work immediately. I set up a new not-for-profit institute, The Center for the Advancement of Genomics (TCAG), and applied for tax-exempt status. With funding from J. Craig Venter Science Foundation (JCVSF) and an endowment that resulted from selling founder's stock of Human Genome Sciences, the enzyme company Diversa, and Celera, I could now get started. I hired Heather, Lynn, and Chris out of Celera, and we began work in my basement in Potomac, Maryland, until we could lease a new workplace. I had several science projects that I wanted to establish as soon as possible.

My first concern was to get my environment project up and running. The scientific evidence is now overwhelming that the 3.5 billion tons of carbon dioxide that we are pumping into the atmosphere each year are altering global climate patterns—modern life, in short, is unsustainable. But I wanted to do more than just using less oil and gas or installing a solar panel. I felt that genomics had something unique to offer. Shotgun sequencing of the oceans could provide a snapshot of ocean health today, help to monitor its health tomorrow, and reveal the microbes that are responsible for creating much of our atmosphere. The metabolic machinery of ocean microbes might also teach us new ways to make alternative sources of fuel such as hydrogen, methane, or ethanol.

I formed the Institute for Biological Energy Alternatives (IBEA), and recruited Ham Smith to be its scientific director. My vision of carrying out environmental genomics would require a substantial DNA sequencing facility, and I had to convince my foundation's board to take a $40 million risk and to build a new facility to rival what we had had at Celera. We established a new not-for-profit organization called the JCVSF–Joint Technology Center

(JTC), which would do the sequencing for TIGR as well. While Heather and the team moved into a temporary building in Rockville, Maryland, construction of the new 120,000-square-foot research building could begin on land that I had purchased with endowment funds a few years earlier. Employees were hired at an aggressive rate, and the enterprise grew even more rapidly as a result of massive layoffs at Celera. Many of my former friends and coworkers came to join my new research organizations.

While the environment was to be the first focus of my efforts, there was still some unfinished business to attend to on the genome front. With my new genomics lab I wanted to follow up on my human genome sequencing to take into account its ethical implications as well as to advance the efforts to turn this understanding into medicine. Of equal importance and as a matter of personal pride, I also wanted to deal once and for all with the continuing attacks and criticisms made by government-funded genome scientists. Instead of dying down, as people got on with their lives after the genome race, they had became even more intense as the struggle continued for sole credit for sequencing the human genome.

Perhaps the most notorious example appeared in April 2002 when the *Journal of Molecular Biology*[1] published "The Human Genome Project: A Player's Perspective." The article was written by Maynard Olson of the University of Washington, who had been described by his peers as the "conscience of the genome project."[2] In a thinly veiled attempt to shift credit away from me to his colleagues, he returned to the old issue of whether my approach was truly new: "Venter's claim to having 'invented' whole-genome sequencing is based on his leadership of a project to sequence a tiny bacterial genome that was nearly devoid of repeats." Then followed the claim that I had lied: "In contradiction to Venter's sworn testimony in June, 1998, Celera had kept its data entirely secret." Olson did give me credit for the fact that the "Celera initiative undoubtedly accelerated the availability of an initial human-genome sequence by approximately two years."

By that time we had written a rebuttal[3] to the Lander, Sulston, and Waterston *PNAS* paper, which argued that they should receive sole credit for sequencing the human genome because my whole-genome shotgun method had failed.[4] This, in turn, prompted Lander et al. to make the same claims using some sleight-of-hand arguments in a nonscientifically reviewed article.[5] Granger Sutton, the computer scientist who had worked at TIGR and Celera on assembling genomes, was particularly angry with Lander, because he believed that even if Sulston and Waterston did not understand what Celera had accomplished (both of them are biologists, not experts in mathematics and computation, after all), Lander certainly should have. Not only did he have an impressive mathematical background, but his own people were developing their version of a whole-genome shotgun assembler, called Arachne, based on our work.

Granger thought that by then the hatchet had been long buried. In June 2001, at prompting from President Clinton and long before the *PNAS* paper, the computational biologists from Celera and the International Human Genome Consortium met on neutral ground (the Howard Hughes Medical Institute in Chevy Chase, Maryland) to discuss sequencing and assembly strategies. As *The New York Times* reported of the gathering: "The meeting, from which the teams' leaders were absent, was conducted by the computational biologists on each side in a cordial atmosphere."[6] Granger told the meeting about how it had been possible to reassemble the Celera human genome without drawing upon the public data at all, and with far better results. In the wake of the subsequent attacks, Granger was understandably furious: The conclusions of the *PNAS* paper had already been refuted by the data presented the year before in Chevy Chase: "We had presented some pretty convincing stuff on how good the whole-genome shotgun approach was."

In fact, by now there were significant figures on the public side who were convinced. One of them was Jim Kent of the University of California, Santa Cruz, a burly, bearded figure who was considered a star player, having single-handedly put together the code—GigAssembler—that would run on 100 Pentium III computers to assemble the public genome in only four weeks, in time for the White House announcement, and all while still a graduate student.[7] I was impressed with that achievement.

Kent did not agree with the conclusions of the *PNAS* paper, because it was clear to him from discrepancies between the public and Celera data[8]—and even before we had sequenced the mouse—"that the reconstruction of public data envisioned in the Lander et al. *PNAS* article could not be completely true." He concluded: "I do think it fair to say that the Celera assembly was better overall than ours. (One would hope so since they always had access to all of our data in addition to what all they were generating.)" He added that Lander's own Arachne assembler, which is remarkably similar to the one we used, is "another indication that the Celera technique, with some limitations basically works very well."

Recall that when we followed up the human genome with the mouse sequence, we ignored the limited public program data in GenBank and used only our own shotgun data, so we would not run into another barrage of sophistry and distortion about what we had really accomplished. With an enhanced assembler we ended up with a better result than we had achieved with the human. Ari Patrinos, then one of the G5, looked back on the *PNAS* paper from Lander et al. and concluded that "it was sad, frankly. The method worked, and there was no better confirmation of the fact that it was used to do the mouse." Even my old foe Michael Morgan conceded that "you have to be very certain to try and shoot someone down, because it usually backfires. All of those articles backfired in one way or another."

While we published a second rebuttal,[9] it was clear to me, as it always has been, that data is the only way to win scientific arguments. I contacted Mike Hunkapiller at ABI; he was also upset by the sustained criticism and wanted to help set the record straight. Given all I had been through with Applera and Tony White in wrangles over data release, I added a legal binding agreement to our purchase of $30 million of new DNA sequencers from ABI that we could publish the Celera data and put it in the public domain without any restrictions. (In 2005, Celera would abandon selling genome information and put it all in the public domain.) In addition, my institute would get a complete copy of the human genome data for research purposes. Once the agreement was signed, we undertook a major collaboration with the remaining scientists at Celera to compare Celera's whole-genome assembly to the other versions of the genome, including the "finished" public version.

Sorin Istrail, one of the senior members of the team led by Gene Myers, was by now the head of bioinformatics at Celera, and he headed the new collaborative effort. The analysis required more than a year and the development of many new computational tools, which for the first time allowed the comparison of whole human genomes. I shared our plans with the *PNAS* editor, who was pleased that we were making all the data available and said that he would like to publish our paper when it was ready, to help end the controversy.

The data were powerful and demonstrated that the whole genome shotgun assembly was accurate. When this work was eventually published at the beginning of 2004, it enabled me to compare accurately the public and private genomes: The Celera results provided more order and orientation, while the public consortium sequence provided better coverage of repeats. The government/Wellcome Trust–funded labs had continued to polish their version of the genome sequence over a four-year period at a cost of about $100 million (perhaps more accurate numbers are not available). The comparisons demonstrated that with each improvement in the quality of the public data, they more closely approached the Celera-only assembly in quality and accuracy (order and orientation). The Celera assembly, in fact, closed many of the remaining gaps in the "finished" genome that was announced with much fanfare in the journal *Nature* by the public program in 2004.[10] We published our paper with no press announcements. The data were powerful enough in their own right.[11]

With the open warfare finally at an end, I was preparing to begin a new phase of the human genome effort. Since the publication of the human genome assembly analysis paper, the team at TCAG (now the Venter Institute, after three of the five not-for-profit institutes were merged together) had been at work on the sequencing and analysis of the genome of a single individual. That individual was me. The reason for this choice was a matter of science, not ego or hubris. The earlier composite versions of the genome, including Celera's, had dramatically underestimated the amount of human variation. The government genome relied on a patchwork of pieces (clones) from a limited number of individuals, so genetic variations were invisible. The Celera genome was a consensus sequence formed from the genomes of five individuals, including my own. We used a winner-takes-all philosophy: Our genome used the parts that were shared by most of the five people. This also lost the variations caused by indels (insertion deletion polymorphisms), places in the genetic code where more than a single letter is changed. When there was a major insertion or deletion of DNA in one individual, the assembly program did not register that change unless it occurred in the majority of the sequences.

In other words, both of the genomes that were unveiled with a great fanfare in June 2000 did not account for one of the key reasons that we want to read them in the first place: a blend or mosaic of DNA from different people erases the individual differences that make some of us predisposed to cancer, heart disease, or whatever (although there had been efforts of course, to chart single-letter variations, SNPs). The earlier genomes had focused on only one copy of a person's genetic code when in fact we inherit two, one from each parent. In some places one parent's gene is dominant; in others, the other one is. We needed to look at all 6 billion base pairs of code, not 3 billion, to have the most accurate representation of what a human sequence really looks like.

While for obvious reasons we never announced that Ham and I were two of the donors in the original sequencing, we never worked to keep our role secret. When the investigative TV news magazine *60 Minutes* covered the genome race, it revealed that I was one of the DNA donors; however, it was not until later, when Nicholas Wade of *The New York Times* was interviewing me about my new institutes, that my genome really became news. I did not think anything of our conversation until the following Saturday morning when *The New York Times* was delivered to my home with the front page story "Scientist Reveals Secret of the Genome: It's His."[12] The headline was inaccurate, but I guess this helps prove that it is not news until *The New York Times* says it is.

Blinding Discovery

The media has reported some of the more depressing discoveries made in my genome with some relish. One front-page story described how, "at the request of *The Wall Street Journal*, Dr. Venter's associates checked his versions of specific genes that have been linked to health risks. During a conference call to discuss the results, Dr. Venter learned his genes place him in greater danger of going blind. Life can be that way when you study your own DNA."[13]

The newspaper was referring to the fact that one of my genes, known as Complement Factor H (CFH), has a variation in a single letter (a SNP called rs1061170) that has been linked by several studies to a "very high" risk for macular degeneration, a common disease that causes degeneration of the center of the retina, with devastating effects on central vision.

Of the two copies of my CFH gene, only one copy is mutated this way, raising my risk of the disease by about three to four times. If both had been mutated, that risk would jump to more than ten times.

Earlier studies had suggested that CFH may play a role in protecting blood vessels from inflammation and damage, so a mutated form may allow inflammation and thus blindness. One of the known properties of Factor H is that it regulates the activation of the complement system, a collection of related proteins that are the body's front-line defense system—the innate system—that attacks foreign invaders while usually avoiding any attacks against healthy cells, the "self."

It can be said that my genome dominated the Celera sequence. As mentioned in Chapter 14, the genome assembly team wanted more coverage of one of the five genomes in the group to ensure that we had an accurate assembly. Ham's DNA had ended up in the highest-quality 50 kb libraries, but because early sequencing libraries in the 2 kb and 10 kb range worked the best from my genome, these libraries were selected for the three-times sequencing coverage. Overall, my DNA had contributed 60 percent of the final Celera genome.

Cancer and My Genome

Many people are born with a mutation that puts them one step closer to developing tumors. In general, some SNPs—one-letter spelling mistakes—can dramatically alter a gene's behavior, while others can have more subtle functional effects that predispose an individual to disease in concert with the individual's genetic background or environment (for instance, there are some genes that raise the risk of lung cancer in a smoker, leaving nonsmokers unaffected), and others that have no effect at all (nonfunctional SNPs).

Genes code for proteins, and of these three SNP types, the ones that change protein structure, and thus function by changing one of the amino acid protein building blocks, are of most interest. These are called "nonsynonymous SNPs." The good news, as far as it goes, is that a search through my genome for mutations in four genes that have already been linked with cancer—Her2, Tp53, PIK3CA, and RBL2—revealed two nonsynonymous SNPs, with no known link to the disease, and two novel ones of unknown effect. Of these novel SNPs one occurs in PIK3CA in what we call a conserved position, that is, a part of the protein that usually varies little because, presumably, it is important.

There is no data on whether this particular change puts me at greater risk. But PIK3CA is one of a gene family that encodes proteins called lipid kinases, enzymes that modify fatty molecules and direct cells to grow, change shape, and move around. PIK3CA mutations are also known to occur in up to 30 percent of colorectal and gastric cancers and glioblastomas, and they are also present, to a lesser extent, in breast and lung cancer. Mutations in PIK3CA can also occur spontaneously in brain tumors. I may look into this more deeply.

Obtaining the sequence of one person's genome has triggered debate, like so much else in genomics. Members of my Celera scientific advisory board were uneasy about any donors being identified. Art Caplan likened the project to the tomb of the unknown soldier, where it was sacred not to know. But the entire point of modern military DNA forensic science, in fact, is to never have future "unknowns." As with many previous controversies in medical science, from heart transplants to test-tube babies, attitudes have changed dramatically with time. This is best illustrated by the announcement that Jim Watson is now having his own genome sequenced by the 454 Life Sciences Company, a new commercial venture that has created a sequencer based on the pioneering work of Mathias Uhlen in Stockholm, who developed piro-sequencing.

Since the revelation of my involvement in the project, I have been asked at every turn what I learned from my genome sequence. (In fact, it was only in 2006 that the reading of all 6 billion base pairs of my code was finished.) On September 4, 2007, we published the first diploid genome sequence for *Homo sapiens* in the free-access journal *PLoS Biology*.[14] Did this incredible knowledge bother me? Was I afraid to have it posted for the world to see on the Internet? I have always argued, and illustrated throughout this book, that there will be very few unequivocal answers revealed by our genomes, and what they will tell us will probably be best expressed in terms of probabilities. Only when we get the big picture of what our genes mean—and that will take decades—will we be able to figure out whether they are telling us that we have a 35 percent chance of getting breast or colon cancer, or whatever.

When it comes to my genome, one of my biggest disappointments came in 2005, when I was diagnosed with two kinds of skin cancer: melanoma and basal cell carcinoma. Fortunately, both were detected early. However, I did not have enough tissue to analyze the genetic changes that caused the tumors, and it would have been fascinating to see how my genome had spun out of control within these cells, how my DNA had let me down so that my cells started to multiply without regard for the health of the whole body.

Still, in broad outline I know what I would have seen. Cancer is thought to arise from an accumulation of genetic defects, with one fashionable idea being that their effects are most felt in the stem cells that supply cell types to given tissues and organs. In colon cancer the first step is a defect in a growth gene called ras, which makes a cell proliferate to form a polyp, a premalignant growth. Typically, another growth-controlling gene in a polyp cell suffers damage, and as the tumor enlarges, more mutations arise as a result of sheer probability and because rapidly proliferating cells are more likely to carry mutations and even "mutator" genes that boost the rate of DNA errors. This is the multi-hit model of cancer set out by my former collaborator, Bert Vogelstein of Johns Hopkins, arguably the most important cancer researcher in the world today. At the Venter Institute, under the direction of Bob Strausberg, we have an ongoing major collaboration with several notable research groups, including Vogelstein's, in which we are looking at somatic changes in genes in tumor cells. Somatic changes are those caused by environmental factors such as toxins and radiation that can mutate genes in the nonsex cells. This can then lead to cancer in an individual, but since it is a noninherited form of cancer it would not be passed on to the children of that person.

Only about 3 to 5 percent of cancers are due to genetic defects inherited from our parents; the other 95 to 97 percent are due to the somatic gene changes. While many research groups are looking for gene changes associated with the cause of cancer, we are primarily focused on gene changes that can predict effective treatment for the tumor. Tyrosine kinase receptors are some of the key cell growth regulatory proteins in our cells. Recent highly effective cancer chemotherapeutic agents have been able to block tyrosine kinase receptors, but their effectiveness often depends on the type of mutation that has occurred in the receptor gene. As a result we have been sequencing the genes in the tyrosine kinase receptor family in search of somatic mutations. We did not have to look far: In our first study we examined the genes in brain tumors and quickly found several unique mutations. We have now extended the study to several other tumor types, including breast and colon cancer.

Sequencing and Cancer

A glimpse of a future in which doctors use "personalized medicine" has come from a study in which a new generation of DNA-reading machine has been able to predict which lung cancer patients will respond to a novel class of drugs. Non–small-cell lung cancer is the leading cause of cancer death worldwide, and earlier work revealed that the tumor cells in about one-fourth of sufferers contain extra copies of the epidermal growth factor receptor (EGFR) gene and as a result are more likely to respond to a class of inhibitor drugs such as gefitinib and erlotinib. The 454 Life Sciences Corporation in Connecticut, in collaboration with scientists at Dana-Farber Cancer Center and Broad Institute, near Boston, used the 454 Sequencing method, which is capable of generating hundreds of thousands of DNA sequences in a single pass, to analyze EGFR gene mutations in tumor samples from twenty-two patients with lung cancer, allowing EGFR inhibitor therapy to target the patients who will benefit the most from it.

As the sequencers in my new facility were churning out more reads of my genome, I turned my attention to a project that combined two great loves of my life: science and sailing. The concept was simple: Take seawater and capture all the microorganisms swimming in it on filters with microscopic pores, isolate the DNA from all the captured organisms simultaneously, make shotgun sequencing libraries from the DNA, sequence thousands to millions of sequences in a single pass, assemble the sequences back into chromosomes and chromosome fragments, and finally analyze the sequences for genes and metabolic pathways to understand what, precisely, is living in that part of the ocean. Rather than focusing on a hunt for one particular type of life, we would obtain a snapshot of the microbial diversity in a single drop of seawater—a genome of the ocean itself.

This was, to me, a straightforward extension of work that had started with the EST method and led to the whole-genome shotgun approach, then the first genome of an organism in history, and then of course to the human genome. As had been the case with every one of its predecessors, this project was met with skepticism. Many people were certain that shotgun sequencing seawater would not work, because we were sequencing a soup of vast numbers of different species. To return to the jigsaw puzzle analogy, this would be like taking thousands of different jigsaw puzzles, mixing up all the pieces, and then trying to solve all the individual puzzles simultaneously.

However, I already knew from some of the earliest genome sequencing experiments at TIGR that our computational tools could accurately assemble more than one complete genome from such a complex mixture. In 1996 we were sent a sample of a presumed *Streptococcus* pneumonia species that had been isolated from a patient. On sequencing the bacterium's genome, the assembler revealed not one but two separate but closely related microbial species' genomes. From this and many more experiences—not least assembling 27 million pieces of human DNA back into the human chromosomes—I became certain that a unique genome sequence would provide a unique mathematical assembly of sequences from that genome.

To help convince a DOE grant review board to fund an ocean pilot project I carried out a simple demonstration: I took every known sequenced microbial genome (approximately one hundred genomes at that time) and fractured their sequences into small stretches with fewer than one thousand base pairs. I then pooled together all these fragments and ran the lot through the genome assembler. On examining the data it became clear that the sequences had been reconstructed correctly by our algorithms into each and every individual genome, with no false assemblies. The review committee was impressed but still not convinced that the procedure would work in the ocean.

I decided to design a pilot experiment to conduct a real test, funding it from my foundation once again. I contacted Anthony H. Knapp, the head of the Bermuda Biological Station for Research, who arranged for Jeff Hoffman from my lab to take some water samples in the Sargasso Sea. We had deliberately chosen the Sargasso because it was thought of as an ocean desert, one starved of nutrients and therefore devoid of much microbial life. Only a few microorganisms had ever been characterized from those waters.

Once the filters containing the microorganisms from the first samples were processed in the sequencing lab and the data examined, I knew we had a winner. We had opened the doors to a world that has been mostly unknown to modern science. From the sunlit surface to the darkest submarine canyons stretches an ocean of life that is beyond human imagination, containing on the order of 10^{30} single-cell organisms and 10^{31} viruses. That amounts to 1,000,000,000,000,000,000,000,000,000,000 organisms, representing millions of unique species, or a billion trillion organisms for every human on the planet.

In much of science we come to know only what we are able to see or measure. In the case of microorganisms, we know a lot about those we have been able to cultivate. The problem is that relatively few organisms—less than 1 percent—have been isolated and grown in culture; when it comes to the 99 percent that do not grow in the laboratory we know virtually nothing. In fact, to all intents and purposes they do not exist. I was thrilled by the idea that my shotgun technique could help reveal the 99 percent of life that had been missed. We could now crack the ocean code, which varies from sea to sea, whether it comes from sulfur vents on the sea bed, near a garden of soft coral, or the top of a submarine volcano.

We have since discovered tens of thousands of new species, many of them strange and exotic. In all we discovered more than 1.3 million new genes in only 200 liters of surface seawater. To put this number in context, the first samples analyzed doubled the number of known genes on the planet. To find such an enormous quantity in what had been regarded as one of the world's most nutrient-impoverished bodies of water posed huge challenges to evolutionary biology.

This research had practical implications, too. Some twenty thousand of the proteins we isolated were involved in processing hydrogen. Another eight hundred new genes harnessed the energy of light. That figure quadruples the number of photoreceptors (such as the ones found in the back of your eye) known to science and suggests that some new type of light-driven biology may explain the Sargasso Sea's unexpectedly high diversity.

We could have continued to sample in the Sargasso Sea, but I wanted to see if there was greater diversity at different sites around the world. So began the *Sorcerer II* Expedition, with the backing of the Venter Institute endowment, the Gordon and Betty Moore Foundation, the DOE, and the Discovery Channel. My yacht was specially refitted for a circumnavigation of our planet so she could crisscross the oceans and collect samples day and night. This voyage of discovery—and self-discovery—has led to a new field, environmental genomics,[15] a field that has been hailed as novel and exciting. I feel that this endeavor could match the long-term impact of sequencing the human genome, if not exceed it.

For two years, I flew back and forth to join the crew of *Sorcerer II* as she sampled the seas from Halifax, Nova Scotia, to the Eastern Tropical Pacific. I found one passage in particular—through the Panama Canal to Cocos Island and then down the Galapagos—to be a particularly transforming experience as I combined genomics with writing this book and diving with sharks, all under the gaze of the TV cameras. It was thrilling to be at the heart of a venture inspired in part by the journeys of the HMS *Beagle* and the HMS *Challenger* in the nineteenth century.

To capture DNA, we took a sample of water every two hundred nautical miles and then filtered it through progressively smaller filters to collect bacteria and then viruses. The filters were stored in refrigerators on board before being airlifted to be sequenced in Rockville. There, a team led by Shibu Yooseph used phenomenal computing power—including supercomputers that had also been used to animate *Shrek* and to simulate H-bomb blasts—to reconstruct and analyze the vast amount of shotgunned microbial DNA data. He compared each DNA fragment with every other to produce clusters of related sequences and predict the proteins in the data. At the Salk Institute, in La Jolla, Gerard Manning also compared the data against Pfam, a collection of signature profiles for all known protein families, with the help of "accelerator" hardware from Time Logic, a company in Carlsbad, California. His team did almost 350 million comparisons, an order of magnitude or two more than had been achieved before. The final computation took two weeks but would have run for well more than a century on a standard computer. The trove of data was breathtaking. In a trio of papers published in 2007 in the journal *PLoS Biology* my team, led by Doug Rusch, described four hundred newly discovered microbes and 6 million new genes, doubling the number then known to science.[16]

The biggest impact of the expedition has been on established ideas about the tree of life. It used to be thought that the light-detecting protein pigment in our own eyes was relatively rare. But our gene trawl revealed that all surface marine organisms make proteorhodopsins that detect colored light. These proteins help microbes to use sunlight, as plants do, but without photosynthesis. Instead they use their "light-harvesting" machinery to pump charged atoms into their equivalent of solar batteries. Blue and green battery variants are found in different environments—blue-light varieties dominate in the open ocean, such as the indigo Sargasso Sea, while green-light varieties live near coasts.

During the circumnavigation my team discovered some new proteins that protect microbes from ultraviolet rays and others that are involved in repairing the damage caused by UV light. We found that certain protein profiles are more popular in the ocean than on land. For example, land-loving Gram-positive bacteria are best known for their hardy spores, but this feature is absent in their marine relatives. Flagella, whiplike extensions propelling bacteria forward, and pili, short extensions used to exchange genetic material between bacteria (the microbial equivalent of sex), are also less frequent in the oceans.

We were also surprised to discover that many kinds of protein that were thought to be specific to one kingdom of life were more widespread. Take glutamine synthetase (GS), the protein that plays a key role in nitrogen metabolism. More than 9,000 GS or GS-like sequences were uncovered. Many were typical of what we call type II GS (one of the three basic types of this protein) This was an unexpected finding, because type II GS is associated more with eukaryotes, such as our own cells, and not with the mostly "simpler" life—bacteria and viruses—in the filters that we analyzed.

Of all the families of proteins we studied, the kinase families are particularly interesting. Protein kinases are enzymes that regulate many of the most basic cellular functions in our bodies. They control the activity of proteins and small molecules in the cells by attaching phosphate chemical groups to them Because of their importance, they are key targets for treating cancer and other diseases. Previously, it was thought that different families of kinases were found in different kingdoms of life. Our cells were believed to use eukaryotic protein kinases (ePK), while bacteria relied on histidine kinases. We found however, that ePK-like kinases are common in bacteria, as well, and in fact, more common than histidine kinases. It also emerged that ten key protein features were the same in almost all kinase families, revealing them to be at the core of what determines a kinase. In this way, data on the genes shared by vast numbers of organisms can be used as a kind of a time machine: We can work out the kinases that must have been functioning in a common ancestor and, in this particular case, deduce that several of these protein families must have existed before the divergence of the three domains of life billions of years ago.

When it comes to climate change, *Sorcerer II* has given rise to another intriguing insight. Some patches of the ocean have more carbon-hungry organisms than others. Traditionally it has been thought that these marine populations tracked local nutrient levels, so that high levels of organisms indicated nutrient-rich waters. But the reality may not be so simple. Bacterial viruses—phages—may actually be responsible for keeping microbe levels low in some seas. If we can understand this relationship better, and learn how to inhibit the viruses, or make the bacteria resistant to phage attack, a lot more organisms could be capturing carbon dioxide and damping down climate change. The new understanding raises even more striking possibilities.

Based on the discoveries of millions of new genes, we are beginning to assemble a toolkit to begin a new phase of evolution. Microbes play a crucial role in the Earth's atmosphere. Trees breathe in carbon dioxide, thanks to photosynthesis. The same is true of the oceans, but with the involvement of additional mechanisms. Could we design new organisms to live in the emission-control system of a coal-fired plant to soak up its carbon dioxide? Could we harness microbes and their extraordinary biochemistry to alter the atmosphere? Could we persuade the planet's microbial lungs to take deeper breaths? This is not as crazy as it sounds. After all, we owe the oxygen in the air we breathe to a microbial population change that took place more than 2 billion years ago. These microbes had to rid themselves of oxygen to avoid being poisoned, and their "oxygen waste" became part of the atmosphere. To counter the effects of burning fossil fuels, perhaps soil microbes could be persuaded to hang on to more carbon. The communities that make up the Earth's expanded lungs could be concentrated in mines, deep aquifers, or deserts.

The first step is to read the genomes of microbes, trees, and the thousands of other creatures that deal with pollution, whether carbon dioxide, radionuclides, or heavy metals. Many such genomes have now been sequenced, mostly with my methods, and many by my own teams. I have also extended the approach I used in the Sargasso Sea to study the air that New Yorkers breathe in Manhattan. This city is now the test bed for my Air Genome Project in which I hope to identify the bacteria, fungi, and viruses that enter our lungs with each breath. Many more microbes are being sequenced as I write. With the information this collected on myriad microorganisms, I can not only study new ways to monitor air quality and monitor bioterrorism but also see if there is a way to harness these organisms and their clever chemistry in cleanups.

We already have a long list of promising organisms to study when it comes to ameliorating the effects of global warming. Peat bogs contain the *Methylococcus capsulatus,* which cycles the greenhouse gas methane. *Rhodopseudomonas palustris* is a soil bug that converts carbon dioxide into cell material and nitrogen gas into ammonia, and can generate hydrogen. *Nitrosomonas europaea* and *Nostoc punctiforme* also take part in nitrogen fixation. Of the many marine microbes, a diatom called *Thalassiosira pseudonana* plays a role in transferring carbon to the ocean depths. All could play a role in engineering our sick atmosphere.

We could go further. Was it possible to use our current understanding to design and chemically build the chromosome of a novel species to create the first self-replicating artificial life form to tap new alternative energy sources? This proposal is bound to trigger unease from biofundamentalists, but it is a natural extension of previous efforts of many individuals over the millennia to make useful products from biological processes. Biotechnology dates back thousands of years, to the first fermenting of wine to make alcohol, the first biofuel.

There is already evidence that these microorganisms could revolutionize the petrochemical industry. DuPont is one of the many companies that has traditionally relied on a cheap oil supply, converting it into polymers used in clothing, carpets, ropes, and bulletproof vests. Now they are at the vanguard of commercial experiments to switch from oil to using engineered bacteria fed on sugar, a renewable source of carbon that depends on the way plants fix carbon dioxide from the air.

DuPont scientists working with Genencor in Palo Alto have modified E. coli bacteria to turn glucose into a compound called propanediol. In a Tennessee plant, tons of the bacteria make the compound from corn sugar to manufacture their Sorona polymer, which the company uses to make stain-resistant carpets and clothing. This is only the beginning. What if we could engineer bacteria to produce fuels such as butane, propane, or even octane, all from sugar? Better still, what if we could engineer them to use cellulose, the sugar polymer that helps give plants and trees their structure? This is the kind of visionary technology that could change the world. The limited sources of oil on our planet have led to a massively disproportionate distribution of wealth, caused and fueled wars, challenged our national security, emitted a pall of pollution, and driven climate change, from hurricanes to floods and droughts.

My own attempts to create designer genomes date in spirit back to 1995. After sequencing the first two genomes in history at TIGR, we undertook a major study to attempt to determine the minimal gene set required for a single cell to live. This was only a step away from creating a synthetic chromosome containing only the genes predicted to be required for life. We hoped that understanding life in its most basic form would pave the way to a new level of control over the genetic architecture of an organism.

Before undertaking such a bold project, I commissioned an ethical review of the idea of making a genome from scratch. The exercise required more than eighteen months and the solicitation of opinions from most major religions. While our approach proved to be scientifically valid, it did raise a number of concerns, ranging from the potential dangers of the technology (biological weapons, unanticipated environmental effects) to the challenge it posed to our conception of the meaning of life. By the time the review was finished, however, I had already started Celera and begun sequencing the human genome. The quest for the artificial genome would have to wait.

After Celera I returned to the problem of synthetic life with a vengeance. The next milestone is perhaps best summed up by what happened after my car pulled up to the curb in front of the Oval Room Restaurant at 800 Connecticut Avenue, a few blocks from the Oval Office on Pennsylvania Avenue. It was Friday, September 3, 2003, and I had been summoned to an urgent lunchtime meeting by Ari Patrinos, the man who brought about the truce in the genome war, who was then still working at the Department of Energy's biology directorate. Joining us was his boss, Raymond Lee Orbach, director of the Office of Science at the DOE; John H. Marburger III, the science advisor to the president and director of the Office of Science and Technology Policy; Lawrence Kerr, director of Bioterrorism, Research, and Development for the Office of Homeland Security at the White House; a third DOE official; and me. What was remarkable was how this high-powered gathering had been arranged only two hours beforehand.

The group was keen to discuss a breakthrough in the synthetic genome project that DOE was funding at my Institute for Biological Energy Alternatives (in 2004 IBEA merged into the Venter Institute)—a $3 million effort to "develop a synthetic chromosome," the first step toward making a self-replicating organism with a completely artificial genome. The day before, I had called Ari to inform him that our team, principally Ham Smith and Clyde Hutchison, had made a leap forward in our ability to synthesize the DNA for a small genome as a step toward our synthetic species project. We had finally achieved the synthesis of biologically active Phi-X174, a bacteriophage that infects the bacterium E. coli, which we had been attempting for more than five years, without any success.

I had always viewed the Phi-X174 synthesis as an essential step in our much larger goal of making a chromosome for a synthetic species. For Phi-X174 to thrive in a bacterium requires essentially every base pair of its DNA code to be correct: There is no room for error. I figured that unless we could assemble the five thousand or so base pairs of Phi-X174 successfully, we could never synthesize a minimal bacterial chromosome consisting of around 500,000 base pairs. On several occasions we had produced molecules of the correct size, but because they were not infective, we knew that the DNA contained errors. The project moved slowly as we built the new institute and the scientific team, and several strategies were devised for the synthesis. In my own mind I remained confident that we would be successful when we could dissect the problem in a systematic way. I insisted, for example, that we sequence the DNA at every stage in order to determine where the errors crept in and to figure out how to overcome them. This methodical approach helped take the guesswork out of the science.

Ham and Clyde took this principle even further by coming to understand the details of every chemical and enzymatic reaction. In a marathon session they worked out the final problems. While they had yet to test the infectivity of this assembly of the artificial virus, they were confident in the synthesis and scheduled a dinner with me to discuss the next steps. The three of us met at Jean Michel, a French restaurant not far from my house. They were both like young postdoctoral fellows, brimming with the kind of excitement that comes only when you know that you have solved a problem. But there were still two crucial steps left to prove they had succeeded: They had to demonstrate that the synthetic virus was infective, which proved we had something that really worked, and they had to sequence the genome of the synthesized phage to prove we did not have an artifact—surplus base pairs in the code of Phi-X174 or contamination by another virus. Ham said that he would contact me the second the infectivity was tested.

Our re-created virus did indeed kill bacteria, just as the real thing did, and now I found myself back in a restaurant again, this time to discuss the implications of that achievement with top people from the government. We were shown to our table, and I immediately began to outline what we had done, from the minimal genome project up through the Phi-X174 synthesis. Marburger interrupted me constantly with questions, revealing a good understanding of where this was all headed. When it became clear that we could probably synthesize any virus under ten thousand base pairs in less than a week in the lab, while larger viruses such as Marburg or Ebola (both about eighteen thousand base pairs, and both very unpleasant) could be done in a month or so, Kerr just sat there, silently mouthing *wow* over and over. I informed them that I had already contacted Bruce Alberts, the president of the National Academy of Sciences, and Don Kennedy, the editor of *Science.* Data release was, as ever, an issue: We were prepared to censor our methods, if necessary, in order to limit what ill-intentioned people could do with this powerful technology.

The issue eventually went all the way up to the White House. Would I be willing to have my work reviewed by a new committee if it were established to investigate "dual-use" research—that is, work that could cause as much harm as good? I thought it was a good idea. Leaving aside our own synthetic genome work, I believed there was an urgent need for this kind of review; as only one example, teams had been attempting to make the H5N1 bird flu more infective to humans to see what would turn it into a pandemic strain. What finally evolved out of these kinds of concerns was a large committee with representatives from across government called the National Science Advisory Board for Biosecurity. We were urged to publish the Phi-X174 paper without any self-censorship, and it appeared in the *PNAS* on December 23, 2003.[17] We decided in the end to publish in the *Proceedings* because three of the paper's authors—Ham, Clyde, and I—are members of the academy, and we knew that we would have the space for what we wanted to say and that the journal would publish it quickly.

Because of the implications of our science, and because it had been backed by the DOE, the secretary of energy, Spencer Abraham, agreed to participate in a press conference in downtown Washington, D.C. Before a roomful of reporters he called the work "nothing short of amazing" and predicted that it could lead to the creation of designer microbes tailored to deal with pollution or excess carbon dioxide or even to meet future fuel needs.

Long Life?

The team combing my genome for DNA patterns linked with disease, disability, and decay does not always deliver bad news. When Jiaqi Huang told me last Christmas that I was "V/V homozygous for I405V" in a gene called CETP (Cholesteryl ester transfer protein), she meant that I had a variant linked to long life—to ninety and beyond—which should also help me remain lucid and retain memories into old age. The significance of this variant was recognized by a team at the Albert Einstein College of Medicine in New York, led by Nir Barzilai, which had previously linked the gene with longevity. The team examined 158 people of Ashkenazi Jewish (Eastern European) descent who were ninety-five or older. Compared with elderly people lacking the gene variant, those who possessed it were twice as likely to have good brain function. The researchers also validated their findings independently in a younger group. The protein made by my gene variant alters another, called the Cholesterol Ester Protein, to affect the size of "good" HDL and "bad" LDL cholesterol packaged into particles of fat and protein (lipoproteins). Centenarians were three times likelier to possess CETP VV compared with the general population and also had significantly larger HDL and LDL lipoproteins than people in the control group. It is thought that larger cholesterol particles are less likely to wedge themselves in blood vessels, putting me at a lower risk of heart attacks and strokes—at least from the very limited perspective of this gene. And, of course, if we could mimic the protective effects of the CETP VV variant with a therapy, we might be able to improve the quality of life of the West's graying population.

Ham joined us at the podium to field some of the questions. Although we had rehearsed several times what he should and should not say, he seemed to forget all that when he was asked by one reporter about the possibility of making deadly pathogens. After Ham blurted out that "we could make the smallpox genome," I interrupted to point out that while that was indeed possible, it was known that smallpox DNA is not infective on its own, attempting to pour at least a little cold water on Ham's speculation. Ham interjected, "But you and I discussed ways to get around that," and then turned toward me and said with a sheepish grin, "I probably shouldn't have said that, huh?" Fortunately, our exchange did not go further than a paragraph in *The New York Times,* and the coverage was mostly favorable. My old collaborator in the fly genome project, Gerry Rubin, told *USA Today:* "It's a very important technical advance. You can envision the day when one could sit down at a computer, design a genome, and then build it."[18]

Some media chose not to do a story on the announcement or felt let down that we were only announcing a virus and not the long-anticipated synthetic living cell. (I find it amusing that my press coverage had gone from skepticism in the past to ennui that I was "only" announcing a synthetic virus, not a new life form.) Scientists were also divided. Eckard Wimmer of Stony Brook University, who had taken three years to make a poliovirus, called it "a very smart piece of work," but to others our virus was no big deal.[19]

Nonetheless, the advance was sufficiently exciting for Ari Patrinos to become president of my new company, Synthetic Genomics, dedicated to pushing forward with the work. Ham Smith and I also convinced Clyde Hutchison to join the Venter Institute full-time to participate in the project to construct a synthetic genome based on *Mycoplasma genitalium,* a hugely ambitious project compared with rebuilding the phage. *M. genitalium* was the organism we had used with *Haemophilus influenzae* to demonstrate the viability of the shotgun method in 1995—the very year that saw the birth of the synthetic genome project, at least in concept.

Before we became distracted by the human genome, Ham and I had asked simple questions: If one species needed 1,800 genes for life (*Haemophilus influenzae*) and another needed only 482 (*Mycoplasma genitalium*), was there a minimal operating system for life? And could we define that operating system? In other words we could ask the old question "What is life?" in genetic terms.

Not only were these simple questions, but they were naïve, a fact that became clear when we sequenced our third genome, that of the archaea *Methanococcus jannaschii,* a so-called autroph that was dependent only on inorganic chemicals to survive. In place of the sugar metabolism used by the other microbes, *Methanococcus* converts carbon dioxide to methane to generate cellular energy. It began to dawn on me that there are different cassettes of genes that can be substituted in microbial cells depending on the environment they live in. Cells such as *Methanococcus* survive where there is little or no sugar and thus lack the genes that provide the ability to metabolize it. We therefore could not define *the* minimal operating system for life because it depended on exactly where that life had to thrive. We could at best define *a* minimal genome, a concept that would evolve much further as we generated more data.

We undertook a series of studies to knock genes out of *Mycoplasma genitalium* to see which ones it could live without. Clyde and his postdoctoral fellow, Scott N. Peterson, devised a novel approach that we named whole-genome transposon mutagenesis, which involved randomly inserting unrelated DNA into the middle of genes, thereby disrupting their function, to see what effect it had on the organism. (The unrelated DNA was in the form of transposons, small segments of DNA that contain the necessary genetic elements to randomly insert anywhere into the genome. A significant portion of our genome is composed of such DNA parasites, more than actually codes for genes themselves.)

In our experiments the membranes of the microbes were made leaky for a while so that transposon DNA could enter them and find a new home in the genome at random. When a transposon inserted itself into a gene sequence, it silenced that particular gene. In order to track exactly what we had done, we added an antibiotic resistance gene to the transposon. That way we knew that any cells that survived in the presence of an antibiotic had the resistance gene and, as a consequence, contained the transposon that delivered it, too. It was easy to design a way to start reading the genetic code from the end of the transposon into the genetic code of *Mycoplasma genitalium* colonies that had survived. We had the complete genome sequence of *Mycoplasma genitalium,* and this sequence revealed exactly where in the genome the transposon had inserted. If it lay in the middle of a gene and the cells lived, we could define that gene as nonessential for the life of that cell under those growth conditions. Without our defining the environment, the data gave new insights into why the gene functions necessary for life are elusive.

A simple example involves two genes in *Mycoplasma genitalium,* one that codes for a protein involved in bringing the sugar glucose into the cell, and the second for a protein that imports the sugar fructose. *Mycoplasma genitalium* can survive on either kind of sugar. If only glucose is provided, transposons can insert into the fructose transporter gene without any consequence to the cell. From these experiments you might conclude that the fructose transporter gene was a nonessential gene, which is correct for these conditions. However, if only fructose is available to the cell, the fructose transporter becomes essential. Context is critical if we are to understand gene function.

Another complication was that our colonies of *Mycoplasma* were not clones, and it was possible that one variant of *Mycoplasma* with a survival gene was supporting its brothers and sisters in which a gene had been knocked out with a transposon. Over the years a team headed by John I. Glass did careful experiments on clones to ensure this was not happening.

With computational analysis of what the genes in the organism did, carried out by comparing thirteen related sequenced genomes, we ended up with a set of about ninety-nine genes that we thought could be dispensed with in the *Mycoplasma genitalium* genome. One-fifth of its genome was therefore redundant, and we now had a glimpse of life at its bare genetic minimum.

With the new techniques in hand from the Phi-X174 effort, Ham, Clyde, and I launched an attempt to construct the entire *Mycoplasma genitalium* genome from lab-made chemicals. As I write, this work is now completed at the hands of a twenty-person team. At each and every stage we have had to develop new approaches to deal with the tremendous technical challenges that faced us.

Aware of how even a single spelling mistake can be lethal, we have had to resequence the 580,000 bases of *Mycoplasma* at an unprecedented level of accuracy: A decade ago the standard was about one error in 10,000 bases, but with new machines we have cut it to less than one in half a million. The result could well be the only bacterial sequence in existence that is essentially correct: no one else has *had* to get it 100 percent right before, even our most purist critics.

Now we had to rebuild a stripped-down version. The team used a standard laboratory machine to make small structures of DNA, known as oligonucleotides, or "oligos." These are the building blocks of our artificial genome. With clever chemistry Ham and his team are painstakingly stitching myriad tiny blocks of fifty or so base pairs into fewer small pieces, growing them in E. coli, and then turning these many small pieces into a handful of bigger ones—cassettes of genes—until they eventually get two large pieces that can be assembled into the circular genome of the new life form. Overall, we have had to make and manipulate synthetic DNA on a scale ten to twenty times bigger than has been accomplished before.

We have now made the circular genome, and are inserting the synthetic DNA into bacteria. We are holding our breath to see whether one or more microbes among the 100 billion in the test tube "boots up" with a strand of our man-made DNA and a daughter cell starts metabolizing and multiplying according to our version of life's recipe. Already, we have succeeded in transplanting the genome of one bacterium into another, marking the first example of species transmutation and generating headlines worldwide.[20] In readiness for experiments to transplant a synthetic genome, we have also applied for patents on how to create what we call "*Mycoplasma laboratorium.*"

If our plan succeeds, a new creature will have entered the world, albeit one that relies on an existing bug's cellular machinery to read its artificial DNA. We have often been asked if this will be a step too far. I always reply that—so far at least—we are only reconstructing a diminished version of what is already out there in nature. I add that we have conducted a major ethical review of what we are doing, and we feel that this is good science. With a synthetic genome we can insert and remove single genes or sets of genes to test in an unequivocal manner the hypotheses that we generated from the gene knockout experiments and really figure out how life works.

As I move past my sixtieth birthday, passing my father's sadly missed milestone, I do so with my life turning in a very positive new direction. Although Claire and I divorced, she remarried and seems happy with her life. It was she who suggested merging TIGR with the Venter Institute, and on September 12, 2006,* the board of trustees of my three organizations, TIGR, The J. Craig Venter Science Foundation, and the Venter Institute, voted unanimously to merge all three into one organization called the J. Craig Venter Institute. This action united all the organizations that I founded more than fourteen years ago into one of the largest private research institutes in the world, one with more than five hundred scientists and staff, more than 250,000 square feet of laboratory space, more than $200 million in combined assets, and an annual budget well north of $70 million. Our scientific publications, dating back to the first genome and continuing annually, make the team at the Venter Institute one of the most cited in modern science. My board also voted to allow me to open a West Coast Venter Institute in La Jolla, California. A new building slated to be completed in 2009 will be located on the University of California, San Diego, campus between the Scripps Institute of Oceanography and the Medical School. Perhaps the best change in my lif

* Seven months later Claire followed her new husband to the University of Maryland.

took place when Heather and I began dating in early 2006 and became engaged in July of that year.

Now that I am the first chemical machine to gaze upon his own sequence, I am still struggling to make sense of what it all means, an endeavor that will probably take decades. Over that time I am sure that millions more people will have the opportunity to do the same as the cost of sequencing drops to the point where we can read a human genome for about $1,000. There is still an ocean of great science left for me to explore.

The first synthetic genome, a stripped-down version of a natural organism, is only the beginning. I now want to go further. My company, Synthetic Genomics Inc., is already trying to develop cassettes—modules of genes—to turn an organism into a biofactory that could make clean hydrogen fuel from sunlight and water or soak up more carbon dioxide. From there I want to take us far from shore into unknown waters, to a new phase of evolution, to the day when one DNA-based species can sit down at a computer to design another. I plan to show that we understand the software of life by creating true artificial life. And in this way I want to discover whether a life decoded is truly a life understood.

ACKNOWLEDGMENTS

A Life Decoded has been many years in the making. I first started thinking about writing about my experiences in the early 1990s, when many had encouraged me to do so, as a way to relate my unusual background and my adventures in the laboratory and on the open ocean. After leaving TIGR to form Celera to sequence the human genome I thought it was also worth documenting that experience, but two events postponed the project—lack of time and an expressed interest by journalist James Shreeve to write or coauthor a book on reading the human genetic code. We ultimately decided that a coauthored book would detract from Jamie's ability to make an independent assessment of the genome race and would also conflict with my desire to write my own book in my own voice. I agreed to give Jamie unfettered access to Celera, and to not start my autobiography for at least two years. Four years later, after the completion of the first sequencing of the human genome and after having being fired from Celera, I felt the time was right for me to attempt such a demanding undertaking.

As I was beginning to look for an agent to help assess the viability of such a book and to help in the process, I was approached by John Brockman, who also thought it was time for me to write my version of events. From the beginning John was very encouraging and urged me to write the book myself. Over the years, John has been not only a great agent but a good friend and sounding board. I thank him for that and for his part of the process of making this book a reality.

I had the opportunity to interview several publishers interested in my story and was impressed with the teams at Penguin in the UK and Viking in the USA. I don't think I could have been assigned a better editor than Rick Kot at Viking. There is no question in my mind that Rick's edits and enthusiasm improved the readability of every page and the quality of the book.

As I assume is the case with many book projects, *A Life Decoded* was not linear in its development. While running my research programs full time, I needed significant discipline to try to make progress. I did most of my writing on airplanes and at sea on the *Sorcerer II* Expedition. After four years of effort I had written more than 240,000 words. I hired Roger Highfield of *The Daily Telegraph* in London to help trim and reorganize my text. In addition to his edits Roger also conducted some key interviews that have provided unique points of view, which have widened the perspective and context of this story. Following Roger's edits I spent an additional six months rewriting the manuscript. Some areas underwent several rounds, with Roger providing valuable feedback on scientific readability. I thank him for his efforts.

I have asked and in some cases challenged friendships by asking for feedback on interest, readability, science, breadth of coverage (from sailing to science), and accuracy. A few individuals have had a disproportionate influence on me and this project. From the first outlines to the finished book, my fiancée, Heather, has given me unrivaled help by providing her usual unfiltered feedback and encouragement on every one of the multiple drafts, from the roughest early version to the final proofreads. It is clear to me that the book would not have happened without her encouragement and assistance. My friend and colleague Ham Smith has likewise read virtually every draft of the book, offering enthusiastic feedback, encouragement, and suggestions, all while continuing to be a wonderful partner from the first genome sequenced to the new field of synthetic genomics. Special thanks also go to Erling Norrby and Juan Enriquez, Venter Institute Trustees, friends, and frequent members of the *Sorcerer II* crew for their reading several versions of the book.

During the process of writing *A Life Decoded* I have conducted numerous taped interviews with family members, including my mother, Elizabeth Venter; her brother, David Wisdom; and my late father's sister, Marge Hurlow, and her husband, Robert (Bud) Hurlow, who has also joined or provided several sailing adventures. David Wisdom has also provided a thoroughly researched family genealogy and history that I relied on for parts of this book.

Others who have either read the manuscript or checked sections for accuracy or provided interviews include but are not limited to: Ari Patrinos, Clyde Hutchinson, Ken Nealson, my brother and sister-in-law Keith and Laurel Venter, my brother Gary, my mother, Elizabeth Venter, Bruce Cameron, Ronald (Ron) Nadel, Jack Dixon, Dave Kiernan, Mala Htun, Ashley Myler Klick, Tim Friend, Rich Bourke, Claire Fraser-Liggett, Charles Howard and the crew of *Sorcerer II,* Olivia Judson, Joe Kowalski, Julie Gross Adelson, and Reid Adler. On the analysis of my genetic code I have had the privilege of working with a very dedicated team of scientists at the Venter Institute, including Robert (Bob) Strausberg, Samuel (Sam) Levi, Jaqui Huangl, and Pauline Ng.

I have attempted at every turn to make *A Life Decoded* as accurate as possible by referring to news articles, scientific articles, depositions, interviews, other genome books (James Shreeve's *Genome Wars,* and John Sulston's and Georgina Ferry's *The Common Thread* were paricularly helpful) as well as having it fact-checked by a variety of readers, but there will be errors. There can only be one person responsible and that, of course, is me. I have also set up a Web site for the book, *www.ALifeDecoded.org,* where I will post supplementary material, scientific papers, my genetic code, key links, and any corrections.

NOTES

1. WRITING MY CODE

1. James Shreeve, *The Genome War: How Craig Venter Tried to Capture the Code of Life and Save the World* (New York: Ballantine, 2005), p. 6.
2. Fortunately, Lish went on to great things at *Esquire* magazine, then as an editor at *The Quarterly* and at Knopf, founding literary magazines and writing a number of novels and collections of short stories. In 1994, he was named one of the two hundred major writers of our time by the French periodical *Le Nouvel Observateur.* "Lish is our Joyce, our Beckett, our most true modernist," said *Kirkus Reviews* when reviewing Lish's story collection *Krupp's Lulu.*
3. Daniel Max, "Gordon Lish: An Editor Who Attracts Controversy," *St. Petersburg Times,* May 3, 1987, p. 7D.
4. Leah Garchik, "News Personals," *San Francisco Chronicle,* March 1, 1991, p. A8.

2. UNIVERSITY OF DEATH

1. A taste for alcohol has also been linked with one variant of a gene called COMT (Catechol-O-methyltransferase), responsible for an enzyme that breaks down dopamine. However, my genome contains a variant that is linked with a lower risk of alcoholism. Although I like to drink, it seems that I prefer to light up my pleasure centers by exciting stimuli and experiences.

3. ADRENALINE JUNKIE

1. One of the few original copies of *Honest Jim* resides at the Venter Institute.
2. James D. Watson, *A Passion for DNA: Genes, Genomes and Society* (New York: Oxford University Press, 2000), p. 97.
3. Francis Crick, *What Mad Pursuit: A Personal View of Scientific Discovery* (London: Weidenfeld & Nicolson, 1988), p. 64.
4. Now part of the Venter Institute collection.
5. James D. Watson, *The Double Helix* (London: Weidenfeld & Nicolson, 1981), p. 98.
6. James D. Watson, *Genes, Girls and Gamow* (Oxford: Oxford University Press, 2001), p. 5.
7. Matt Ridley, *Francis Crick: Discoverer of the Genetic Code* (London: Harper Press, 2006), p. 77.
8. Watson, *A Passion for DNA,* p. 120.

9. Venter, J.C., Dixon, J.E., Maroko, P.R. and Kaplan, N.O. "Biologically Active Catecholamines Covalently Bound to Glass Beads," *Proc. Natl. Acad. Sci.*, USA *69,* 1141–45, 1972.

5. SCIENTIFIC HEAVEN, BUREAUCRATIC HELL

1. "Apart from my work my main interests are gardening and what can best be described as 'messing about in boats.' " Fred Sanger, autobiography, Nobelprize.org. He would, for example, go sailing with César Milstein.
2. Chung, F.Z., Lentes, K.U., Gocayne, J.D., FitzGerald, M.G., Robinson, D., Kerlavage, A.R., Fraser, C.M., and Venter, J.C. "Cloning and Sequence Analysis of the Human Brain Beta-Adrenergic Receptor: Evolutionary Relationship to Rodent and Avian Beta-Receptors and Porcine Muscarinic Receptors." *FEBS Lett. 211,* 200–6, 1987.
3. Gocayne, J.D., Robinson, D.A., FitzGerald, M.G., Chung, F.-Z., Kerlavage, A.R., Lentes, K.-U., Lai, J.-Y., Wang, C.D., Fraser, C.M., and Venter, J.C., "Primary Structure of Rat Cardiac Beta-Adrenergic and Muscarinic Cholinergic Receptors Obtained by Automated DNA Sequence Analysis: Further Evidence for a Multigene Family." *Proc. Natl. Acad. Sci.,* USA *84,* 8296–8300, 1987.
4. Cook-Deegan, *The Gene Wars,* p. 139.
5. A so-called restriction map is made by cutting the DNA with different enzymes, then determining the size of the fragments produced by each enzyme. By using a variety of enzymes, a "digest map" can be constructed giving the order and size of the fragments. When the fragments were sequenced, they revealed the sites snipped by the enzymes and these could be lined up into the right order on the restriction map.
6. Cook-Deegan, *The Gene Wars,* p. 184.
7. James Shreeve, *The Genome War: How Craig Venter Tried to Capture the Code of Life and Save the World* (New York: Ballantine, 2005), p. 79. Watson was talking to Gerry Rubin at the time.
8. Cook-Deegan, *The Gene Wars,* pp. 313–14.
9. Ibid., pp. 226, 220.

6. BIG BIOLOGY

1. James D. Watson, *DNA: The Secret of Life* (New York: Knopf, 2003), p. 180.
2. Christopher Anderson and Peter Aldhous, *Nature 354,* November 14, 1991.
3. Watson, *DNA: The Secret of Life,* p. 280.
4. Adams, M.D., Kelley, J.M., Gocayne, J.D., Dubnick, M., Polymeropoulos, M.H., Xiao, H., Merril, C.R., Wu, A., Olde, B., Moreno, R., Kerlavage, A.R., McCombie, W.R., and Venter, J.C., "Complementary DNA Sequencing: 'Expressed Sequence Tags' and the Human Genome Project," *Science 252,* 1651–56, 1991.
5. Leslie Roberts, *Science 252,* 1991.
6. John Sulston and Georgina Ferry, *The Common Thread* (London: Corgi, 2003), p. 9.
7. Ibid., p. 125.
8. Leslie Roberts, "Genome Patent Fight Erupts," *Science 184,* 184–86, October 11, 1991.
9. Robert Cook-Deegan, *The Gene Wars: Science, Politics and the Human Genome* (New York: Norton, 1994), p. 311.
10. James Shreeve, *The Genome War: How Craig Venter Tried to Capture the Code of Life and Save the World* (New York: Ballantine, 2005), p. 85.

11. Although it depended on my DNA (60 percent), it was also built with the DNA of four other people by a "majority rules" approach.
12. There were indeed enough pieces of DNA from the shotgun method tiling this region to be sure.
13. Roberts, "Genome Patent Fight Erupts."
14. Cook-Deegan, *The Gene Wars,* p. 208. A name coined by Brenner, who joked that he personally preferred THUG.
15. Peter Aldhous, *Nature 353,* 785, 1991.
16. Letter from Jan Witkowski to Craig Venter, October 30, 1991.
17. Christopher Anderson, *Nature 353,* 485–86, 1991.
18. Alex Barnum, *San Francisco Chronicle,* December 2, 1991.
19. Sulston and Ferry, *The Common Thread,* p. 103.
20. Robin McKie, "Scandal of U.S. Bid to Buy Vital UK Research," *Observer,* January 26, 1992, p. 2.
21. Ibid.
22. Ibid., p. 3.
23. Sulston and Ferry, *The Common Thread,* p. 115.
24. Cook-Deegan, *The Gene Wars,* p. 333.
25. Cook-Deegan, *The Gene Wars,* p. 336.
26. Ibid., p. 328.
27. Victor McElheny, *Watson and DNA: Making a Scientific Revolution* (New York: John Wiley, 2003), p. 266.
28. Christopher Anderson, *Nature 358,* July 9, 1992.
29. Michael Gottesman, "Purely Academic" *Molecular Interventions 4,* 10-15, 2004.
30. Gina Kolata, "Biologist's Speedy Gene Method Scares Peers But Gains Backer," *The New York Times,* July 28, 1992, p. C1.
31. Ibid.
32. Cook-Deegan, *The Gene Wars,* p. 325.

7. TIGR CUB

1. John Sulston and Georgina Ferry, *The Common Thread* (London: Corgi, 2003), p. 127.
2. See chapter 2, note 1.
3. Gina Kolata, "Biologist's Speedy Gene Method Scares Peers But Gains Backer," *The New York Times,* July 28, 1992, p. C1.
4. Robert Cook-Deegan, *The Gene Wars: Science, Politics and the Human Genome* (New York: Norton, 1994), p. 327.
5. Francis Collins, *The Language of God: A Scientist Presents Evidence for Belief* (New York: Free Press, 2006), p. 36.
6. Cook-Deegan, *The Gene Wars,* p. 341.
7. Editorial, "Venter's Venture," *Nature 362,* 575-76, 1993.

8. GENE WARS

1. Daniel S. Greenberg, "Clinton Goes Slow on Health Research," *The Baltimore Sun,* August 10, 1993, p. 11A.
2. Eliot Marshall, "Varmus: The View from Bethesda," *Science 262,* 1364, 1993.
3. "NIH Shakeup Continues," *Science 262,* 643, 1993.
4. James Shreeve, *The Genome War: How Craig Venter Tried to Capture the Code of Life and Save the World* (New York: Ballantine, 2005), p. 90.

5. Sandra Sugawara, "A Healthy Vision," *The Washington Post*, November 16, 1992.
6. Robert F. Massung*, Joseph J. Esposito, Li-ing Liu, Jin Qi, Theresa R. Utterback, Janice C. Knight, Lisa Aubin, Thomas E. Yuran, Joseph M. Parsons, Vladimir N. Loparev, Nickolay A. Selivanov, Kathleen F. Cavallaro*, Anthony R. Kerlavage, Brian W. J. Mahy and J. Craig Venter, "Potential Virulence Determinants in Terminal Regions of Variola Smallpox Virus Genome," *Nature 366,* 748–51, December 30, 1993.
7. "Gone but Not Forgotten," *The Economist*, August 14, 1993.
8. Betsy Wagner, "Smallpox is Now a Hostage in the Lab," *The Washington Post,* January 4, 1994.
9. Christopher Anderson, "NIH Drops Bid for Gene Patents," *Science 263,* 909–10, February 18, 1994.
10. However, genes are by no means the whole story: in *Science* in 2002, Terrie Moffitt of King's College London found an intriguing nature-nurture effect in which only maltreated children with this less active version are more likely to develop behavioral problems. A. Caspi, A., McClay, J., Moffitt, T., Mill, J., Martin, J., Craig, I., Taylor, A., and Poulton, R. "Evidence that the Cycle of Violence in Maltreated Children Depends on Genotype," *Science 297,* 851–54, 2002.
11. Eliot Marshall, "HGS Opens Its Databanks—For a Price," *Science 266,* 25, October 7, 1994; and David Dickson, "HGS Seeks Exclusive Option on All Patents Using Its cDNA Sequences," *Nature 371,* 463, October 6, 1994.
12. "Breast Cancer Discovery Sparks New Debate on Patenting Human Genes," *Nature 371,* 271–72, September 22, 1994.
13. "Ownership and the Human Genome," *Nature 371,* 363–364, September 29, 1994.
14. Jerry Bishop, "Merck's Plan for Public-Domain Gene Data Could Blow Lid Off Secret Genetic Research," *The Wall Street Journal Europe*, September 30, 1994.
15. Eliot Marshall, "A Showdown Over Gene Fragments," *Science 266,* 208–10, October 14, 1994.
16. John Sulston and Georgina Ferry, *The Common Thread* (London: Corgi, 2003), p. 139.
17. Eliot Marshall, "The Company That Genome Researchers Love to Hate," *Science 266,* 1800–02, December 16, 1994.

9. SHOTGUN SEQUENCING

1. Ashburner, M., *Won for All: How the Drosophila Genome Was Sequenced* (Cold Spring Harbor Laboratory Press, 2006), p. 7.
2. Rachel Nowak, "Venter Wins Sequence Race—Twice," *Science 268,* 1273, June 2, 1995.
3. *Time,* June 5, 1995, p. 21.
4. Nicholas Wade, "Bacterium's Full Gene Makeup Is Decoded," *The New York Times,* May 26, 1995, p. A16.
5. Fleischmann, R.D., Adams, M.D., White, O., Clayton, R.A., Kirkness, E.F., Kerlavage, A.R., Bult, C.J., Tomb, J.-F., Doughtery, B.A., Merrick, J.M., McKenney, K., Sutton, G., FitzHugh, W., Fields, C., Gocayne, J.D., Scott, J., Shirley, R., Liu, L.-I., Glodek, A., Kelley, J.M., Weidman, J.F., Phillips, C.A., Spriggs, T., Hedblom, E., Cotton, M.D., Utterback, T.R., Hanna, M.C., Nguyen, D.T., Saudek, D.M., Brandon, R.C., Fine, L.D., Fritchman, J.L., Fuhrmann, J.L., Geoghagen, N.S.M., Gnehm, C.L., McDonald, L.A., Small, K.V., Fraser, C.M., Smith, H.O., Venter, J.C. "Whole-Genome Random Sequencing and Assembly of *Haemophilus influenzae* Rd," *Science 269,* 496–512, 1995.
6. Smith, H.O., Tomb, J.-F., Doughtery, B.A., Fleischmann, R.D. and Venter, J.C., "Frequency and Distribution of DNA Uptake Signal Sequences in the *Haemophilus influenzae* Rd Genome," *Science 269,* 538–40, 1995.

7. James Shreeve, *The Genome War: How Craig Venter Tried to Capture the Code of Life and Save the World* (New York: Ballantine, 2005), p. 110.
8. Nicholas Wade, "First Sequencing of Cell's DNA Defines Basis of Life," *The New York Times*, August 1, 1995, p. C1.
9. Rachel Nowak, "Homing In on the Human Genome," *Science 269*, 469, July 28, 1995.
10. Fraser, C.M., Gocayne, J.D., White, O., Adams, M.D., Clayton, R.A., Fleischmann, R., Bult, C.J., Kerlavage, A.R., Sutton, G., Kelley, J.M., Fritchman, J.L., Weidman, J.F., Small, K.V., Sandusky, M., Fuhrmann, J., Nguyen, D., Utterback,T.R., Saudek, D.M., Phillips, C.A., Merrick, J.M., Tomb, J., Dougherty, B.A., Bott, K.F., Hu, P., Lucier, T.S., Peterson, S.N., Smith, H.O., Hutchison, C.A., Venter, J.C., "The Minimal Gene Complement of *Mycoplasma genitalium*," *Science 270*, 397–403, 1995.
11. Andre Goffeau, "Life with 482 Genes," *Science 270*, October 20, 1995.
12. Karen Young Kreeger, "First Completed Microbial Genomes Signal Birth of New Area of Study," *The Scientist*, November 27, 1995.
13. Adams, M.D., Kerlavage, A.R., Fleischmann, R.D., Fuldner, R.A., Bult, C.J., Lee, N.H., Kirkness, E.F., Weinstock, K.G., Gocayne, J.D., White, O., Sutton, G., Blake, J.A., Brandon, R.C., Man-Wai, C., Clayton, R.A., Cline, R.T., Cotton, M.D., Earle-Hughes, J., Fine, L.D., FitzGerald, L.M., FitzHugh, W.M., Fritchman, J.L., Geoghagen, N.S., Glodek, A., Gnehm, C.L., Hanna, M.C. , Hedbloom, E., Hinkle Jr., P.S., Kelley, J.M., Kelley, J.C., Liu, L.I., Marmaros, S.M., Merrick, J.M., Moreno-Palanques, R.F., McDonald, L.A., Nguyen, D.T., Pelligrino, S.M., Phillips, C.A., Ryder, S.E., Scott, J.L., Saudek, D.M., Shirley, R. Small, K.V., Spriggs, T.A., Utterback, T.R., Weidman, J.F., Li, Y., Bednarik, D.P., Cao, L., Cepeda, M.A., Coleman, T.A., Collins, E.J., Dimke, D., Feng, P., Ferrie, A., Fischer, C., Hastings, G.A., He, W.W., Hu, J.S., Greene, J.M., Gruber, J., Hudson, P., Kim, A., Kozak, D.L., Kunsch, C., Hungjun, J., Li, H., Meissner, P.S., Olsen, H., Raymond, L., Wei, Y.F., Wing, J., Xu, C., Yu, G.L., Ruben, S.M., Dillon, P.J., Fannon, M.R., Rosen, C.A., Haseltine, W.A., Fields, C., Fraser, C.M., Venter, J.C. "Initial Assessment of Human Gene Diversity and Expression Patterns Based Upon 52 Million Basepairs of cDNA Sequence." *Nature* 377, Suppl., 3–174, 1995.
14. John Maddox, "Directory to the Human Genome," *Nature 376*, 459–60, August 10, 1995.
15. Elyse Tanouye, *The Wall Street Journal*, September 28, 1995.
16. Tim Friend, *USA Today*, September 28, 1995.
17. Nicholas Wade, *The New York Times*, September 28, 1995.
18. David Brown and Rick Weiss, *The Washington Post*, September 28, 1995.
19. Sue Goetinck, *The Dallas Morning News*, September 28, 1995.
20. Ibid.
21. John Carey, "The Gene Kings," *Business Week*, May 8, 1995.
22. Richard Jerome, "The Gene Hunter," *People*, June 12, 1995.
23. Troy Goodman, *U.S. News and World Report*, October 9, 1995.
24. Bult, C.J., White, O., Olsen, G.J., Zhou, L., Fleischmann, R.D., Sutton, G.G., Blake, J.A., FitzGerald, L. M., Clayton, R.A., Gocayne, J.D., Kerlavage, A.R., Dougherty, B.A., Tomb, J.-F., Adams, M.D., Reich, C.I., Overbeek, R., Kirkness, E.F., Weinstock, K. G., Merrick, J.M., Glodek, A., Scott, J.L., Geoghagen, S.M., Weidman, J.F., Fuhrmann, J.L., Nguyen, D., Utterback, T.R., Kelley, J.M., Peterson, J.D., Sadow, P.W., Hanna, M.C., Cotton, M.D., Roberts, K.M. Hurst, M.A., Kaine, B.P., Borodovsky, M., Klenk, H.-P., Fraser, C.M., Smith, H.O., Woese, C.R and Venter, J.C. "Complete Genome Sequence of the Methanogenic Archaeon, *Methanococcus jannaschii*," *Science 372*, 1058–73, 1996.
25. Tim Friend, *USA Today*, August 23–25, 1996.

26. *The Christian Science Monitor*, August 23, 1996.
27. *The Economist*, August 24, 1996.
28. Jim Wilson, *Popular Mechanics*, December 1996.
29. *San Jose Mercury News*, August 23, 1996.
30. Curt Suplee, *The Washington Post*, September 30, 1996.
31. Nicholas Wade, "Thinking Small Paying Off Big in Gene Quest," *The New York Times*, February 3, 1997.

10. INSTITUTIONAL DIVORCE

1. Gina Kolata, "Wallace Steinberg Dies at 61; Backed Health Care Ventures," *The New York Times*, July 29, 1995.
2. Angus Phillips, "He Leaves His Body to Science, His Heart to Sailing." *The Washington Post*, November 24, 1996.
3. Nicholas Wade, *The New York Times*, June 24, 1997.
4. Beth Berselli, "Gene Split: Research Partners Human Genome and TIGR Are Ending Their Marriage of Convenience," *The Washington Post*, July 7, 1997.
5. *The* [*Memphis*] *Commercial Appeal*, July 4, 1997.
6. Tim Friend, "20,000 New Genes Boon to Research," *USA Today*, June 25, 1997.
7. Ibid.

11. SEQUENCING THE HUMAN

1. Maurice Wilkins, *The Third Man of the Double Helix: The Autobiography of Maurice Wilkins* (Oxford: Oxford University Press, 2003), p. 206.
2. James Shreeve, *The Genome War: How Craig Venter Tried to Capture the Code of Life and Save the World* (New York: Ballantine Books, 2005), p. 19.
3. Elizabeth Pennisi, "DNA Sequencers' Trial by Fire," *Science 280*, 814–17, May 8, 1998.
4. John Sulston and Georgina Ferry, *The Common Thread* (London: Corgi, 2003), p. 172.
5. Shreeve, *The Genome War*, p. 163.
6. Ibid., p. 21.
7. Nicholas Wade, "Scientist's Plan: Map All DNA Within 3 Years," *The New York Times*, May 10, 1998, p. 1, 20.
8. Ibid.
9. Ibid.
10. Nicholas Wade, "Beyond Sequencing of Human DNA," *The New York Times*, May 12, 1998.
11. Sulston and Ferry, *The Common Thread*, p. 172.
12. Ibid., p. 174.
13. Justin Gillis and Rick Weiss, "Private Firm Aims to Beat Government to Gene Map," *The Washington Post*, May 12, 1998, p. A1.
14. Wade, "Beyond Sequencing of Human DNA."
15. Gillis and Weiss, "Private Firm."
16. Ibid.
17. Pennisi, "DNA Sequencers' Trial by Fire."
18. Sulston and Ferry, *The Common Thread*, p. 171.
19. Shreeve, *The Genome War*, p. 23.
20. Ibid., p. 51.
21. Sulston and Ferry, *The Common Thread*, p. 180.

22. Ashburner, M., *Won for All: How the Drosophila Genome Was Sequenced* (Cold Spring Harbor Laboratory Press, 2006), p. 1.
23. Ibid., p. 15.
24. Sulston and Ferry, *The Common Thread,* p. 176.
25. Ibid.
26. Shreeve, *The Genome War,* p. 48.
27. Ibid., p. 53.
28. Sulston and Ferry, *The Common Thread,* p. 188.
29. Shreeve, *The Genome War,* p. 53.

12. *MAD* MAGAZINE AND DESTRUCTIVE BUSINESSMEN

1. John Sulston and Georgina Ferry, *The Common Thread* (London: Corgi, 2003), p. 190.
2. James Shreeve, *The Genome War: How Craig Venter Tried to Capture the Code of Life and Save the World* (New York: Ballantine, 2005), p. 125.
3. Ibid., p. 93.
4. Ibid., p. 226.
5. Maynard Olson, "The Human Genome Project: A Player's Perspective," *Journal of Molecular Biology 319,* 931–42, 2002.

13. FLYING FORWARD

1. James Shreeve, *The Genome War: How Craig Venter Tried to Capture the Code of Life and Save the World* (New York: Ballantine, 2005), p. 285.
2. Ashburner, M., *Won for All: How the Drosophila Genome Was Sequenced* (Cold Spring Harbor Laboratory Press, 2006), p. 45.
3. Shreeve, *The Genome War,* p. 300.
4. Ashburner, *Won for All,* p. 55.
5. John Sulston and Georgina Ferry, *The Common Thread* (London: Corgi, 2003), p. 232.
6. Mark D. Adams, Susan E. Celniker, Robert A. Holt, Cheryl A. Evans, Jeannine D. Gocayne, Peter G. Amanatides, Steven E. Scherer, Peter W. Li, Roger A. Hoskins, Richard F. Galle, Reed A. George, Suzanna E. Lewis, Stephen Richards, Michael Ashburner, Scott N. Henderson, Granger G. Sutton, Jennifer R. Wortman, Mark D. Yandell, Qing Zhang, Lin X. Chen, Rhonda C. Brandon, Yu-Hui C. Rogers, Robert G. Blazej, Mark Champe, Barret D. Pfeiffer, Kenneth H. Wan, Clare Doyle, Evan G. Baxter, Gregg Helt, Catherine R. Nelson, George L. Gabor Miklos, Josep F. Abril, Anna Agbayani, Hui-Jin An, Cynthia Andrews-Pfannkoch, Danita Baldwin, Richard M. Ballew, Anand Basu, James Baxendale, Leyla Bayraktaroglu, Ellen M. Beasley, Karen Y. Beeson, P. V. Benos, Benjamin P. Berman, Deepali Bhandari, Slava Bolshakov, Dana Borkova, Michael R. Botchan, John Bouck, Peter Brokstein, Phillipe Brottier, Kenneth C. Burtis, Dana A. Busam, Heather Butler, Edouard Cadieu, Angela Center, Ishwar Chandra, J. Michael Cherry, Simon Cawley, Carl Dahlke, Lionel B. Davenport, Peter Davies, Beatriz de Pablos, Arthur Delcher, Zuoming Deng, Anne Deslattes Mays, Ian Dew, Suzanne M. Dietz, Kristina Dodson, Lisa E. Doup, Michael Downes, Shannon Dugan-Rocha, Boris C. Dunkov, Patrick Dunn, Kenneth J. Durbin, Carlos C. Evangelista, Concepcion Ferraz, Steven Ferriera, Wolfgang Fleischmann, Carl Fosler, Andrei E. Gabrielian, Neha S. Garg, William M. Gelbart, Ken Glasser, Anna Glodek, Fangcheng Gong, J. Harley Gorrell, Zhiping Gu, Ping Guan, Michael Harris, Nomi L. Harris, Damon Harvey, Thomas J. Heiman, Judith R. Hernandez, Jarrett Houck, Damon Hostin, Kathryn A. Houston, Timothy J. Howland, Ming-Hui Wei, Chinyere Ibegwam, Mena Jalali, Francis Kalush, Gary H. Karpen, Zhaoxi Ke,

James A. Kennison, Karen A. Ketchum, Bruce E. Kimmel, Chinnappa D. Kodira, Cheryl Kraft, Saul Kravitz, David Kulp, Zhongwu Lai, Paul Lasko, Yiding Lei, Alexander A. Levitsky, Jiayin Li, Zhenya Li, Yong Liang, Xiaoying Lin, Xiangjun Liu, Bettina Mattei, Tina C. McIntosh, Michael P. McLeod, Duncan McPherson, Gennady Merkulov, Natalia V. Milshina, Clark Mobarry, Joe Morris, Ali Moshrefi , Stephen M. Mount, Mee Moy, Brian Murphy, Lee Murphy, Donna M. Muzny, David L. Nelson, David R. Nelson, Keith A. Nelson, Katherine Nixon, Deborah R. Nusskern, Joanne M. Pacleb, Michael Palazzolo, Gjange S. Pittman, Sue Pan, John Pollard, Vinita Puri, Martin G. Reese, Knut Reinert, Karin Remington, Robert D. C. Saunders, Frederick Scheeler, Hua Shen, Bixiang Christopher Shue, Inga Sidén-Kiamos, Michael Simpson, Marian P. Skupski, Tom Smith, Eugene Spier, Allan C. Spradling, Mark Stapleton, Renee Strong, Eric Sun, Robert Svirskas, Cyndee Tector, Russell Turner, Eli Venter, Aihui H. Wang, Xin Wang, Zhen-Yuan Wang, David A. Wassarman, George M. Weinstock, Jean Weissenbach, Sherita M. Williams, Trevor Woodage, Kim C. Worley, David Wu, Song Yang, Q. Alison Yao, Jane Ye, Ru-Fang Yeh, Jayshree S. Zaveri, Ming Zhan, Guangren Zhang, Qi Zhao, Liansheng Zheng, Xiangqun H. Zheng, Fei N. Zhong, Wenyan Zhong, Xiaojun Zhou, Shiaoping Zhu, Xiaohong Zhu, Hamilton O. Smith, Richard A. Gibbs, Eugene W. Myers, Gerald M. Rubin, and J. Craig Venter, "The Genome Sequence of *Drosophila Melanogaster,*" *Science 287,* 2185–95, March 24, 2000.

7. Justin Gillis, "Will this MAVERICK Unlock the Greatest Scientific Discovery of His Age? Copernicus, Newton, Einstein and VENTER?," *USA Weekend,* January 29–31, 1999.
8. Philip E. Ross, "Gene Machine," *Forbes,* February 21, 2000.

14. THE FIRST HUMAN GENOME

1. Brown eyes have the same number of melanocytes as blue eyes, but they produce a relative abundance of melanin. The blue look is due not to the color of the melanin pigment itself but rather to a light scattering effect (like the one that makes the sky blue) of the packaged melanin. Newborn babies have blue eyes because they have not yet made much melanin.
2. David Ewing Duncan, *The Geneticist Who Played Hoops with My DNA : . . . and Other Masterminds from the Frontiers of Biotech* (London: Fourth Estate, 2005).
3. James C. Mullikin and Amanda A. McMurray, "Sequencing the Genome, Fast," *Science 283,* 1867–68, March 19, 1999.
4. Monday, March 15, 1999, NHGRI Release: Human Genome Project Announces Successful Completion of Pilot Project, Launches Large-Scale Effort to Sequence the Human Genome with New Awards, Accelerated Timetable.

> The international Human Genome Project today announced the successful completion of the pilot phase of sequencing the human genome and the launch of the full-scale effort to sequence all 3 billion letters (referred to as bases) that make up the human DNA instruction book. Based on experience gained from the pilot projects, an international consortium now predicts they will produce at least 90 percent of the human genome sequence in a "working draft" form by the spring of 2000, considerably earlier than expected. "I am extremely pleased that the Human Genome Project has accelerated efforts to complete one of the most important scientific projects in human history — unlocking the secrets of the genetic code. The Project will forever change how we understand the human body and disease, leading to improved prevention, treatments, and cures for what are currently medical mysteries," said Vice President Al Gore. "Specifically, I am thrilled that we are moving

into full-scale sequencing and are on track to complete a working draft of the human genome a year and half ahead of schedule."

5. Francis Collins, *The Language of God: A Scientist Presents Evidence for Belief* (New York: Free Press, 2006), p. 119.
6. James Shreeve, *The Genome War: How Craig Venter Tried to Capture the Code of Life and Save the World* (New York: Ballantine, 2005), p. 186.
7. Collins, *The Language of God,* p. 120.
8. Tim Friend, "Feds May Have Tried to Bend Law for Gene Map," *USA Today,* March 13, 2000.
9. Shreeve, *The Genome War,* p. 321.
10. John Sulston and Georgina Ferry, *The Common Thread* (London: Corgi, 2003), p. 182.
11. Ibid., p. 240.
12. Ibid., p. 228.
13. Collins, *The Language of God,* p. 121.
14. Ibid.
15. Sulston and Ferry, *The Common Thread,* p. 185.
16. Ibid., p. 241.
17. Ibid., p. 265.
18. Ibid., p. 277.
19. David Whitehouse, "Gene Firm Labeled a 'Con Job' " *BBC News Online,* March 6, 2000.

    Dr. John Sulston, director of the Sanger Centre, Britain's leading gene-sequencing lab, has attacked the American company Celera Genomics and its director Dr. Craig Venter over their intention to make money selling a combination of public and private DNA data. In an escalation of the international war of words over the use of genetic information from the project to read the human genome, our DNA blueprint, Dr. Sulston said that the public has got to know what is really going on. "It would be hilarious if it wasn't so serious for all of us," he told the BBC. Dr. Sulston said that Celera "hoover up all the public data, add a bit of their own and sell it as a packaged product. It is fair enough if people want to buy it. That's up to them." But he added that the Celera data is something of a "con job. . . . The emerging truth is absolutely extraordinary. They really intend to establish a complete monopoly position on the human genome for a period of at least five years." As well as the ethical considerations of 'owning' human genes, Dr. Sulston said that the danger was that Celera Genomics will get its way and persuade politicians to reduce public funding for genome studies in the belief that it can all be done by private companies.

20. Sorin Istrail, Granger G. Sutton, Liliana Florea, Aaron L.Halpern, Clark M. Mobarry, Ross Lippert, Brian Walenz, Hagit Shatkay, Ian Dew, Jason R. Miller, Michael J. Flanigan, Nathan J. Edwards, Randall Bolanos, Daniel Fasulo, Bjarni V. Halldorsson, Sridhar Hannenhalli, Russell Turner, Shibu Yooseph, Fu Lu, Deborah R. Nusskern, Bixiong Chris Shue, Xiangqun Holly Zheng, Fei Zhong, Arthur L. Delcher, Daniel H. Huson, Saul A. Kravitz, Laurent Mouchard, Knut Reinert, Karin A. Remington, Andrew G. Clark, Michael S. Waterman, Evan E. Eichler, Mark D. Adams, Michael W. Hunkapiller, Eugene W.Myers, and J. Craig Venter, "Whole Genome Shotgun Assembly and Comparison of Human Genome Assemblies," *Proc. Natl. Acad. Sci.* USA, published online, February 9, 2004, 10.1073.

21. Memo from Lynn Holland to Celera senior staff; containing Rich Roberts e-mail and Eric Lander e-mail.
22. Shreeve, *The Genome War,* p. 314.
23. Sulston and Ferry, *The Common Thread,* p. 244.
24. Peter G. Gosselin and Paul Jacobs, "Rush to Crack Genetic Code Breeds Trouble Science: Public-Private Rift Arises After Company Seeks Exclusive Rights in Exchange for Sharing Expanding Data," *Los Angeles Times,* March 6, 2000.
25. Justin Gillis, "Gene-Mapping Controversy Escalates; Rockville Firm Says Government Officials Seek to Undercut Its Effort," *The Washington Post,* March 7, 2000.
26. Nicholas Wade, "Genome Decoding Plan Is Derailed by Conflicts," *The New York Times,* March 9, 2000.
27. Gillis, "Gene-Mapping Controversy Escalates."
28. Sulston and Ferry, *The Common Thread,* p. 246.
29. Transcript of Briefing by Directors of Office on Science and Technology Policy and the Human Genome Project, *U.S. Newswire,* March 14, 2000.
30. Ibid.
31. Frederick Goolden and Michael Lemonick, "The Race Is Over," *Time,* July 3, 2000.
32. Sulston and Ferry, *The Common Thread,* p. 250.
33. Shreeve, *The Genome War,* p. 296.
34. Bill Clinton, *My Life* (London: Hutchinson, 2004), p. 889.
35. Deb Reichmann, "A Blue Dress and a Presidential Blood Sample," *Pittsburgh Post-Gazette* (Associated Press), September 22, 1998. Charles B. Babcock, "The DNA Test," *The Washington Post,* September 22, 1998.
36. Collins, *The Language of God,* p. 122.
37. Bob Davis and Ron Winslow, "Joint Release of DNA Drafts is Planned," *The Wall Street Journal,* June 20, 2000.
38. Ibid.

15. THE WHITE HOUSE, JUNE 26, 2000

1. Matt Ridley. *Genome: The Autobiography of a Species* (New York: Harper Perennial, 2000), p. 5.
2. Bill Clinton, *My Life* (London: Hutchinson, 2004), p. 910.
3. John Sulston and Georgina Ferry, *The Common Thread* (London: Corgi, 2003), p. 258.
4. Ibid., p. 252.
5. Francis Collins, *The Language of God: A Scientist Presents Evidence for Belief* (New York: Free Press, 2006), p. 2.
6. Ibid., p. 3.
7. Ibid.

16. PUBLISH AND BE DAMNED

1. Michael Ashburner: *See* James Shreeve, *The Genome War: How Craig Venter Tried to Capture the Code of Life and Save the World* (New York: Ballantine, 2005), p. 361.
2. J. Craig Venter, Mark D. Adams, Eugene W. Myers, Peter W. Li, Richard J. Mural, Granger G. Sutton, Hamilton O. Smith, Mark Yandell, Cheryl A. Evans, Robert A. Holt, Jeannine D. Gocayne, Peter Amanatides, Richard M. Ballew, Daniel H. Huson, Jennifer Russo Wortman, Qing Zhang, Chinnappa D. Kodira, Xiangqun H. Zheng, Lin Chen, Marian Skupski, Gangadharan Subramanian, Paul D. Thomas, Jinghui Zhang, George L. Gabor Miklos, Catherine Nelson, Samuel Broder, Andrew G. Clark, Joe Nadeau, Victor

A. McKusick, Norton Zinder, Arnold J. Levine, Richard J. Roberts, Mel Simon, Carolyn Slayman, Michael Hunkapiller, Randall Bolanos, Arthur Delcher, Ian Dew, Daniel Fasulo, Michael Flanigan, Liliana Florea, Aaron Halpern, Sridhar Hannenhalli, Saul Kravitz, Samuel Levy, Clark Mobarry, Knut Reinert, Karin Remington, Jane Abu-Threideh, Ellen Beasley, Kendra Biddick, Vivien Bonazzi, Rhonda Brandon, Michele Cargill, Ishwar Chandramouliswaran, Rosane Charlab, Kabir Chaturvedi, Zuoming Deng, Valentina Di Francesco, Patrick Dunn, Karen Eilbeck, Carlos Evangelista, Andrei E. Gabrielian, Weiniu Gan, Wangmao Ge, Fangcheng Gong, Zhiping Gu, Ping Guan, Thomas J. Heiman, Maureen E. Higgins, Rui-Ru Ji, Zhaoxi Ke, Karen A. Ketchum, Zhongwu Lai, Yiding Lei, Zhenya Li, Jiayin Li, Yong Liang, Xiaoying Lin, Fu Lu, Gennady V. Merkulov, Natalia Milshina, Helen M. Moore, Ashwinikumar K Naik, Vaibhav A. Narayan, Beena Neelam, Deborah Nusskern, Douglas B. Rusch, Steven Salzberg, Wei Shao, Bixiong Shue, Jingtao Sun, Zhen Yuan Wang, Aihui Wang, Xin Wang, Jian Wang, Ming-Hui Wei, Ron Wides, Chunlin Xiao, Chunhua Yan, Alison Yao, Jane Ye, Ming Zhan, Weiqing Zhang, Hongyu Zhang, Qi Zhao, Lian sheng Zheng, Fei Zhong, Wenyan Zhong, Shiaoping C. Zhu, Shaying Zhao, Dennis Gilbert, Suzanna Baumhueter, Gene Spier, Christine Carter, Anibal Cravchik, Trevor Woodage, Feroze Ali, Huijin An, Aderonke Awe, Danita Baldwin, Holly Baden, Mary Barnstead, Ian Barrow, Karen Beeson, Dana Busam, Amy Carver, Angela Center, Ming Lai Cheng, Liz Curry, Steve Danaher, Lionel Davenport, Raymond Desilets, Susanne Dietz, Kristina Dodson, Lisa Doup, Steven Ferriera, Neha Garg, Andres Gluecksmann, Brit Hart, Jason Haynes, Charles Haynes, Cheryl Heiner, Suzanne Hladun, Damon Hostin, Jarrett Houck, Timothy Howland, Chinyere Ibegwam, Jeffery Johnson, Francis Kalush, Lesley Kline, Shashi Koduru, Amy Love, Felecia Mann, David May, Steven McCawley, Tina McIntosh, Ivy McMullen, Mee Moy, Linda Moy, Brian Murphy, Keith Nelson, Cynthia Pfannkoch, Eric Pratts, Vinita Puri, Hina Qureshi, Matthew Reardon, Robert Rodriguez, Yu-Hui Rogers, Deanna Romblad, Bob Ruhfel, Richard Scott, Cynthia Sitter, Michelle Smallwood, Erin Stewart, Renee Strong, Ellen Suh, Reginald Thomas, Ni Ni Tint, Sukyee Tse, Claire Vech, Gary Wang, Jeremy Wetter, Sherita Williams, Monica Williams, Sandra Windsor, Emily Winn-Deen, Keriellen Wolfe, Jayshree Zaveri, Karena Zaveri, Josep F. Abril, Roderic Guigo, Michael J. Campbell, Kimmen V. Sjolander, Brian Karlak, Anish Kejariwal, Huaiyu Mi, Betty Lazareva, Thomas Hatton, Apurva Narechania, Karen Diemer, Anushya Muruganujan, Nan Guo, Shinji Sato, Vineet Bafna, Sorin Istrail, Ross Lippert, Russell Schwartz, Brian Walenz, Shibu Yooseph, David Allen, Anand Basu, James Baxendale, Louis Blick, Marcelo Caminha, John Carnes-Stine, Parris Caulk, Yen-Hui Chiang, My Coyne, Carl Dahlke, Anne Deslattes Mays, Maria Dombroski, Michael Donnelly, Dale Ely, Shiva Esparham, Carl Fosler, Harold Gire, Stephen Glanowski, Kenneth Glasser, Anna Glodek, Mark Gorokhov, Ken Graham, Barry Gropman, Michael Harris, Jeremy Heil, Scott Henderson, Jeffrey Hoover, Donald Jennings, Catherine Jordan, James Jordan, John Kasha, Leonid Kagan, Cheryl Kraft, Alexander Levitsky, Mark Lewis, Xiangjun Liu, John Lopez, Daniel Ma, William Majoros, Joe McDaniel, Sean Murphy, Matthew Newman, Trung Nguyen, Ngoc Nguyen, Marc Nodell, Sue Pan, Jim Peck, Marshall Peterson, William Rowe, Robert Sanders, John Scott, Michael Simpson, Thomas Smith, Arlan Sprague, Timothy Stockwell, Russell Turner, Eli Venter, Mei Wang, Meiyuan Wen, David Wu, Mitchell Wu, Ashley Xia, Ali Zandieh, Xiaohong Zhu1, "The Sequence of the Human Genome," *Science 291,* 1304–51, February 16, 2001.

3. R. Waterston, E. Lander, and J. Sulston, "On the Sequencing of the Human Genome," *Proc. Natl. Acad. Sci.* USA, *99,* 3712–16, 2002.
4. John Sulston and Georgina Ferry, *The Common Thread* (London: Corgi, 2003), p. 271.
5. James Shreeve, *The Genome War: How Craig Venter Tried to Capture the Code of Life and Save the World* (Ballantine, New York, 2005) p. 364.

6. David Ewing Duncan, *The Geneticist Who Played Hoops with My DNA : . . . and Other Masterminds from the Frontiers of Biotech* (London: Fourth Estate, 2005), p. 134.
7. Mural, R.J., Adams, M.D., Myers, E. W., Smith, H.O., Miklos, G.L., Wides, R., Halpern, A., Li, P.W., Sutton, G., Nadeau, J., Salzbert, S.L., Holt, R., Kodira, C.D., Lu, F., Evangelista, C.C., Gan, W., Heiman, T.J., Li, J., Merkulov, G.V., Naik, A.K., Qi, R., Wang, A., Wang, X., Yan, X., Yooseph, S., Zheng, L., Zhu, S.C., Biddick, K., Bolanos, R., Delcher, A., Dew, I., Fasulo, D., Flanigan, M., Huson, D., Kravitz, S., Miller, J.R., Mobarry, C., Reinert, K., Remington, K., Zhang, Q., Zheng, X.H., Nusskern, D., Lai, Z., Lei, Y., Zhong, W., Yao, A., Guan, P., Ji, R., Gu, Z., Wang, Z., Zhong, F., Ziao, C., Chiang, C., Yandell, M., Wortman, J., Amanatides, P., Hladun, S., Pratts, E., Johnson, J., Dodson, K., Woodford, K., Evans, J.C., Gropman, B., Rusch, D., Venter, E., Wang, M., Smith, T., Houck, Tompkins, D.E., Haynes, C., Jacob, D., Chin, S. Allen, D., Dahlke, C., Sanders, B., Li, K., Liu, F., Levitsky, A., Majoros, W., Chen, Q., Xia, A., Lopez, J., Donnelly, M., Newman, M., Glodek, A., Kraft, C., Nodell, M., Beeson, K., Cai, S., Caulk, P., Chen, Y., Coyne, M., Dietz, S., Dullaghan, P., Fosler, C., Gire, C., Gocayne, J.D., Hoover, J., Howland, T., Ma, D., McIntosh, T., Murphy, B., Murphy, S., Nelson, K., Parker, K., Prudhomme, A., Puri, Vinita, Qureshi, H., Raley, J.C., Reardon, M., Regier, M., Rogers, Y., Romblad, D., Scott, J., Scott, R., Sitter, C., Sprague, A., Stewart, E., Strong, R., Suh, E., Sylvester, K., Tint, N.N., Tsonis, C., Wang, G., Wang, G., Williams, M., Williams, S., Windsor, S. Wolfe, K., Wu, M., Zaveri, J., Zubeda, N., Subramanian, G., Venter, J.C. "A Comparison of Whole-Genome Shotgun-Derived Mouse Chromosome 16 and the Human Genome," *Science, 296,* 1661–71, May 19, 2002.
8. Sulston and Ferry, *The Common Thread,* p. 261.
9. Meredith Wadman, "Biology's Bad Boy Is Back," *Fortune,* March 8, 2004.
10. Another possible mechanism of this effect lies in a peculiarity at the twilight of life. In the first twenty-four hours of life, parental DNA has not even been mixed to form an individual at this stage, and the DNA is not even being used. Parental genetic messages scurry hither and thither to control and influence the early stages of development. These messages are written in RNA, the more ancient genetic code found in our cells alongside DNA, and are bequeathed by the egg and sperm. One can think of this as an RNA operating system that enables the development software to run in the cell.

17. BLUE PLANET AND NEW LIFE

1. Maynard Olson, "The Human Genome Project: A Player's Perspective." *J. Mol. Biol. 319,* 931–42, 2002.
2. John Sulston and Georgina Ferry, *The Common Thread* (London: Corgi, 2003), p. 192.
3. Myers, E.W., Sutton, G.G., Smith, H.O., Adams, M.D., Venter, J.C., "On the Sequencing and Assembly of the Human Genome," *Proc. Natl. Acad. Sci.* USA, 99, 7, 4145–46, 2002.
4. R. Waterston, E. Lander, and J. Sulston, "On the Sequencing of the Human Genome," *Proc. Natl. Acad. Sci.* USA, *99,* 3712–16, 2002.
5. Ibid.
6. Nicholas Wade, "Genome Project Rivals Trade Notes, Cordially," *The New York Times,* June 12, 2001, p. 2.
7. Nicholas Wade, "Grad Student Becomes Gene Effort's Unlikely Hero," *The New York Times,* February 13, 2001, p. 1.
8. Particularly on chromosome 22.
9. Mark D. Adams, Granger G. Sutton, Hamilton O. Smith, Eugene W. Myers, and J. Craig

Venter, "The Independence of Our Genome Assemblies," *Proc. Natl. Acad. Sci.* USA, *100,* 3025-26, 2003.

10. International Human Genome Sequencing Consortium. "Finishing the Euchromatic Sequence of the Human Genome," *Nature 431,* 931-45, October 21, 2004.
11. Sorin Istrail, Granger G. Sutton, Liliana Florea, Aaron L.Halpern, Clark M. Mobarry, Ross Lippert, Brian Walenz, Hagit Shatkay, Ian Dew, Jason R. Miller, Michael J. Flanigan, Nathan J. Edwards, Randall Bolanos, Daniel Fasulo, Bjarni V. Halldorsson, Sridhar Hannenhalli, Russell Turner, Shibu Yooseph, Fu Lu, Deborah R. Nusskern, Bixiong Chris Shue, Xiangqun Holly Zheng, Fei Zhong, Arthur L. Delcher, Daniel H. Huson, Saul A. Kravitz, Laurent Mouchard, Knut Reinert, Karin A. Remington, Andrew G. Clark, Michael S. Waterman, Evan E. Eichler, Mark D. Adams, Michael W. Hunkapiller, Eugene W.Myers, and J. Craig Venter, "Whole Genome Shotgun Assembly and Comparison of Human Genome Assemblies," *Proc. Natl. Acad. Sci.* USA, published online, February 9, 2004, 10.1073.
12. Nicholas Wade, "Scientist Reveals Secret of the Genome: It's His," *The New York Times,* April 27, 2002.
13. Antonio Regalado, "Entrepreneur Puts Himself Up for Study in Genetic 'Tell-All,' " *The Wall Street Journal,* October 18, 2006.
14. Samuel Levy, Granger Sutton, Pauline Ng, Lars Feuk, Aaron L. Halpern, Brian Walenz, Nelson Axelrod, Jiaqi Huang, Ewen Kirkness, Gennady Denisov, Yuan Lin, Jeffrey R. MacDonald, Andy Wing Chun Pang, Mary Shago, Tim Stockwell, Alexia Tsiamouri, Vineet Bafna, Vikas Bansal, Saul Kravitz, Dana Busam, Karen Beeson, Tina McIntosh, John Gill, Jon Borman, Yu-Hui Rogers, Marvin Frazier, Stephen Scherer, Robert L. Strausberg, J. Craig Venter, "The Diploid Genome Sequence of an Individual Human," *PLoS Biology,* 5: September 4, 2007.
15. Venter, J.C., Remington, K., Heidelberg, J., Halpern, A., Rusch, D., Eisen, J., Wu, D., Paulsen, I., Nelson, K., Nelson, W., Fouts, D., Levy, S., Knap, A., Lomas, M., Nealson, K., White, O., Peterson, J., Hoffman, J., Parsons, R., Baden-Tillson, H., Pfannkoch, C., Rogers, Y.H., and Smith, H., "Environmental Genome Shotgun Sequencing of the Sargasso Sea," *Science 304,* 66–74, 2004.
16. Douglas B. Rusch, Aaron L. Halpern, Granger Sutton, Karla B. Heidelberg, Shannon Williamson, Shibu Yooseph, Dongying Wu, Jonathan A. Eisen, Jeff M. Hoffman, Karin Remington, Karen Beeson, Bao Tran, Hamilton Smith, Holly Baden-Tillson, Clare Stewart, Joyce Thorpe, Jason Freeman, Cynthia Andrews-Pfannkoch, Joseph E. Venter, Kelvin Li, Saul Kravitz, John F. Heidelberg, Terry Utterback, Yu-Hui Rogers, Luisa I. Falco, Valeria Souza, German Bonilla-Rosso, Luis E. Eguiarte, David M. Karl, Shubha Sathyendranath, Trevor Platt, Eldredge Bermingham, Victor Gallardo, Giselle Tamayo-Castillo, Michael R. Ferrari, Robert L. Strausberg, Kenneth Nealson, Robert Friedman, Marvin Frazier, J. Craig Venter, "The Sorcerer II Global Ocean Sampling Expedition: Northwest Atlantic through Eastern Tropical Pacific," *PLoS Biology,* 398-431, 2007; Shibu Yooseph, Granger Sutton, Douglas B. Rusch, Aaron L. Halpern, Shannon J. Williamson, Karin Remington, Jonathan A. Eisen, Karla B. Heidelberg, Gerard Manning, Weizhong Li, Lukasz Jaroszewski, Piotr Cieplak, Christopher S. Miller, Huiying Li, Susan T. Mashiyama, Marcin P. Joachimiak, Christopher van Belle, John-Marc Chandonia, David A. Soergel, Yufeng Zhai, Kannan Natarajan, Shaun Lee, Benjamin J. Raphael, Vineet Bafna, Robert Friedman, Steven E. Brenner, Adam Godzik, David Eisenberg, Jack E. Dixon, Susan S. Taylor, Robert L. Strausberg, Marvin Frazier, J. Craig Venter, "The Sorcerer II Global Ocean Sampling Expedition: Expanding the Universe of Protein Families," *PLoS Biology,* 432-66, 2007; Natarajan Kannan, Susan S. Taylor, Yufeng Zhai, J. Craig Venter, Gerard Manning,

"Structural and Functional Diversity of the Microbial Kinome," *PLoS Biology*, 467–78, 2007.

17. Smith H.O., Hutchison C.A., III, Pfannkoch, C, and Venter, J.C., "Generating a Synthetic Genome by Whole Genome Assembly: _X174 Bacteriophage from Synthetic Oligonucleotides," *Proc. Natl. Acad. Sci.* USA, *100,* 15440–445, 2003.
18. Elizabeth Weise, "Scientists Create a Virus That Reproduces," *USA Today*, November 14, 2003.
19. Elizabeth Pennisi, "Venter Cooks Up a Synthetic Genome in Record Time," *Science 302,* 1307, November 21, 2003.
20. Carole Lartigue, John I. Glass, Nina Alperovich, Rembert Pieper, Prashanth P. Parmar, Clyde A. Hutchison III, Hamilton O. Smith, and J. Craig Venter, "Genome Transplantation in Bacteria: Changing One Species to Another," *Science*, June 28, 2007.

INDEX